Student Companion and Problem-Solving Guide

FOR

GIORDANO'S

COLLEGE PHYSICS

REASONING AND RELATIONSHIPS

Second Edition
VOLUME 1

Richard Grant
Roanoke College

BROOKS/COLE
CENGAGE Learning

Australia • Brazil • Japan • Korea • Mexico • Singapore • Spain • United Kingdom • United States

BROOKS/COLE
CENGAGE Learning

© 2013 Brooks/Cole, Cengage Learning

ISBN-13: 978-1-111-57102-3

ISBN-10: 1-111-57102-3

Brooks/Cole
20 Channel Center Street
Boston, MA 02210
USA

Cengage Learning is a leading provider of customized learning solutions with office locations around the globe, including Singapore, the United Kingdom, Australia, Mexico, Brazil, and Japan. Locate your local office at: **www.cengage.com/global**

Cengage Learning products are represented in Canada by Nelson Education, Ltd.

To learn more about Brooks/Cole, visit **www.cengage.com/brookscole**

Purchase any of our products at your local college store or at our preferred online store **www.cengagebrain.com**

READ IMPORTANT LICENSE INFORMATION

Printed in the United States of America
1 2 3 4 5 6 7 15 14 13 12 11

PREFACE

INTRODUCTION

This is the second edition of the *Student Companion and Problem Solving Guide* to accompany Giordano's *College Physics: Reasoning and Relationships, Volume 1, Second Edition*. Many of you taking an introductory physics course are not physics majors, and this may be the only physics course you take. The approach of Giordano's *College Physics* is to teach physics in the broader science context, rather than require that you learn a collection of quantitative problems that don't seem to connect to the real world. I have written the *Student Companion and Problem-Solving Guide* with the same emphasis on reasoning and relationships as the main text and have added some additional content to help you if you're preparing for the MCAT exam.

SIGNIFICANT FIGURES

The approach to significant figures in the main text is reflected in this guide. However, remember that significant figure rules should not be applied until the last calculation is done. You should leave all the information in your calculator until the final answer, since removing and re-entering intermediate results will often lead to pre-mature rounding. Also, for reasoning problems it is most beneficial to use your calculator very little. By sketching the problem, creating diagrams, writing out a formula, and thinking about its meaning, you can learn much about the physical behavior of a system without actually performing any numerical calculations. This skill is emphasized throughout the main text and within this student companion.

KEY FEATURES OF THIS STUDENT COMPANION

Summary of Key Concepts and Problem-Solving Strategies

Each chapter of this guide contains a "Summary of Key Concepts and Problem-Solving Strategies" which highlights the core ideas from the main text chapter and explains how you use these concepts to solve problems, either for homework or for an exam. This is a good place to look if you need help with some of the mathematical material in the book or if you would just like a quick review of a chapter.

Frequently Asked Questions

"Frequently Asked Questions" are collected from my experience with students in the classroom and the kinds of questions they ask about homework assignments or material from the main text. This section is helpful if you're having trouble solving problems or if you would like to make sure that you avoid some common pitfalls in understanding the chapter material.

Selection of End-of-Chapter Answers and Solutions

The "Selection of End-of-Chapter Answers and Solutions" has expanded explanations of specific end-of-chapter questions and problems that are representative of important concepts in the book. These answers and solutions carefully explain how to reach the answer and what the underlying assumptions are when you are working on a solution.

III

Additional Worked Examples and Capstone Problems

In addition to the Worked Examples in the text, which follow the five-step problem-solving methodology, I have included "Additional Worked Examples and Capstone Problems." The Capstone Problems typically combine concepts from several chapters in the text, and therefore serve as a good review as you prepare for exams.

MCAT Review Problems and Solutions

In the physical science section of the MCAT exam there are two types of questions, discrete questions and those questions associated with a passage. Discrete questions can usually be done quickly and should therefore be done first. This ensures you'll answer as many questions in the allotted time as possible and helps to build your confidence during the test resulting in a calmer, logical approach to the passages. Though many rather expensive MCAT preparatory courses would have you believe that there is something unique about their problem-solving strategy, there are only minor differences between them and the approach described in this textbook. A comparison of the problem-solving strategy employed in this textbook with a typical strategy used by MCAT preparatory courses is illustrated below. You can see that, the content is the same but the packaging is different.

Typical MCAT Strategy	General Problem-Solving Strategy
Read the problem: You should read the problem carefully and highlight the points which seem important. Don't just gloss over the problem.	Recognize the principle
Think about the problem: Determine what the problem is asking, what physical relationships are important, and identify and eliminate any unnecessary information.	
Predict an answer: Quickly and logically pull together the appropriate givens and relationships to solve for an answer. A sketch often helps to organize your thoughts.	Sketch the problem Identify the relationships
Scan for the closest match: Look for the answer which most closely matches your answer. Be sure your answer makes physical sense.	Solve What does it mean?

In addition to the general problem-solving approach used in this textbook and guide, there are time-saving tricks specific to tackling multiple choice questions like those found on the MCAT. These strategies can help you quickly eliminate some wrong answers, and perhaps identify the correct answer without fully solving the problem:

First: Check the units of the answers. If the question is asking for the speed of an object, then all answers without the dimension of length/time can be eliminated.

Second: The size and scale of an answer should be realistic. MCAT problems are generally based on real, physical concepts so the answers should be realistic. For example, if a problem asks for the speed of an object just before it hits the ground after being dropped off the top of a three-story building, then an answer of 10,000 m/s is not reasonable and can therefore be eliminated.

Third: Apply basic mathematic principles and common sense. Calculators are not permitted on the MCAT so you should be comfortable with rounding off numbers in your calculations. For example, the acceleration of an object due to gravity, $g = 9.8 \text{ m/s}^2$, should be approximated as 10 m/s^2. Also, you need to know the simple trigonometric relationships including those involving the angles 30°, 45°, and 60°.

$$\sin 30° = \cos 60° = \frac{1}{2}$$

$$\sin 60° = \cos 30° = \frac{\sqrt{3}}{2}$$

$$\sin 45° = \cos 45° = \frac{1}{\sqrt{2}}$$

Additional relationships include the following:

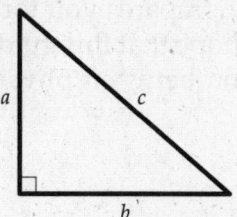

Pythagorean Theorem: $c^2 = a^2 + b^2$

3–4–5 Triangle: If $a = 3$ and $b = 4$, then $c = 5$. This also works for multiples of 3–4–5 (i.e., 6–8–10, 9–12–15, etc.).

Finally, there are specific approaches that can help you tackle passages and their associated questions. When reading a passage you should be highlighting and summarizing the key points for each paragraph so you can avoid having to read the passage several times. The following approach should be used during this process:

Figures, graphs, and tables: Look carefully at the information provided in any figures, graphs, or tables included in the passage. By first understanding this material you can save time not having to look at them while reading the text of the passage. It is much quicker to read a passage that refers to figures, graphs, or tables when you already know what information they contain.

Highlight and summarize: Don't just gloss over each paragraph. As you read each paragraph, underline or circle key words, equations, or data and summarize the purpose of each paragraph with a few words.

Review: Read back over the points you've highlighted, the summaries for each paragraph, and the figures, graphs, and tables before tackling the problems.

Though the word MCAT strikes fear in many (if not most) premed students, performing well on the MCAT can help ensure your acceptance into medical school. Although the physical science portion seems to be the most challenging portion of the exam for most students, this does not need to be the case. To improve your chances of performing well on the MCAT you should focus on understanding the main concepts discussed in the textbook and hone your problem-solving strategies through practice and persistence. The knowledge and skills you develop during this course and by practicing the MCAT problems within this guide should be adequate preparation for your success on the physics portion of the MCAT exam.

ACKNOWLEDGMENTS

I would like to thank Susan English of Durham Technical Community College who provided an essential service by accuracy reviewing this manual. I would also like to thank the team at Cengage for their help, especially Charles Hartford, Susan Pashos, Brandi Kirksey, and Brendan Killion. Finally, I would like to thank my wife Jackie for her patience.

I hope that the material in the *Student Companion and Problem-Solving Guide* compliments what you will learn in the textbook and efficiently prepares you for the homework assignments and exams during the semester. I also hope that this guide supports the main goal of the text: to help you see the connection between physics and the world around you.

TABLE OF CONTENTS

1 Introduction

CONTENTS

Part A. Summary of Key Concepts and Problem-Solving Strategies

KEY CONCEPTS

Scientific notation and significant figures

Scientific notation is used to express very large and very small numbers. We can also use prefixes to write very large or very small numbers; i.e., 1 micrometer = 1 μm = 1×10^{-6} meters, and 1 Megasecond = 1 Ms = 1×10^6 seconds.

The accuracy of a quantity is reflected in the number of *significant figures* used to express its value. The result of a calculation should be expressed with an appropriate number of significant figures.

Units of measure

The *primary units* of mechanics involve length, time, and mass. Most scientific work employs the SI system of units, in which length is measured in *meters*, time is measured in *seconds*, and mass is measured in *kilograms*. The values of the standard meter, the standard second, and the standard kilogram are established by international agreement, and are a foundation for all scientific work. All other units in mechanics can be derived from these three primary units.

1

Physical quantities and dimensions

We cannot give a definition of the concepts of length, time, and mass, but must instead treat these as "givens." The definitions of all other physical quantities encountered in mechanics can be derived, or built upon, these three primary quantities. A total of seven primary physical quantities are needed to describe all of physics—the other four are connected with electricity and magnetism, and heat.

The *dimensions* of all quantities in mechanics can be expressed in terms of length **L**, mass **M**, and time **T**. For example, the dimensions of velocity are length divided by time = **L/T**. In the SI system of units, **L** is measured in meters and **T** in seconds, so the units of velocity are meters/second. Dimensional analysis involves checking that the dimensions of an answer correctly match the dimensions of the quantity being calculated.

The mathematics of physics

Several types of mathematics are used in this book: algebra, trigonometry, and vectors. Algebra is essential for solving systems of equations. When dealing with motion or other problems in physics we often need to express position or movement in terms of a *coordinate system* (usually the *x-y-z* set of coordinate axes). Trigonometry and vectors are extremely useful in such calculations. A *vector* quantity has both a *magnitude* and a *direction*. Vectors can be added graphically or in terms of their *components*. Appendix B in your textbook contains a quick review of algebra, trigonometry, and vectors.

PROBLEM-SOLVING STRATEGIES

Problem solving is an essential part of physics. Problems may be quantitative with a precise answer or they may be conceptual. For some problems you may need to use common sense reasoning to estimate the values of important quantities. We will encounter problems involving different laws of physics, including Newton's laws of mechanics, the laws of electricity and magnetism, and quantum theory. Although these problems involve many different situations, they can all be attacked using the same basic problem-solving strategy.

Problem Solving: General Plan of Attack

Recognize the principle

For example, one problem might involve the principle of conservation of energy, whereas another might require Newton's action–reaction principle. The ability to recognize the central principles requires a conceptual understanding of the laws of physics, how they are applied, and how they are interrelated. Such knowledge and skill are obtained from experience, practice, and careful study.

Sketch the problem

A diagram showing all the given information, the directions of any forces, and so forth is valuable for organizing your thoughts. A good diagram will usually contain a coordinate system to be used in measuring the position of an object and other important quantities.

Identify the relationships

For example, Newton's second law gives a relationship between force and motion, and is thus the key to analyzing the motion of an object. This step in the problem-solving process may involve several parts (substeps), depending on the nature of the problem. For example, problems involving collisions may involve steps that aren't needed or necessary for a problem in magnetism. When dealing with a reasoning and relationship problem, one of these substeps will involve identifying the "missing" information or quantities and then estimating their values.

Solve

Using the relationships you've identified as important in the previous step, solve for the unknown quantities.

What does it mean?

Does your answer make sense? Take a moment to think about your answer and reflect on the general lessons to be learned from the problem.

Part B. Frequently Asked Questions

1.   *Can I multiply and divide quantities with different dimensions?*

Yes, you can multiply and divide quantities with different dimensions to derive other quantities. For example, speed is defined as a distance divided by a time. Distance has the dimension of length (**L**) with the units in meters and time (dimension **T**) has the units of seconds. This means that speed has the dimensions of **L/T** and the units of meters/second, abbreviated as m/s.

2.   *Can I add and subtract quantities with different units?*

No, it makes no physical sense to add or subtract quantities with different units. For example, it is meaningless to add a length of 10 m to a time of 4 s. However, you can add or subtract quantities with the same units such as 10 m minus 4 m and obtain a meaningful result, 6 m.

3.   *When do I round up an answer?*

Generally, if the first insignificant figure is a 5 or more you should round up your answer, whereas if the first insignificant figure is less than 5 you don't round up. For example, if your answer is 2.015 m then rounding off to three significant figures would give 2.02 m. On the other hand, if your answer is 2.014 m then rounding off would give 2.01 m.

4.   *Why is it important to include the units with my answer? Aren't the numbers what we're really looking for?*

In physics, most numbers are meaningless unless they have an associated unit. While the numbers give us something to compare, the units provide us a physical context for the comparison. For example, if someone stated that the mass of an object was 147 without including the units, we would have no way of knowing whether that was a large mass or a small mass: 147 g is much different than 147 kg! Units provide us with essential information. Also, including the units in our calculations provides a way to check our answer. If the units of our final answer are different than expected, then we've probably made a mistake somewhere in our calculations.

5.   *How can I solve a problem if there seem to be too many unknowns?*

You'll be able to solve most of the problems in your text with the information provided, using expressions relating the various quantities. The key is to identify the expressions relating the quantities and use them appropriately. If there are two unknowns then you need to identify two expressions and if there are three unknowns then you need three expressions. Most real-world problems and those in your text identified as reasoning problems require you to estimate or otherwise find a particular quantity before you can solve the problem.

6.   *Every time I try to solve three equations for the three unknowns, I keep ending up back where I started! What am I doing wrong?*

This is a very common problem students have until they gain practice in solving simultaneous equations. The key is to reduce the number of equations and unknowns by combining equations to express one unknown in terms of the others.

Once you've reduced the number of equations and unknowns to two, you can combine these equations to obtain one equation with one unknown, then solve this single equation. For example, suppose you wish to solve the following three equations for x, y, and z:

$$3x - 5y + z = -11 \qquad (\#1)$$
$$-4x + 3y + 9z = 82 \qquad (\#2)$$
$$x - 4y - 3z = -50 \qquad (\#3)$$

We see that multiplying (#1) by 4, then multiplying (#2) by 3, then adding the results will produce an equation without x:

$$-20y + 4z + 9y + 27z = -44 + 246$$
Therefore, $-11y + 31z = 202$ \qquad (\#4)

Similarly, we see that multiplying (#3) by 4, then adding the result to (#2) will produce another equation without x:

$$3y + 9z - 16y - 12z = 82 - 200$$
Therefore, $-13y - 3z = -118$ \qquad (\#5)

Now we have two equations, (#4) and (#5), and two unknowns. Combine them by multiplying (#4) by 13, then multiplying (#5) by -11, then adding the results:

$$13(31z) + 11(3z) = 13(202) + 11(118)$$

We have produced an equation with only z as an unknown.

Solving for z: $\boxed{z = 9}$

Substituting back into equations (#5) and (#1) we find:

$\boxed{y = 7}$ and $\boxed{x = 5}$

7. *Is the sine of an angle always equal to y/r and the cosine equal to x/r?*

No, certainly not. It all depends on how you define your coordinate system and which angle you are referring to. For example, consider the following figure:

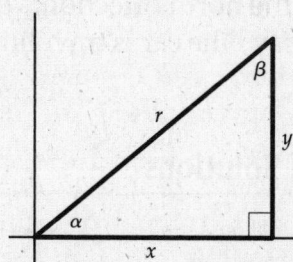

If you are interested in finding the sine of the angle α, then $\sin \alpha = y/r$. However, if you are interested in finding the sine of the angle β, then $\sin \beta = x/r$. Similarly you can define the cosine in terms of x or y depending on the angle of interest in a given coordinate system. In general, it's best not to memorize one way or the other but to tackle each situation separately.

8.   *Is the x-component of a vector always associated with the cosine of the angle and the y-component associated with the sine?*

No. This question is similar to the previous question in that it all depends on how you define your coordinate system and which angle you are referring to. Consider the following figure:

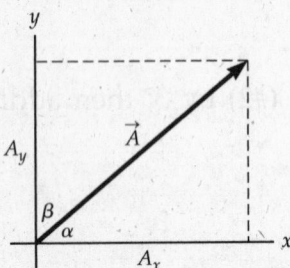

Depending on which angle you use, you can express the components of vector $\vec{A}$ in terms of the angle α or β.

$$A_x = |\vec{A}| \cos \alpha = |\vec{A}| \sin \beta$$
$$A_y = |\vec{A}| \sin \alpha = |\vec{A}| \cos \beta$$

9.   *What is the difference between a "dimension" and a "unit"?*

Dimensions refer the physical nature of a quantity (such as length, time, or mass) whereas the units refer to standards of measure for the physical quantity such as meters, seconds, or kilograms. For example, if you measure a distance as 10 m, then you would say that the distance has the dimension of length and the units of meters.

10.   *How do I know if a physical quantity is a vector?*

Vectors have both a magnitude and a direction and are typically indicated with an arrow above the symbol (e.g., $\vec{A}$). However, if the vector nature of a physical quantity is not given, then a common sense test usually works. Ask yourself if it makes sense to include a direction while describing the physical quantity. For example, it makes no sense to say "the color of that shirt is blue in the north direction," or "she is 9 years old, 30° south of east." It does make sense to say "the car is traveling with a velocity of 10 m/s due west."

Part C. Selection of End-of-Chapter Answers and Solutions

QUESTIONS

Q1.3 Which of the following are units of volume?

cubic meters $acres/m^2$

$mm \times mi^2$ $hours \times mm^3/s$

$kiloseconds \times ft^2$ $mm \times cm \times mm^2/ft$

$kg^2 \times cm$

Answer

Using dimensional analysis we can determine which are units of volume, since volume has dimensions of cubic length (L^3).

cubic meters has the dimensions of L^3, so this is a volume.

mm × mi^2 has the dimensions of $L \times L^2$ or L^3, so this is a volume.

kiloseconds × ft^2 has the dimensions of $T \times L^2$, so this is *not* a volume.

kg × cm has the dimensions of $M^2 \times L$, so this is *not* a volume.

acres/m^2 has the dimensions of L^2/L^2. This is dimensionless, so it is *not* a volume.

hours × mm^3/s has the dimensions of $T \times L^3/T$ or L^3, so this is a volume.

mm × cm × mm^2/ft has the dimensions of $L \times L \times L^2/L$ or L^3, so this is a volume.

Q1.6 Which of the following quantities have the properties of a vector and which have the properties of a scalar?

mass   density

velocity   temperature

displacement (change in position)

Answer

If a quantity has both a magnitude (a number) and a direction, then it is a vector. Quantities without a direction are scalars. With this is mind we see that displacement and velocity are both vectors and mass, density, and temperature are all scalars.

PROBLEMS

P1.12 For each of the following formulas and associated values, find and express the solution with the appropriate number of significant figures and units. (a) The area A of a rectangle of length $L = 2.34$ m and width $w = 1.874$ m. (b) The area A of a circle of radius $r = 0.0034$ m. (c) The volume V of a cylinder with height $h = 1.94 \times 10^{-2}$ m and radius $r = 1.878 \times 10^{-4}$ m. (d) The perimeter P of a rectangle of length $L = 207.1$ m and width $w = 28.07$ m.

Solution

Recognize the principle

We will need to keep track of our significant figures and units using the rules described in the chapter.

Sketch the problem

None required.

Identify the relationships and Solve

(a) $A = L \times w$ where $L = 2.34$ m and $w = 1.874$ m.

Since the less accurate of the two quantities is L, we round the answer to three significant figures:

$$A = 4.38516 \text{ m}^2 = \boxed{4.39 \text{ m}^2}$$

(b) $A = \pi r^2$ where $r = 0.0034$ m.

Since r has two significant figures we round our answer to two significant figures and use scientific notation:

$$A = \boxed{3.6 \times 10^{-5} \text{ m}^2}$$

(c) $V = \pi r^2 h$ where $h = 1.94 \times 10^{-2}$ m and $r = 1.878 \times 10^{-4}$ m.

Since the less accurate of the two quantities is h, we round the answer to three significant figures and use scientific notation:

$$V = \boxed{2.15 \times 10^{-9} \text{ m}^3}$$

(d) $P = 2L + 2w$ where $L = 207.1$ m and $w = 28.07$ m.

Since the less accurate of the two quantities is L, we round the answer to one digit to the right of the decimal place:

$$P = \boxed{470.3 \text{ m}}$$

What does it mean?

Following rules for significant figures when combining quantities is important to avoid overstating the accuracy of the result.

P1.19   A U.S. football field is 120 yards long (including the end zones). How long is the field in (a) meters, (b) millimeters, (c) feet, and (d) inches?

Solution

Recognize the principle

This problem requires keeping track of significant figures and conversion from yards to meters, to millimeters, to feet, and then to inches.

Sketch the problem

None required.

Identify the relationships

The conversion factor for converting yards to meters is 0.9144 m/yard, yards to millimeters is 914.4 m/yard, yards to feet is 3 feet/yard, then yards to inches is 36 inches/yard.

Solve

A U.S. football field is 120 yards long. We assume this has three significant figures. That is, it is accurate to within 0.5 yards. Therefore, using the conversion factors:

(a) 120 yards $\times$ (0.9144 m/yard) = 109.7 m. Rounded to three significant figures, this is $\boxed{1.10 \times 10^2 \text{m}}$.

(b) 120 yards $\times$ (914.4 mm/yard) = 109,700 mm. With three significant figures, this is $\boxed{1.10 \times 10^5 \text{ mm}}$.

(c) 120 yards $\times$ (3 ft/yard) = 360 ft, or with three significant figures, $\boxed{3.60 \times 10^2 \text{ ft}}$.

(d) 120 yards $\times$ (36 in/yard) = 4320 in. or with three significant figures, $\boxed{4.32 \times 10^3 \text{ in.}}$

What does it mean?

There are many ways to express the length of a football field, that's why including a unit with an answer is so important. Also, keeping track of significant figures is quite easy as long as you remember the simple rules.

P1.34 Consider the equations

$$5x + 2y = 13 \qquad -3x + 7y = 25$$

(a) Put each equation into the standard format for a straight line ($y = mx + b$) and plot both lines. Use your graph to find the point (x, y) where these lines intersect. (b) Use algebra to find the values of x and y that simultaneously satisfy these two equations. (c) How do your answers to parts (a) and (b) compare?

Solution

Recognize the principles

We can use algebra to get the equations into the desired format. If we then graph the lines, the point of intersection (on both lines) is the single solution to both equations.

Sketch the problem

We will plot the needed graphs of the two equations as we answer part (a).

Identify the relationships

We can solve each of the given equations for y by subtracting the x term from both sides of the equation and dividing by the factor multiplying the y.

Solve

(a) Solving for y in each of the equations, we find

$$y = -2.5x + 6.5$$
$$y = 0.43x + 3.6$$

Plotting these two expressions we obtain the following graph,

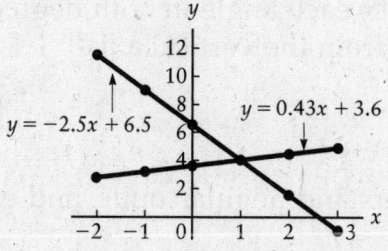

The figure shows the plot of these two lines. Reading from the graph we find that the two lines cross at approximately

$$(x, y) = (1, 4)$$

(b) Applying techniques for solving simultaneous equations as described in the text, we find a common value for the x term,

$$(5x + 2y = 13)(3) \Rightarrow 15x + 6y = 39$$
$$(-3x + 7y = 25)(5) \Rightarrow -15x + 35y = 125$$

Then adding these two equations to eliminate x we find

$$\begin{array}{r} 15x + 6y = 39 \\ -15x + 35y = 125 \\ \hline 0x + 41y = 164 \end{array}$$

Solving for y,

$$y = \frac{164}{41} = 4$$

Finally, solve one of the equations for x and substitute in $y = 4$:

$$15x + 6y = 39$$

$$\Rightarrow x = \frac{39 - 6y}{15} = \frac{39 - 6(4)}{15} = 1$$

Therefore,

$$(x, y) = (1, 4)$$

(c) From the graph in part (a), the lines cross at approximately (1, 4). This is the same point found in part (b), so the methods agree on this single point that solves both equations.

What does it mean?

By knowing how to solve simultaneous equations we can quickly identify the solution to a problem without the need to graph the equations and approximate the solution from the point at which the lines cross. This could easily be extended to three equations.

P1.36 Consider the motion of the hour hand of a clock. What angle does the hour hand make with respect to the vertical 12 o'clock position when it is (a) 3:00, (b) 6:00, (c) 6:30, (d) 9:00, and (e) 11:10? Express each angle in both degrees and radians, and measure angles going clockwise from the vertical axis.

Solution

Recognize the principle

We will need to use some trigonometry, understand angular units, and employ some conversion factors.

Sketch the problem

The 3 o'clock configuration is shown below.

Identify the relationships

We can take measurements directly off the clock face. Note that at 3 o'clock the hour hand makes a right angle with the vertical 12 o'clock position, therefore we can derive a conversion factor:

$$\frac{90°}{3\,h} = 30°/h$$

Using radians instead of degrees, this conversion factor is written as

$$\frac{\pi/2}{3\,h} = \frac{\pi}{6}\,h^{-1}$$

Solve

(a) 3 o'clock → 3(30°) = $\boxed{90° \text{ or } \pi/2 = 1.57 \text{ rad}}$

(b) 6:00 → 6(30°) = $\boxed{180° \text{ or } \pi = 3.14 \text{ rad}}$

(c) 6:30 → 6(30°) + 30°/2 = $\boxed{195°}$ or $\pi + (1/2)(\pi/6)$ = $\boxed{13\pi/12 = 3.40 \text{ rad}}$. (Remember that the hour hand would be halfway between the six and the seven at 6:30.)

(d) 9:00 → 9(30°) = $\boxed{270° \text{ or } 3\pi/2 = 4.71 \text{ rad}}$

(e) 11:10 → 11(30°) + (10/60)(30°) = $\boxed{335°}$ or $335°\left(\dfrac{2\pi}{360°}\right) = (1.86)\pi = \boxed{5.85 \text{ rad}}$

What does it mean?

By first determining a conversion factor, we can easily convert from one unit of measure to another.

P1.60 A mountaineer uses a global positioning system receiver to measure his displacement from base camp to the top of Mount McKinley. The coordinates of base camp are $x = 0$, $y = 0$, and $z = 4300$ m (here z denotes altitude), and those of the top of Mount McKinley are $x = 1600$ m, $y = 4200$ m, and $z = 6200$ m. What is the magnitude of the displacement in going from base camp to the top?

Solution

Recognize the principle

We will need to add the components of vectors in three dimensions. This involves the same algebra we've used for two-dimensional problems.

Sketch the problem
Position vectors to the base camp and to Mount McKinley are shown below.

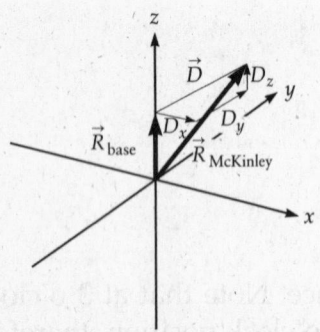

Identify the relationships
The three-dimensional displacement vector has a magnitude given by $D = \sqrt{D_x^2 + D_y^2 + D_z^2}$, where the displacement vector $\vec{D}$ is equal to the difference in the position vector to the top of the mountain, $\vec{R}_{McKinley}$, to that of the position vector to the base camp, $\vec{R}_{base}$. This difference can be written as the vector equation:

$$\vec{D} = -\vec{R}_{base} + \vec{R}_{McKinley} = \vec{R}_{McKinley} - \vec{R}_{base}$$

As in two dimensions, the components of the displacement vector from the base camp to the top of Mount McKinley can be found by subtracting each component of the Mount McKinley position vector from that of the base camp position vector. The third (z) coordinate represents the altitude change.

Solve
We subtract each dimension:

$$D_x = R_{McKinley,x} - R_{base,x} = 1600 \text{ m} - 0 \text{ m} = 1600 \text{ m}$$
$$D_y = R_{McKinley,y} - R_{base,y} = 4200 \text{ m} - 0 \text{ m} = 4200 \text{ m}$$
$$D_z = R_{McKinley,z} - R_{base,z} = 6200 \text{ m} - 4300 \text{ m} = 1900 \text{ m}$$

The magnitude of the displacement vector is then,

$$D = \sqrt{D_x^2 + D_y^2 + D_z^2} = \sqrt{(1600 \text{ m})^2 + (4200 \text{ m})^2 + (1900 \text{ m})^2} = \boxed{4900 \text{ m}}$$

What does it mean?
This means that if you ran a straight cable from the top of Mount McKinley directly to base camp, it would be 4.9 km long. Based on our sketch of the problem, this result seems reasonable.

P1.62 The solar system in perspective. The diameters of the Earth and the Sun are approximately 1.3×10^7 m and 1.4×10^9 m respectively, and the average Sun-Earth distance is 1.5×10^{11} m. Consider a scale model of the solar system, where the Earth is represented by a peppercorn 3.7 mm in diameter. (a) Calculate the diameter of the Sun in this model. Name an object of this size that could be used in the model. (b) To keep the model in scale, how far would the peppercorn Earth be from the center of the model Sun? (c) Pluto's orbit can take it as far as 5.9×10^{12} m away from the Sun. A grain of sand could represent Pluto in our model. How far away from the model Sun would this grain of sand be?

Solution

Recognize the principle

We will need to use scientific notation, keep track of the significant figures, and use some conversion factors. To make the calculations easier, we can first determine a conversion factor for our model.

Sketch the problem

None required.

Identify the relationships

We can use the ratio of the peppercorn diameter to the Earth diameter to find a conversion factor for our model. This conversion factor can then be used to bring the other diameters and orbit distances to scale.

$$\frac{3.7 \text{ mm}}{1.3 \times 10^7 \text{ m}} = 2.846 \times 10^{-7} \text{ mm/m}$$

Therefore 2.846×10^{-7} mm in our model represents 1 m in the solar system. We will keep several significant figures in for this conversion factor and then round off our final answers. We also need some real distances:

> Diameter of the Sun $\approx 1.4 \times 10^9$ m
> Typical Earth–Sun distance $\approx 1.5 \times 10^{11}$ m
> Typical Pluto–Sun distance $\approx 5.9 \times 10^{12}$ m

Solve

We can apply this conversion factor to determine distances in our model.

(a) $(1.4 \times 10^9 \text{ m})(2.846 \times 10^{-7} \text{ mm/m}) = 398 \text{ mm} = \boxed{40 \text{ cm}}$

This is approximately the size of a large beach ball.

(b) $(1.5 \times 10^{11} \text{ m})(2.846 \times 10^{-7} \text{ mm/m}) = 4.3 \times 10^4 \text{ mm} = \boxed{43 \text{ m}}$

The peppercorn Earth would orbit 43 m away from the center of the beach ball Sun.

(c) $(5.9 \times 10^{12} \text{ m})(2.846 \times 10^{-7} \text{ mm/m}) = 1.7 \times 10^6 \text{ mm} = \boxed{1.7 \text{ km} \approx 1 \text{ mi}}$

The grain of sand representing Pluto would orbit approximately 1 mile from the center of the beach ball Sun.

What does it mean?

It is a challenge to depict a scale model of the solar system where the planet diameters and orbit distances are both to scale simultaneously. Such an accurate model will not fit on a page in a book, or in a classroom!

P1.70 At room temperature, 1.0 g of water has a volume of 1.0 cm^3. (a) What is the approximate volume of one water molecule? (b) The human body is composed mainly of water. Assuming for simplicity your body is only water (and nothing else), find the approximate number of water molecules in your body.

Solution

Recognize the principle

We will need to use some basic algebra skills and some information from elementary chemistry. We will also need to estimate the volume of a typical human body.

Sketch the problem

None required.

Identify the relationships

For part (a), the molar mass of water is about 18 g/mol. An estimate for the number of water molecules in 1.0 g can be found by dividing 1.0 g by the molar mass to get the number of moles and multiplying by Avogadro's number to get the number of molecules. Then, 1.0 cm^3 divided by the number of molecules gives the volume of one molecule. For part (b), if we assume an average person is approximately 1.5 m tall, 0.5 m wide, and 0.2 m thick, we can come up with an approximate volume for the human body.

Solve

(a) The number of molecules in 1.0 cm^3 of water is

$$\frac{1.0\ \text{g}}{18\ \text{g/mol}} \times 6.022 \times 10^{23}\ \frac{\text{molecules}}{\text{mol}} = 3.3 \times 10^{22}\ \text{molecules}$$

By dividing 1.0 cm^3 by the number of molecules, the approximate volume of one water molecule can be found.

$$\boxed{\frac{1.0\ \text{cm}^3}{3.3 \times 10^{22}\ \text{molecules}} \approx 3 \times 10^{-23}\ \text{cm}^3/\text{molecule}}$$

(b) From the estimate of the size of a human body and assuming for the sake of the estimation that the human body is box-like, the volume will be the height times the width times the thickness,

$$V = (1.5\ \text{m}) \times (0.5\ \text{m}) \times (0.2\ \text{m})$$
$$V = 0.15\ \text{m}^3$$

Dividing this volume by the volume of a water molecule, an approximate value for the number of water molecules in a human body can be found.

$$\frac{V_{\text{body}}}{V_{\text{water molecule}}} = \frac{0.15\ \text{m}^3}{3 \times 10^{-23}\ \text{cm}^3} \times \frac{(100\ \text{cm})^3}{\text{m}^3} \boxed{\approx 5 \times 10^{27}\ \text{water molecules}}$$

What does it mean?

There are roughly 5 billion billion billion molecules in the human body. It's pretty amazing that we can calculate this with only a few simple assumptions!

Part D. Additional Worked Examples and Capstone Problems

The following worked example provides you with practice in working with vectors. The capstone problems are intended to strengthen your understanding of some of the key concepts in this chapter and pull together these concepts. In later chapters, capstone problems will often incorporate concepts from multiple chapters, much like problems you would encounter on a final exam. The first capstone problem provides additional practice in the use of vectors while the second problem illustrates the importance of keeping track of significant figures. Use these three problems as a test of your understanding of the chapter material.

WE 1.1 Working with Vectors

In the United States, a football field has the shape and dimensions shown in the figure. Suppose a football player starts at point A (at the end of one goal line) and runs in a straight line to point B (at the opposite end of the other goal line). (a) How far did he run? (b) At what angle θ did he run relative to the sideline?

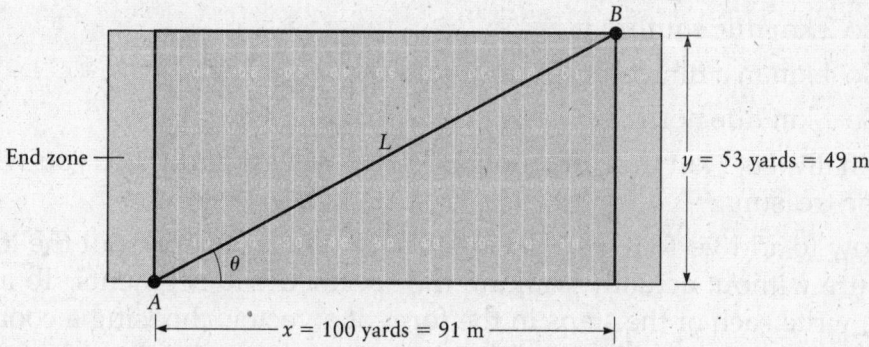

Recognize the principle

The path followed by the football player forms the hypotenuse of a right triangle. We can therefore use the Pythagorean theorem to find L and use other trigonometric relations to find θ.

Sketch the problem

The figure provided in the problem shows the path taken by the football player and the angle θ.

Identify the relationships and Solve

(a) To find L, we apply the Pythagorean theorem to the right triangle with sides L, x, and y. We get

$$L = \sqrt{x^2 + y^2}$$

Inserting the values of x and y from the figure gives

$$L = \sqrt{(91 \text{ m})^2 + (49 \text{ m})^2} = \boxed{100 \text{ m}}$$

(b) To find θ, we use the inverse tangent function:

$$\theta = \tan^{-1}\left(\frac{y}{x}\right) = \tan^{-1}\left(\frac{49 \text{ m}}{91 \text{ m}}\right)$$
$$= \boxed{28°}$$

What does it mean?

Because the values of x and y were given to two significant figures, we gave the values of L and θ to two significant figures.

CP 1.1 Treasure Hunt: Navigating with Vectors

You are given a treasure map with directions on how to navigate your way around a city to a final destination where the treasure is located. The map provides the following instructions:

Start at the corner of 1st Street and Elm Avenue.

- Go 3 km due east.
- Go 5 km in a direction 25° north of west.
- Go 2 km due south.
- Go 4 km in a direction 30° north of east.
- Go 1 km due north.
- Finally, go 3 km in a direction 15° south of east and there you will find the treasure.

Knowing how to add vectors, you decide to save time and work out the location of the treasure without actually walking the specified trip segments. To achieve this goal (a) write each of the steps in the form of a vector, choosing a coordinate system with east as the positive x axis and north as the positive y axis. (b) Next, determine the location of the treasure relative to the corner of 1st Street and Elm Avenue. Express your answer in terms of a distance and a direction.

Solution

Recognize the principle

We will need to write vectors in component form and add those vectors to find the resultant vector.

Sketch the problem

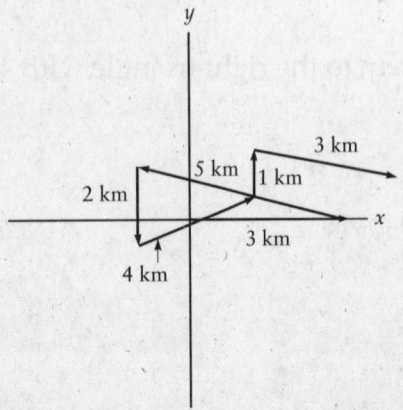

Identify the relationships

We can label each of the steps with the vectors $\vec{A}$ through $\vec{F}$ and write each of these vectors in component form.

$A_x = 3.00$ km　　　　　　　　　　$A_y = 0$
$B_x = -5.00$ km $(\cos 25°) = -4.53$ km　$B_y = 5.00$ km $(\sin 25°) = 2.11$ km
$C_x = 0$　　　　　　　　　　　　　$C_y = -2.00$ km
$D_x = 4.00$ km $(\cos 30°) = 3.46$ km　$D_y = 4.00$ km $(\sin 30°) = 2.00$ km
$E_x = 0$　　　　　　　　　　　　　$E_y = 1.00$ km
$F_x = 3.00$ km $(\cos 15°) = 2.90$ km　$F_y = -3.00$ km $(\sin 15°) = -0.78$ km

Solve

We add up the x-components of the vectors to obtain the x-component of the total vector and repeat the process for the y-components. Therefore, the resultant vector from the origin (1st Street and Elm Avenue) to where the treasure is buried is given in component form by

$$R_x = (3.00 - 4.53 + 0 + 3.46 + 0 + 2.90) \text{ km} = 4.83 \text{ km}$$
$$R_y = (0 + 2.11 - 2.00 + 2.00 + 1.00 - 0.78) \text{ km} = 2.33 \text{ km}$$

The distance is given by

$$R = |\vec{R}| = \sqrt{R_x^2 + R_y^2} = \sqrt{(4.83 \text{ km})^2 + (2.33 \text{ km})^2} = \boxed{5.36 \text{ km}}$$

and the direction is given by

$$\theta = \tan^{-1}\left(\frac{R_y}{R_x}\right) = \tan^{-1}\left(\frac{2.33 \text{ km}}{4.83 \text{ km}}\right) = \boxed{25.8° \text{ km north of east}}$$

What does it mean?

Comparing our answer with the sketch of the problem, this result seems reasonable.

CP 1.2 When Precision Counts

Suppose the king of your country has commissioned you and another scientist to independently determine whether or not his royal crown is pure gold. To determine this you are to calculate the density of the crown and compare the value with the known density of gold, $\rho_{gold} = 1.93 \times 10^4$ kg/m³. You and the other scientist independently measure the mass and volume of the crown and obtain the following data:

	Mass of the crown	Volume of the crown
Your data	2.17 kg	112.52 cm³
The other scientist's data	2.2 kg	112.5 cm³

Based on the precision of your measurements (i.e., keeping track of your significant figures), what conclusion can you and the other scientist draw about the purity of the king's crown?

Solution

Recognize the principle

You will need to keep track of the significant figures in your calculations and properly interpret the meaning of those significant figures when comparing the results with the known density of gold.

Sketch the problem

No sketch needed.

Identify the relationships

The density is given by the mass divided by the volume.

$$\rho = \frac{m}{V}$$

Also, the volume is given in cm^3 so you'll need to convert this to m^3 using the following conversion:

$$1 \text{ cm}^3 = \left(\frac{1 \text{ m}}{100 \text{ cm}}\right)^3 (1 \text{ cm}^3) = 1.0 \times 10^{-6} \text{ m}^3$$

Finally, you must keep track of the significant figures.

Solve

Your data:

$$\rho = \frac{m}{V} = \frac{2.17 \text{ kg}}{112.52(1.00 \times 10^{-6}) \text{ m}^3} = \boxed{1.93 \times 10^4 \text{ kg/m}^3}$$

The other scientist's data:

$$\rho = \frac{m}{V} = \frac{2.2 \text{ kg}}{112.5(1.0 \times 10^{-6}) \text{ m}^3} = \boxed{2.0 \times 10^4 \text{ kg/m}^3}$$

Based on your more precise measurements, you can confidently conclude that the king's crown is made of pure gold. However, the other scientist's lack of precision leads to a less confident conclusion.

What does it mean?

Precision matters and by keeping track of significant figures we can ensure that the conclusions drawn are based on the precision of the data.

Part E. MCAT Review Problems and Solutions

PROBLEMS

1. A car travels due east for a distance of 3 miles and then due north for an additional 4 miles before stopping. What is the shortest straight-line distance between the starting and ending points of the trip?

    (a) 3 mi

    (b) 4 mi

    (c) 5 mi

    (d) 7 mi

2. In Problem 1, what is the angle α of the shortest path relative to due north?

   (a) $\alpha = \cos^{-1}\left(\dfrac{3}{5}\right)$

   (b) $\alpha = \sin^{-1}\left(\dfrac{5}{3}\right)$

   (c) $\alpha = \sin^{-1}\left(\dfrac{4}{3}\right)$

   (d) $\alpha = \tan^{-1}\left(\dfrac{3}{4}\right)$

3. A vector $\vec{A}$ makes an angle of 60° with the x axis of a Cartesian coordinate system. Which of the following statements is true of the indicated magnitude?

   (a) A_x is greater than A_y.

   (b) A_y is greater than A_x.

   (c) A_y is greater than A.

   (d) A_x is greater than A.

4. Force is a vector quantity measured in units of newtons, N. What must be the angle between two forces of 5 N and 3 N acting at the same point if the resultant vector has a magnitude of 8 N?

   (a) 0°

   (b) 45°

   (c) 90°

   (d) 180°

5. Two forces, $\vec{A}$ and $\vec{B}$, both act on a point C. Both vectors have the same magnitude of 10 N and act at right angles to each other. What is the closest estimate of the magnitude of their resultant?

   (a) 0 N

   (b) 14 N

   (c) 20 N

   (d) 100 N

6. A vector drawn in an x–y coordinate system has x- and y-components which are equal in magnitude. What can we conclude about this vector?

   (a) Its magnitude is zero.

   (b) It makes a 45° angle with the x axis.

   (c) It is directed along the y axis.

   (d) None of the above.

7. For two vectors to add to zero, which of the following must be true?

   (a) The vectors are perpendicular.

   (b) The vectors point in the same direction.

   (c) The vectors are equal in magnitude but opposite in direction.

   (d) One of the vectors must have a magnitude of zero.

8. In order to add or subtract two quantities, which of the following must be true?
   (a) Both quantities must have the same magnitude.
   (b) Both quantities must have the same direction.
   (c) Both quantities must have the same dimensions.
   (d) None of the above. You can always add or subtract quantities.

9. Which of the following is the sum of two masses, 5.22 kg and 0.347 kg, stated with the correct number of significant figures?
   (a) 5.567 kg
   (b) 5.57 kg
   (c) 5.6 kg
   (d) 6 kg

10. Which of the following is 20 m divided by 0.005 s stated to the correct number of significant figures and the correct units?
    (a) 4×10^3 m/s
    (b) 4.00×10^3 m/s
    (c) 4000 s/m
    (d) None of the above. You cannot divide quantities with different dimensions.

SOLUTIONS

1. *MCAT strategies*

A check of the answers shows that they all have the correct units so none can be eliminated. Since the two legs of the trip are perpendicular and have the values of 3 mi and 4 mi, we recognize this as a 3-4-5 right triangle and conclude that the correct answer is (c) 5 mi. No further analysis is necessary.

2. *MCAT strategies*

Answers (b) and (c) are physically impossible since the argument of the inverse sine function cannot be greater than 1. Therefore, these answers can be eliminated.

Recognize the principle

We will need to use the correct trigonometric relationships for a right triangle.

Sketch the problem

Note that the problem specifies α relative to due north.

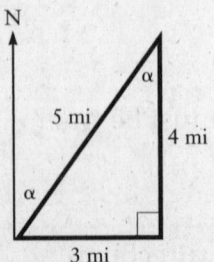

Identify the relationships

We can see from the figure that

$$\sin\alpha = \frac{3}{5}, \cos\alpha = \frac{4}{5}, \text{ and } \tan\alpha = \frac{3}{4}$$

Solve

Inverting these expressions we obtain the following:

$$\alpha = \sin^{-1}\left(\frac{3}{5}\right), \alpha = \cos^{-1}\left(\frac{4}{5}\right), \text{ and } \alpha = \tan^{-1}\left(\frac{3}{4}\right)$$

Therefore, (d) $\alpha = \tan^{-1}\left(\frac{3}{4}\right)$ is the correct answer.

What does it mean?

By understanding some basic trigonometric relationships we were able to eliminate two possible answers and quickly zero in on the correct remaining answer.

3. MCAT strategies

Choices (c) and (d) can be eliminated immediately because the projection or component of a vector can never be greater in magnitude than the vector itself.

Recognize the principle

As the angle between the vector and the axis increases toward 90°, the magnitude of the projection on the x axis (the x-component) decreases toward zero.

Sketch the problem

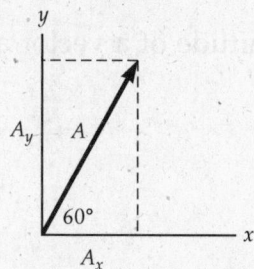

Identify the relationships

The angle between the vector and the x axis is 60°, which means the angle between the vector and the y axis must be 30°.

Solve

Therefore, the projection onto the y axis, A_y, must be greater than the projection onto the x axis, A_x. So, (b) is the correct answer.

What does it mean?

Checking our answer with the sketch confirms we have answered correctly.

4. MCAT strategies

We recognize that the only way the vector sum of a 3 N and a 5 N force can equal 8 N is if both forces act in the same direction. Therefore, the angle between them must be 0°. So, we conclude that (a) is the correct answer without any further analysis.

5. *MCAT strategies*

Answer (a) can be eliminated since the only way two vectors will sum to zero is if they are equal in magnitude and opposite in direction. Also, answer (c) can be eliminated since it requires the two vectors to be along the same direction. Choice (d) can also be eliminated since it is too large to be the hypotenuse. Therefore without out any calculation, we find that the only remaining answer, (b), must be correct. To check we can calculate the hypotenuse of the right triangle using the Pythagorean Theorem:

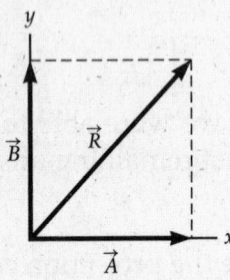

$$R = \sqrt{10^2 + 10^2}\ \text{N} = \sqrt{200}\ \text{N} = 10\sqrt{2}\ \text{N} = 10(1.41)\ \text{N} \approx \boxed{14\ \text{N}}$$

6. *MCAT strategies*

No answers can be easily eliminated so we proceed with our problem-solving strategy.

Recognize the principle

We must understand the relationship between the magnitude of a vector and its components.

Sketch the problem

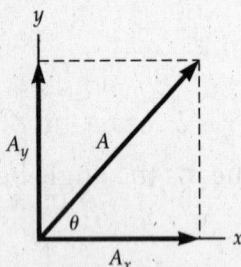

Identify the relationships

Representing the components of the vector in terms of the angle θ we find

$$A_x = A\cos\theta \text{ and } A_y = A\sin\theta$$

Solve

Setting the components equal to each other we find

$$A_x = A_y$$
$$A\cos\theta = A\sin\theta$$
$$\cos\theta = \sin\theta$$

Therefore,

$$\theta = 45°$$

So the correct answer is (b).

What does it mean?

In order for both the x- and y-components of a vector to be equal in magnitude, the projection of the vector along these axes must be the same. This occurs when the angle between the vector and the axes are the same, 45°.

7. MCAT strategies

We recognize that the only way the vector sum of two vectors can be zero is if they are equal in magnitude and opposite in direction. Therefore, we conclude that (c) is the correct answer without any further analysis.

8. MCAT strategies

Since you can add quantities with different magnitudes or different directions, answers (a) and (b) can be quickly eliminated.

Recognize the principle

This problem involves the relationships between quantities with different dimensions.

Sketch the problem

None required.

Identify the relationships and Solve

It makes no physical sense to add or subtract quantities with different dimensions. Therefore, (c) is the correct answer.

What does it mean?

As a quick check, suppose we have a length of 10 m and a time of 4 s. It makes no sense to add these quantities; however, we can add or subtract quantities with the same dimension (i.e., a length and a length) such as 10 m minus 4 m and obtain a meaningful result, 6 m.

9. MCAT strategies

No answers can be easily eliminated so we proceed with our problem-solving strategy.

Recognize the principle

We need to keep track of the significant figures in the calculations.

Sketch the problem

None required.

Identify the relationships

The location of the least significant digit in the answer is determined by the location of the least significant digit in the starting quantity that is known with the least accuracy. For this question, the less accurate of the two numbers is 5.22 kg, so the least significant digit here and in the final answer is two places to the right of the decimal point.

Solve

$$5.22 \text{ kg} + 0.347 \text{ kg} = 5.567 \text{ kg}$$

Rounding off to the second decimal place we get 5.57 kg. Therefore, (b) is the correct answer.

What does it mean?

Keeping track of significant figures is easy as long as we remember to round our answer to the number of significant figures present in the least accurate starting quantity.

10. *MCAT strategies*

Answer (c) can be eliminated since it does not have the correct units of m/s. Answer (d) can be eliminated since it is clearly possible to divide quantities with different units as in the case of speed which is a distance divided by a time.

Recognize the principle

We need to keep track of the significant figures in the calculations.

Sketch the problem

None required.

Identify the relationships

The number of significant digits in the answer is determined by the number of significant digits in the starting quantity that is known with the least accuracy. For this question, both quantities have one significant digit.

Solve

$$\frac{20 \text{ m}}{0.005 \text{ s}} = 4000 \text{ m/s}$$

Writing this in scientific notation and keeping the correct number of significant figures we get, 4×10^3 m/s. Therefore, (a) is the correct answer.

What does it mean?

Keeping track of significant figures is easy as long as we remember to round our answer to the number of significant figures present in the least accurate starting quantity.

2 Motion, Forces, and Newton's Laws

CONTENTS

Part A. Summary of Key Concepts and Problem-Solving Strategies

KEY CONCEPTS

Motion

A complete description of the motion of an object involves its *position*, *velocity*, and *acceleration*. For motion in one dimension (along a line), position can be specified by a single quantity, x. The change in position is given by Δx and a period of time given by Δt. The *instantaneous velocity* is related to changes in position by

$$v = \lim_{\Delta t \to 0} \frac{\Delta x}{\Delta t}$$

and is referred to as simply the "velocity." If we are interested in determining the average velocity of an object over a period of time, we can use the expression

$$v_{\text{ave}} = \frac{\Delta x}{\Delta t}$$

Similarly, the *instantaneous acceleration* is given by

$$a = \lim_{\Delta t \to 0} \frac{\Delta v}{\Delta t}$$

and is usually referred to simply as the "acceleration." The average acceleration is given by

$$a_{\text{ave}} = \frac{\Delta v}{\Delta t}$$

Position, velocity, and acceleration are vector quantities. For motion in two or three dimensions we must consider how the directions of these quantities relate to our chosen coordinate axes. We'll explore this further in Chapter 4.

Newton's laws of motion

The connection between *force* and motion is at the heart of *mechanics*. A force is a push or a pull. Force is a vector. Given the forces acting on an object, we can calculate how it will move using Newton's three laws of motion.

- *Newton's first law:* If the total force acting on an object is zero, the object will move with a constant velocity. Zero total force acting on an object does not necessarily imply that the object is at rest, only that it is moving at a constant speed and direction. Zero velocity is one example of a constant velocity.

- *Newton's second law:* If the total force acting on an object is not zero, then the object will accelerate. In other words, forces cause acceleration. The expression describing this relationship is $\vec{a} = \frac{\Sigma \vec{F}}{m}$, where m is the mass of the object. Note that the acceleration is in the same direction as the total force and the total force is the vector sum of all the forces acting on the object.

- *Newton's third law:* For every action (force) there is a reaction (force) of equal magnitude and opposite direction. All forces come in *action-reaction pairs.* The two forces in an action-reaction pair act on different objects. Only one of a pair of forces is considered when studying a particular object's motion.

PROBLEM-SOLVING STRATEGIES

In this chapter the end-of-chapter questions and problems focus on two main areas. The first is understanding the graphical relationship between the position, velocity, and acceleration of an object. The second involves solving for the position, velocity, or acceleration of an object using the key relationships defined above. The following two schematics will help remind you of the thought process you should use while tackling problems from this chapter. A third focus of the end-of-chapter questions and problems is in understanding and applying Newton's laws of motion. Since these will be dealt with in more detail in the next two chapters we will not highlight the strategies until then.

Problem Solving: Graphically Relating Position, Velocity, and Acceleration

Recognize the principle

To graphically represent the motion of an object, we simply need to understand the definitions of position, velocity, and acceleration and how they are related.

Sketch the problem

If the graph of the motion is given in the problem then we can proceed to the next step. If the graph of the motion is what we are trying to solve for then we can also proceed to the next step in order to help us determine its shape.

Identify the relationships

- The *instantaneous velocity* at a particular time *t* is the slope of the position-time curve at that time.
- The *average velocity* during a particular time interval is equal to the slope of the line connecting the start and end of that interval on the position-time curve.
- The *instantaneous acceleration* at a particular time *t* is the slope of the velocity-time curve at that time.
- The *average acceleration* during a particular time interval is equal to the slope of the line connecting the start and end of that interval on the velocity-time curve.

Solve

Using these relationships we can easily determine the shapes of the position, velocity, and acceleration versus time graphs for objects undergoing many types of motion. The following examples will help illustrate:

- Zero velocity means a horizontal line on the position-time graph.
- Zero acceleration means a horizontal line on the velocity-time graph (constant velocity) and a constant-slope line on the position-time graph.
- Constant acceleration means a constant-slope line on the velocity-time graph and a curved (quadratic) line on the position-time graph.

For other example see the section on Frequently Asked Questions.

What does it mean?

Often objects move in such a way that we cannot easily mathematically describe their motion mathematically; however, a sketch can help us visualize the situation in a qualitative way.

Problem Solving: Calculating Position, Velocity, and Acceleration

Recognize the principle

When we are asked for quantitative solutions to motion problems, we are usually dealing with the motion over a period of time. Therefore, we can use the expressions for the average velocity and average acceleration.

Sketch the problem

Although a sketch is often not necessary it may be useful to either draw the situation (i.e., a car rolling down a hill), or sketch the position, velocity, and acceleration versus time graphs to help us understand the problem.

Identify the relationship

- Velocity: $v = \lim\limits_{\Delta t \to 0} \dfrac{\Delta x}{\Delta t}$. This is the slope of the x-t graph.
- Average velocity: $v_{\text{ave}} = \dfrac{\Delta x}{\Delta t} \to \Delta x = v_{\text{ave}}\Delta t$
- Acceleration: $a = \lim\limits_{\Delta t \to 0} \dfrac{\Delta v}{\Delta t}$. This is the slope of the v-t graph.
- Average acceleration: $a_{\text{ave}} = \dfrac{\Delta v}{\Delta t} \to \Delta v = a_{\text{ave}}\Delta t$

Solve

Using these relationships we can determine the value of an unknown given the other parameters in an expression. For example, suppose we are asked to find the velocity of an object after 10 s which starts from rest and accelerates at 10 m/s^2. We can calculate the change in the velocity using the expression:

$$\Delta v = a_{\text{ave}}\,\Delta t = (10 \text{ m/s}^2)(10 \text{ s}) = 100 \text{ m/s}$$

Since the initial velocity is zero, then the velocity of the object after 10 s is 100 m/s.

What does it mean?

By keeping track of our units we can be sure that the answer agrees with what is being asked. In our example, a check of the answer ensures that the velocity has the correct units of m/s.

Part B. Frequently Asked Questions

1. *Doesn't negative speed imply the object is moving backwards?*

No. Speed is the magnitude of the velocity of an object and therefore cannot be negative. For one-dimensional motion if an object is moving in the $-x$ direction at 10 m/s we would say it has a velocity of -10 m/s. Alternatively we could say that it has a speed of 10 m/s and is traveling in the $-x$ direction.

2. *Can the instantaneous speed of an object at a point in time ever equal the object's average speed over some time interval which contains that point in time?*

Yes. During any interval of time there is at least one point at which the instantaneous speed is equal to the average speed over that time interval. There is a mathematical proof for this statement but it also just makes sense. Suppose you took 1 h to travel a distance of 60 miles. Your average speed during this time interval would be 60 mi/h. In order to do this you would need to spend all of your time going exactly 60 mi/h or some time going faster than and some time slower than 60 mi/h. Either way there is at least one instant when your speed is exactly 60 mi/h.

3. *Given a position-time graph how do I sketch the velocity-time graph?*

The velocity at an instant in time is equal to the slope of the position-time graph at that point in time. Therefore, if the slope of the position-time graph is positive, then the velocity is positive. If the slope is negative, then the velocity is negative. The greater the slope (positive or negative) of a position-time graph then the greater the magnitude of the velocity. Consequently, a straight line on a position-time graph corresponds to a constant velocity. The following figure will help illustrate the relationship between a position-time graph and its corresponding velocity-time graph.

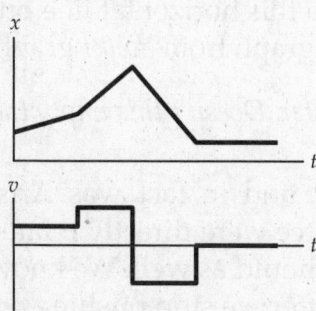

4. *Given a velocity-time graph how do I sketch the position-time graph?*

It's much easier to create a velocity-time (v-t) graph from a position-time (x-t) graph as in the previous FAQ, so if you're given a v-t graph and asked for an x-t graph it helps to think of the reverse process using the logic of the previous FAQ. For example, if you have a straight horizontal line on a v-t graph, you ask yourself what the x-t graph should look like in order to create this horizontal line on a v-t graph. Proceeding in this way, it's easy to derive an x-t graph from a v-t graph.

5.   *Given a velocity-time graph how do I sketch the acceleration-time graph?*

The acceleration at an instant in time is equal to the slope of the velocity-time graph at that point in time. Therefore, if the slope of the velocity-time graph is positive, then the acceleration is positive. If the slope is negative, then the acceleration is negative. The greater the slope (positive or negative) of a velocity-time graph then the greater the magnitude of the acceleration. Consequently, a straight line on a velocity-time graph corresponds to a constant acceleration. Do not let negative velocities confuse your reasoning. It's possible to have either a positive or negative acceleration for a negative velocity since it's the slope of the velocity graph which provides the sign for the acceleration. The following figure will help illustrate the relationship between a velocity-time graph and its corresponding acceleration-time graph.

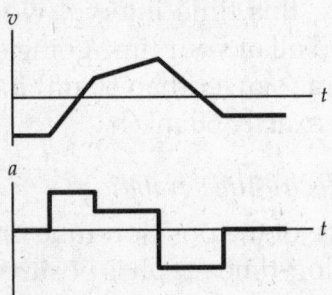

6.   *Given an acceleration-time graph how do I sketch the velocity-time graph?*

Once again it's much easier to create an acceleration-time (*a-t*) graph from a velocity-time (*v-t*) graph as in the previous FAQ, so if you're given an *a-t* graph and asked for a *v-t* graph it helps to think of the reverse process using the logic of the previous FAQ. For example, if you have a straight horizontal line on the *a-t* graph, you ask yourself what the *v-t* graph should look like in order to create this horizontal line on an *a-t* graph. Proceeding in this way, it's easy to derive a *v-t* graph from an *a-t* graph.

7.   *As I push on an object, its velocity gets faster and faster. Doesn't this imply that force and velocity are directly related?*

Although this seems like a reasonable conclusion, and in fact was Aristotle's description of motion, it is wrong. If velocity and force were directly related then as the force on an object becomes zero, the velocity should as well. We know from experience that objects will continue to move even after we stop pushing on them so force and velocity are not proportional. Newton's second law describes the correct relationship between force and motion by directly relating force and acceleration. If a constant force is applied to an object, then it will accelerate at a constant rate resulting in the velocity of the object getting faster and faster as was stated in your question.

8.   *How do I tell if two forces are an action-reaction pair?*

The easiest way to know if two forces form an action-reaction pair is by switching the nouns in the description of a force in order to form its pair. For example, if the action force is a "<u>person</u> pushing on the <u>wall</u>" then the reaction force is the "<u>wall</u> pushing on the <u>person</u>." A second example is illustrated in figure below in which a crate rests on a horizontal surface. In this situation there are actually four forces of interest, two of which act on the crate: the force of gravity on the <u>crate</u> due to the <u>Earth</u> and the contact force of the <u>surface</u> on the <u>crate</u> (also call a normal force). Using the noun-switching technique we find the reaction pairs for these forces: the force of gravity on the <u>Earth</u> due to the <u>crate</u> and the contact force of the <u>crate</u> on the <u>surface</u>. The force of gravity on the Earth due to the crate may seem strange but it does exist and will be discussed further in Chapter 5.

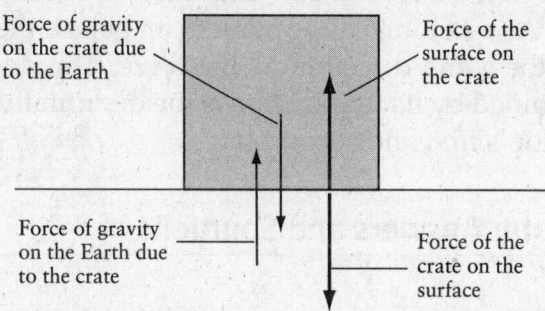

9. *Newton's third law states that for every force there is an equal and opposite force. Since these forces will cancel each other, how does anything ever move?*

This is a good question and gets at the heart of how we interpret and apply Newton's third law. Although there are always two forces during any interaction between two objects (or an object and a surface) and Newton's third law says they are equal in magnitude and opposite in direction, in practice we only concern ourselves with the force (either the action or reaction) which is acting upon the object of interest. For example suppose we are interested in analyzing the motion of a person who is standing on the floor while leaning against a wall. The first of the following figures shows the many forces involved in the scenario. Notice that the forces are paired according to Newton's third law. If we are interested in the person's motion (or lack of motion) we consider only those forces which act upon the person. The second figure is the free-body diagram for the person. Similarly, if we are interested in the wall we would consider only those forces acting upon the wall.

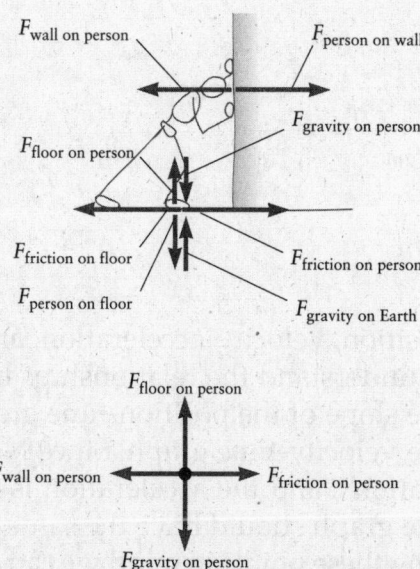

10. *How are Newton's first and second laws related?*

It is a common misconception to think that Newton's first law is simply a special case of Newton's second law. Actually the first law provides us with a deep understanding of the natural state of motion. Unlike Aristotle's notion that objects naturally seek a state of rest, Newton's first law states that objects will maintain their state of motion or rest indefinitely unless acted upon by a force. This resistance to changing motion is called inertia. Newton's second law provides us with the relationship between force and motion. As a net force is applied to an object, the object accelerates proportional to and in the same direction as the force. The degree to which an object accelerates is determined by its mass. Mass is the quantitative measure of inertia thereby linking Newton's first and second laws.

Part C. Selection of End-of-Chapter Answers and Solutions

QUESTIONS

Q2.9   Abracadabra! A magician pulls a tablecloth off of a set table with one swift, graceful motion. Amazingly, the fine china, glassware, and silverware are practically undisturbed. Although amazing, this feat is not an illusion. Describe the behavior of the plates and glasses in terms of the principle of inertia (Newton's first law).

Answer

The place-settings are initially at rest. The principle of inertia (Newton's first law) states that a body will move with a constant velocity unless acted upon by a force. Since there is very little friction between the tablecloth and the place-settings and the tablecloth is pulled out from under the place-settings quickly, the force and thus the acceleration on the place-settings is very small. Therefore, the velocity of the place-settings barely changes, and they remain at rest.

Q2.15   Figure Q2.15 shows a motion diagram for a rocket-powered car. The photos are taken at 1.0-s intervals. Make qualitative plots of the position, velocity, acceleration, and force on the car as functions of time.

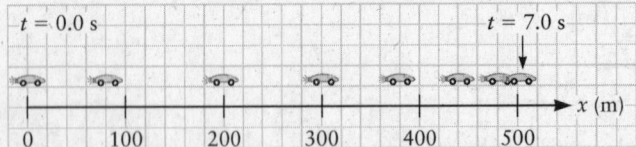

Figure Q2.15

Answer

In order to make qualitative plots of the position, velocity, acceleration, and force on the car as functions of time, we need to understand the relationships between these variables. The velocity of the car is the slope of the position-time graph and the acceleration of the car is the slope of the velocity-time graph. Since Newton's second law states that forces cause accelerations, and the acceleration is directly proportional to the force, then the force-time graph should have the same general shape as the acceleration-time graph. Keeping these points in mind we can analyze the motion of the car. Let's assume that the car starts from rest. It appears from Figure Q2.15 that the car increases its speed at a uniform rate until it reaches the 200-m point. At this point the car begins to slow down at a uniform rate for the remainder of the trip. So, from 0 to 2 s the car has a constant positive acceleration and from 2 to 7 s the car has a constant negative acceleration. Therefore, the graphs for this type of motion would look like the following:

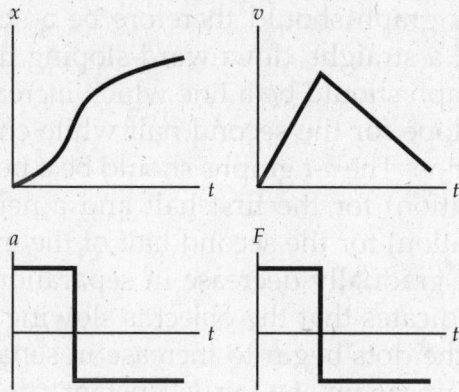

PROBLEMS

P2.13   Figure P2.13 shows three motion diagrams, where the dots indicate the positions of an object after equal time intervals. Assume left-to-right motion. For each motion diagram, sketch the appropriate position-time, velocity-time, and acceleration-time graphs.

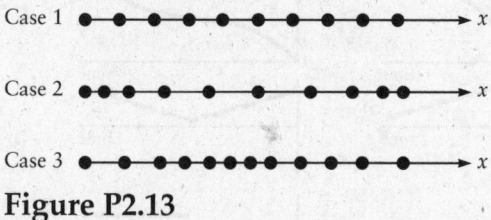

Figure P2.13

Solution

Recognize the principle

Since the dots represent the location of an object between equal time intervals, then as the object's speed increases the spacing between the dots also increases. Similarly, as the object slows down the spacing between the dots decreases. With this in mind we can proceed with our analysis of the problem.

Sketch the problem

Figure P2.13 provides the diagrams to analyze and we will generate sketches of the motion as part of the solution to the problem. So we proceed with the next step in our problem-solving strategy.

Identify the relationships

In Case 1 the dots are a constant distance apart during the motion of the object. This indicates that the velocity of the object is constant, therefore a horizontal line on the *v-t* diagram. The corresponding *x-t* graph should be a straight line with a positive slope and the acceleration of the object is zero throughout. In Case 2 the dots start out close together and gradually increase in separation until the middle of the object's motion. This indicates that the object is speeding up at a fairly constant rate. After this point the dots begin getting closer together, indicating that the object begins slowing down, again at a fairly constant rate. So the velocity of the object increases at a constant rate for the first half of its motion and decreases at a constant rate for the second half of its motion. The *v-t* graph should therefore be a straight, upward-sloping line for the first half and a straight, downward-sloping line for the second half. The corresponding *x-t* graph should be a line which increases in slope for the first half and decreases in slope for the second half while continuing to increase in value over the entire motion. The *a-t* graphs should be a positive, horizontal line (constant positive acceleration) for the first half and a negative, horizontal line (constant negative acceleration) for the second half of the motion. In Case 3 the dots start out far apart and gradually decrease in separation until the middle of the object's motion. This indicates that the object is slowing down at a fairly constant rate. After this point the dots begin to increase in separation indicating that the object begins speeding up, again at a fairly constant rate. Since this is the reverse process to that described by Case 2, the graphs should simply be the inverse as shown in following figures.

Solve

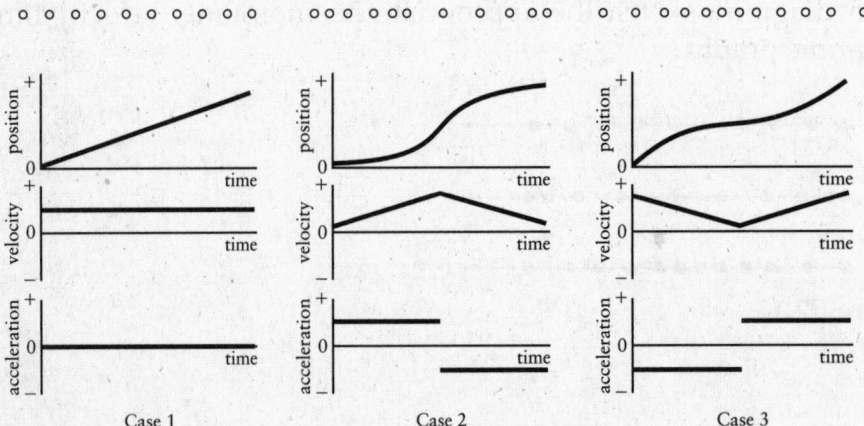

Case 1   Case 2   Case 3

What does it mean?

It's rather impressive to think that from a series of dots representing the positions of an object, we're able to determine how the object's position, velocity, and acceleration vary as functions of time!

P2.33   A squirrel falls from a very tall tree. Initially (at $t = 0$), the squirrel is at the top of the tree, a distance $y = 50$ m above the ground. At $t = 1.0$ s, the squirrel is at $y = 45$ m, and at $t = 2.0$ s, it is at $y = 30$ m. Estimate the average velocity of the squirrel during the intervals from $t = 0.0$ to $t = 1.0$ s and from $t = 1.0$ s to $t = 2.0$ s. Use these results to estimate the average acceleration of the squirrel during this time. (No squirrels were harmed during the writing of this problem.)

Solution

Recognize the principle

The average velocity during a particular time interval is the slope of the position-time graph during that interval. This means that $v_{ave} = \Delta x/\Delta t$. Similarly, the average acceleration during a particular time interval is the slope of the velocity-time graph during that interval. Therefore $a_{ave} = \Delta v/\Delta t$. Since the positions at various points in time are given, we can easily calculate the average velocity and average acceleration for the two time intervals.

Sketch the problem

Our first step is to draw a picture that contains all the relevant information; this is essential for organizing our thoughts and seeing connections. Defining up as the positive x-direction with the $x = 0$ at ground level, we have the following picture:

Identify the relationships

The average velocity is the squirrel's displacement during a particular time interval divided by the length of time of the interval. For each interval,

$$v_{ave} = \frac{\Delta x}{\Delta t} = \frac{x_{final} - x_{initial}}{t_{final} - t_{initial}}$$

Similarly, the average acceleration is the squirrel's change in velocity during a particular interval divided by the length of time of the interval. For each interval,

$$a_{ave} = \frac{\Delta v}{\Delta t} = \frac{v_{final} - v_{initial}}{t_{final} - t_{initial}}$$

Solve

For the interval from $t = 0$ to $t = 1.0$ s,

$$v_{ave} = \frac{\Delta x}{\Delta t} = \frac{(45 \text{ m}) - (50 \text{ m})}{1.0 \text{ s}} = \boxed{-5.0 \text{ m/s}}$$

For the interval from $t = 1.0$ s to $t = 2.0$ s,

$$v_{ave} = \frac{\Delta x}{\Delta t} = \frac{(30 \text{ m}) - (45 \text{ m})}{1.0 \text{ s}} = \boxed{-15 \text{ m/s}}$$

Using the velocities we've just calculated, we can find the average acceleration,

$$a_{ave} = \frac{\Delta v}{\Delta t} = \frac{(-15 \text{ m/s}) - (-5.0 \text{ m/s})}{1.0 \text{ s}} = \boxed{-10 \text{ m/s}^2}$$

What does it mean?

Since the distance between the positions of the squirrel in consecutive "snapshots" increases continuously, the squirrel's velocity continues to increase. In fact, it appears that the squirrel's velocity is increasing at a constant rate indicating that its acceleration is constant. As we shall see in the next chapter, this is an example of free fall.

P2.38 Figure 2.24 shows one of Galileo's experiments in which a ball rolls up an incline. A ball that is initially rolling up the incline will roll up to some maximum height and then roll back down the incline. Draw qualitative plots of the position and velocity as functions of time for the ball. Take $x = 0$ at the bottom of the ramp.

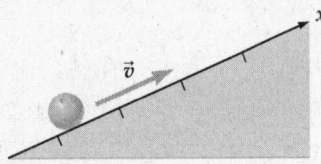

Figure 2.24

Solution

Recognize the principle

As the ball rolls up the incline it will slow down at a constant rate until it momentarily comes to rest. The ball will then begin rolling back down the incline speeding up at a constant rate.

Sketch the problem

The sketch of the problem is provided in Figure 2.24.

Identify the relationships

If we define the positive x-direction to be up the incline then the ball will begin its motion with a positive velocity, reducing to zero at a constant rate. The v-t graph for this portion of the motion should be a straight line sloping downward to zero. After the ball momentarily comes to rest, it will begin moving back down the incline with a negative velocity which increases linearly with time. This portion of the v-t graph should be in the negative portion of the graph and again be a straight line sloping downward. Since the slope of the x-t graph is the velocity of the ball, the x-t graph should form a downward curving arc, starting and ending at $x = 0$. In fact, the shape of the curve should be a parabola.

Solve

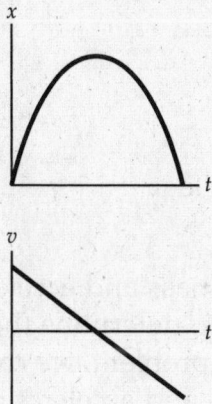

What does it mean?

Since the slope of the v-t graph remains constant throughout the entire process, the acceleration of the ball must remain constant as well. Newton's second law tells us that the force on the ball must also be constant for this situation. As we will see in the next chapter, the force of gravity on the ball is what is responsible for this observed motion and is indeed a constant in this situation.

P2.49   A cannon is fired horizontally from a platform (Fig. P2.49). The platform rests on a flat, icy, frictionless surface. Just after the shell is fired and while it is moving through the barrel of the gun, the shell (mass 3.2 kg) has an acceleration of +2500 m/s². At the same time, the cannon has an acceleration −0.76 m/s². What is the mass of the cannon?

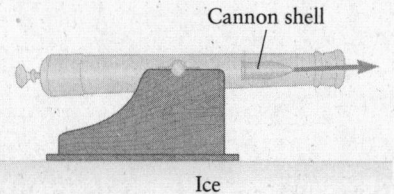

Cannon shell

Ice

Figure P2.49

Solution

Recognize the principle

This is a situation where we need to consider both Newton's second and third laws. Newton's third law tells us that the force of the cannon on the shell is equal in magnitude and opposite in direction to the force of the shell on the cannon. If we can determine the magnitude of this force, then Newton's second law will enable us to determine the unknown mass of the cannon.

Sketch the problem

The following figure illustrates the two forces which form an action-reaction pair.

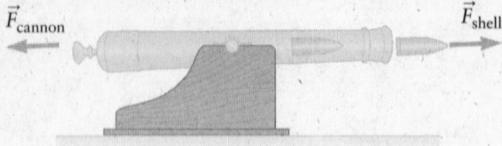

Identify the relationships

Let's begin by studying the motion of the shell. Since the mass and acceleration of the shell are given, we can apply Newton's second law to determine the force on the shell by the cannon. Since this is a one-dimensional problem, we can deal with the magnitudes rather than the vector nature of forces and accelerations, so the mathematics simplifies to the following:

$$a_{shell} = \frac{F_{shell}}{m_{shell}} \rightarrow F_{shell} = m_{shell} \, a_{shell}$$

We can also apply Newton's second law to the cannon. Since the cannon is on a flat, icy, frictionless surface, the only force we need to concern ourselves with is the force due to the interaction with the shell.

$$a_{cannon} = \frac{F_{cannon}}{m_{cannon}} \rightarrow F_{cannon} = m_{cannon} \, a_{cannon}$$

Now, Newton's third law states the force on the cannon and the force on the shell must be equal in magnitude, $F_{shell} = F_{cannon}$, so by equating the two expressions we find,

$$F_{cannon} = F_{shell} \rightarrow m_{cannon} \, a_{cannon} = m_{shell} \, a_{shell}$$

Rearranging to solve for m_{cannon} we find,

$$m_{cannon} = \frac{m_{shell} \, a_{shell}}{a_{cannon}}$$

Solve

Inserting the numerical values which were given and being sure to include only the magnitudes of the accelerations not the directions, we find

$$m_{cannon} = \frac{(3.2 \text{ kg})(2500 \text{ m/s}^2)}{(0.76 \text{ m/s}^2)} = \boxed{11{,}000 \text{ kg}}$$

What does it mean?

Although the force on the cannon and shell are equal in magnitude, the acceleration of the cannon is much smaller than the shell since it has a much greater mass. Consider when you jump up into the air by pushing off of the Earth beneath your feet. During the interaction the force you exert on the Earth is equal in magnitude and opposite in direction to the force the Earth exerts on you. However, the effect this force has on you and the Earth is quite different because of the difference in mass. You accelerate upward while the Earth essentially remains at rest!

P2.52   Throwing heat. In professional baseball, pitchers can throw a fastball at a speed of 90 mi/h. (a) Given that the regulation distance from the pitcher's mound to home plate is 60.5 ft, how long does it take the ball to reach home plate after the ball leaves the pitcher's hand? (b) It takes (on average) 0.20 s for the batter to get the tip of his bat over home plate. How much time does that give him to react? (c) The average fast pitch in professional women's softball is about 60 mi/h, where the regulation distance from the pitchers mound to home plate is 40.0 ft. How do the travel time of a pitched ball and reaction time of a batter in softball compare to those in baseball? For simplicity, assume the pitcher releases the ball just above the center of the pitcher's mound. Keep three significant figures in your calculations.

Solution

Recognize the principle

The main principle we will need to apply in this problem is the concept of average velocity. Also, we'll need to be careful with our units since the problem provides values in mi/h, ft, and s.

Sketch the problem

No sketch is needed.

Identify the relationships

We can begin by rearranging the expression for the average velocity of a moving object,

$$v_{ave} = \frac{\Delta x}{\Delta t} \rightarrow \Delta t = \frac{\Delta x}{v_{ave}}$$

Since the balls are only traveling in one direction (from the pitcher's mound to home plate) the average velocity is the average speed. So, using this expression we can calculate the time it takes the ball to get from the pitcher's mound to home plate. We will also need to convert the average speeds from mi/h to ft/s. Since there are 5280 ft/mi and 3600 s/h, this is an easy conversion.

Solve

(a) The average speed of the baseball in ft/s is given by

$$v_{ave, baseball} = 90 \text{ mi/h} \left(\frac{5280 \text{ ft}}{1 \text{ mi}}\right)\left(\frac{1 \text{ h}}{3600 \text{ s}}\right) = 132 \text{ ft/s}$$

Therefore, the time it takes the ball to travel from the pitcher's mound to home plate is,

$$\Delta t_{baseball} = \frac{\Delta x_{baseball}}{v_{ave, baseball}} = \frac{60.5 \text{ ft}}{132 \text{ ft/s}} = \boxed{0.458 \text{ s}}$$

(b) Since it takes 0.20 s for the batter to get the tip of his bat over home plate, we can determine his reaction time by subtracting 0.20 s from the time it takes the ball to reach home plate. Reaction time in baseball = 0.458 s − 0.20 s = $\boxed{0.258 \text{ s}}$ or about a quarter of a second!

(c) The average speed of the softball in ft/s is given by

$$v_{ave, baseball} = 60.0 \text{ mi/h} \left(\frac{5280 \text{ ft}}{1 \text{ mi}} \right) \left(\frac{1 \text{ h}}{3600 \text{ s}} \right) = 88.0 \text{ ft/s}$$

Therefore, the time it takes the ball to travel from the pitcher's mound to home plate is,

$$\Delta t_{softball} = \frac{\Delta x_{softball}}{v_{ave, softball}} = \frac{40.0 \text{ ft}}{88.0 \text{ ft/s}} = \boxed{0.455 \text{ s}}$$

So, the reaction time for a batter in softball is 0.455 s − 0.20 s = $\boxed{0.255 \text{ s}}$. Again, this is about a quarter of a second.

What does it mean?

Although a softball pitch has only two-thirds the velocity of a baseball pitch, since the distance from the pitcher's mound to home plate in softball is about two-thirds that in baseball, each sport allows the batter about the same time to react.

P2.54   On your vacation, you fly from Atlanta to San Francisco (a total distance of 3400 km) in 4.0 h. (a) Draw a qualitative sketch of how the speed of your airplane varies with time. (b) What is the average speed during your trip? (c) Estimate the top speed during your trip. *Hint*: You reach your top speed about 10 min after taking off. (d) What is your average acceleration during the first 10 min of your trip? (e) What is the average acceleration during the central hour of your trip?

Solution

Recognize the principle

We will need to think about how the speed of a plane varies over the course of a trip and be able to graphically represent that speed as a function of time. Also, we'll need to apply the expression for the average velocity (or speed) of an object.

Sketch the problem

(a) During the first 10 min of the flight the plane accelerates up to its cruising speed which it maintains for the duration of the flight until the last 10 min of the flight, when it reduces its speed, lands, and stops at its final destination. Therefore, the speed would vary with time according to the following graph:

Identify the relationships

We will need to use the expressions for the average speed and average acceleration of an object:

$$v_{ave} = \frac{\Delta x}{\Delta t}, \text{ and } a_{ave} = \frac{\Delta v}{\Delta t}$$

Solve

(b) To calculate the average speed for the entire trip we use the total distance traveled and the total time for the trip and express our answer in m/s.

$$v_{\text{ave, trip}} = \frac{\Delta x_{\text{trip}}}{\Delta t_{\text{trip}}} = \frac{3400 \text{ km}}{4.0 \text{ h}} \left(\frac{1000 \text{ m}}{1 \text{ km}}\right)\left(\frac{1 \text{ h}}{3600 \text{ s}}\right) = \boxed{240 \text{ m/s}}$$

(c) If we assume that the acceleration during the first 10 min and the last 10 min is relatively constant, then the average speed during those times will be half the top speed. Therefore, the plane travels for 3 h and 40 min at its top speed and for 20 min at half its top speed. Using this information and the total distance for the trip we can derive the following expression:

$$v_{\text{top speed}} \times (3.0 \text{ h } 40 \text{ min}) + \frac{v_{\text{top speed}}}{2} \times (20 \text{ min}) = \text{total distance}$$

Converting the times into seconds and inserting the value for the distance,

$$v_{\text{top speed}} \times (13{,}200 \text{ s}) + \frac{v_{\text{top speed}}}{2} \times (1200 \text{ s}) = 3.4 \times 10^6 \text{ m}$$

Solving for the top speed we find,

$$v_{\text{top speed}} = \boxed{250 \text{ m/s}}$$

(d) Using the expression for the average acceleration, we can calculate its value during the first 10 min of the trip.

$$a_{\text{ave}} = \frac{\Delta v}{\Delta t} = \frac{v_{\text{final}} - v_{\text{initial}}}{t_{\text{final}} - t_{\text{initial}}} = \frac{v_{\text{top speed}} - 0}{10 \text{ min}} = \frac{250 \text{ m/s}}{600 \text{ s}} = \boxed{0.42 \text{ m/s}^2}$$

(e) During the central hour of the trip, the top speed has been reached and is maintained at a constant value. Since the speed is constant (and assuming that the direction of the airplane remains constant), the acceleration during this time would be $\boxed{\text{zero}}$.

What does it mean?

It's interested to note that the plane's average speed for the entire trip and its top speed only differ by 10 m/s. This is because for most of the trip the plane is flying at its top speed. This would not be the case for a short trip where the plane only travels for a few minutes at its top speed before beginning its descent into its final destination.

Part D. Additional Worked Examples and Capstone Problems

The following worked example provides you with practice in calculating the average velocity of a moving object. The capstone problems provide additional practice in representing motion on a graph and interpreting graphs. Although these problems do not incorporate all the material discussed in this chapter, they do pull together several of the key concepts. If you can successfully solve these problems then you should feel confident in your understanding of these key concepts, so use these problems as a test of your understanding of the chapter material.

WE 2.1 Understanding the Motion of an Accelerating Car

Consider the following data recorded for an accelerating car:

Time	Position
$t = 0$	$x = 0$
$t = 1.0$ s	$x = 5$ m
$t = 2.0$ s	$x = 20$ m
$t = 3.0$ s	$x = 45$ m
$t = 4.0$ s	$x = 80$ m
$t = 5.0$ s	$x = 125$ m
$t = 6.0$ s	$x = 180$ m

Find the average velocity of the car over the time interval (a) $t = 0$ s to $t = 3.0$ s, (b) $t = 3.0$ s to $t = 6.0$ s, and (c) $t = 0$ s to $t = 6.0$ s. (d) What can you conclude about the car's motion?

Solution

Recognize the principle

The average velocity during a particular time interval is the slope of the position-time graph during that time. Mathematically, this is given by the expression $v_{ave} = \Delta x / \Delta t$. We simply need to calculate the displacement for the various time intervals from the data provided.

Sketch the problem

Since the data is provided in the problem, a sketch is not required.

Identify the relationships

The average velocity is the car's displacement during a particular time interval divided by the length of time of the interval. Therefore, for each interval we simply apply the expression,

$$v_{ave} = \frac{\Delta x}{\Delta t} = \frac{x_{final} - x_{initial}}{t_{final} - t_{initial}}$$

Solve

(a) For this interval $x_{final} = 45$ m, $x_{initial} = 0$ m, $t_{final} = 3.0$ s, and $t_{initial} = 0$ s. Substituting into the expression we find,

$$v_{ave} = \frac{x_{final} - x_{initial}}{t_{final} - t_{initial}} = \frac{45 \text{ m} - 0 \text{ m}}{3.0 \text{ s} - 0 \text{ s}} = \boxed{15 \text{ m/s}}$$

(b) For this interval $x_{final} = 180$ m, $x_{initial} = 45$ m, $t_{final} = 6.0$ s, and $t_{initial} = 3.0$ s. Substituting into the expression we find,

$$v_{ave} = \frac{x_{final} - x_{initial}}{t_{final} - t_{initial}} = \frac{180 \text{ m} - 45 \text{ m}}{6.0 \text{ s} - 3.0 \text{ s}} = \boxed{45 \text{ m/s}}$$

(c) For this interval $x_{final} = 180$ m, $x_{initial} = 0$ m, $t_{final} = 6.0$ s, and $t_{initial} = 0$ s. Substituting into the expression we find,

$$v_{ave} = \frac{x_{final} - x_{initial}}{t_{final} - t_{initial}} = \frac{180 \text{ m} - 0 \text{ m}}{6.0 \text{ s} - 0 \text{ s}} = \boxed{30 \text{ m/s}}$$

(d) Since the velocity of the car is increasing, it must be accelerating.

What does it mean?

We could continue the analysis in a similar way and determine the average acceleration of the car for these time intervals. Try it! You should find that the acceleration remains constant. For each scenario choose the $+x$ axis to be in the direction that the object is moving.

CP 2.1 Representing Motion on a Graph

For the following scenarios qualitatively sketch the position, velocity, and acceleration versus time graphs.

(a) A massive ball rolling along a very smooth horizontal surface.

(b) A massive ball rolling down a smooth inclined surface.

(c) An airplane as it accelerates down a runway.

(d) A car skidding to a stop.

(e) An object under the influence of a force which increases proportionally with time.

Solution

Recognize the principle

We need to understand how to graphically represent position, velocity, and acceleration and how they relate to each other.

Sketch the problem

We will do this as we solve the problem.

Identify the relationships

The key relationships to keep in mind are:

- The slope of a position versus time graph is the velocity of the object.
- The slope of a velocity versus time graph is the acceleration of the object.
- Force and acceleration are directly proportional.

Solve

(a) A massive ball rolling along a very smooth surface would take a great deal of time to slow down and could therefore be considered to be moving at a constant velocity. Constant velocity means the slope of the position-time graph should be positive and constant, the velocity-time graph should be a horizontal line, and the acceleration-time graph should be zero indicating there is zero acceleration.

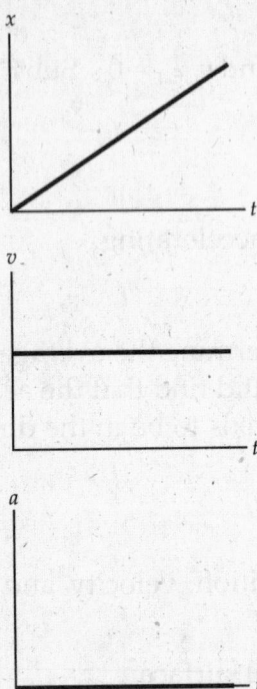

(b) A massive ball rolling down a smooth inclined surface would accelerate at a constant rate due to the force of gravity on the ball. Constant acceleration means the slope of the velocity-time graph should be positive and constant, the acceleration-time graph should be a horizontal line, and the position-time graph should curve upward since its slope (indicating the velocity of the ball) should get steeper and steeper.

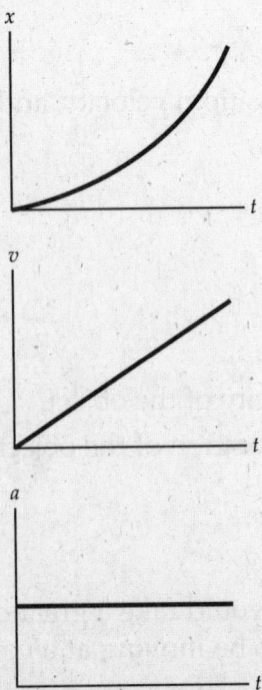

(c) If we assume that the airplane accelerates down the runway at a constant rate (a reasonable assumption), then the graphs for its motion would be qualitatively the same as the ball rolling down the inclined plane.

(d) We can assume that the acceleration of the car is constant as it skids to a stop. Since the acceleration of the car is negative (it's slowing down), the slope of the velocity-time graph should be negative and constant. The position-time graph would then be increasing in value but decreasing in slope, indicating that the car is moving further away but slowing down.

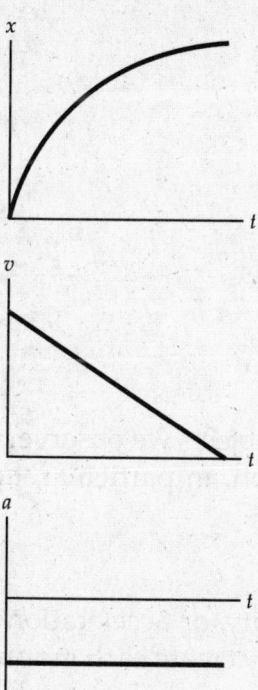

(e) Since the acceleration of an object is directly proportional to the net force acting on the object (Newton's second law), then the acceleration of this object would also increase proportionally with time. Therefore, the acceleration-time graph should have a positive slope. The velocity of this object would increase at an ever increasing rate so the velocity-time graph should be positive and curve upward. The position-time graph for this object should also curve upward at an even greater rate.

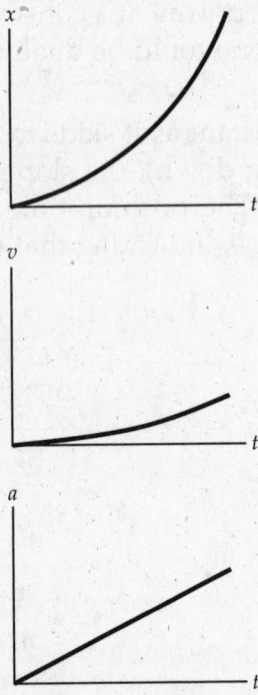

What does it mean?

By practicing qualitative plots of the motion of various objects we observe, we can gain an intuitive understanding of the nature of motion, in particular, how the position, velocity, and acceleration are related.

CP 2.2 Interpreting Motion from a Graph

Match the following graphs ((a)–(d)) of position, velocity, or acceleration versus time with the scenarios ((1)–(4)) which would most likely create each graph. Note: Each graph and scenario should be used only once.

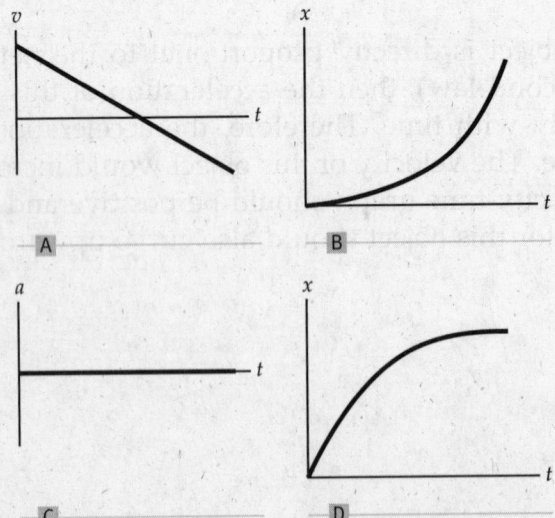

1. A drag racer accelerating down the track at a constant acceleration.

2. A ball thrown vertically upward and returning to your hand.

3.   A car sliding to a stop after slamming on its brakes.

4.   A hockey puck sliding on a smooth icy surface.

Solution

Recognize the principle

The key principle in this problem is to be able to interpret the graphical representations of motion. That is, we need to understand what various lines or curves on a position, velocity, or acceleration versus time graph physically represent.

Sketch the problem

The necessary sketches are given in the problem.

Identify the relationships and Solve

Let's tackle this problem by analyzing what type of motion each graph describes, and then decide which object ((1)–(4)) would move in this way.

(a) This graph represents an object moving in the positive direction (hence the positive velocity) but slowing down at a constant rate illustrated by the constant negative slope. At some point the object momentarily comes to rest (the velocity becomes zero) and begins moving in the negative direction indicated by the negative velocity. The object accelerates in the negative direction at a constant rate illustrated by the constant negative slope on the v-t graph. Since the slope of the line is constant for all points in time we can conclude that the object undergoes a constant negative acceleration for all points in time. Therefore, the force on the object is constant and in the negative direction for all points in time even though the object changes directions during the process. The only object which describes this type of motion is (2) a ball thrown vertically upward and returning to your hand.

(b) This graph represents an object moving in the positive direction and increasing in speed as illustrated by the increasing slope on the x-t graph. Therefore, the object is continually accelerating, perhaps at a constant rate. The only object which describes this type of motion is (1) a drag racer accelerating down the track at a constant acceleration.

(c) This graph represents an object moving with zero acceleration. That is, the object is moving at a constant velocity, not accelerating, with zero net force. The object which best describes this type of motion is (4) a hockey puck sliding on a smooth icy surface.

(d) Finally, this graph represents an object slowing down until it stops. The object which describes this motion is (3) a car sliding to a stop after slamming on its brakes.

What does it mean?

Once again, practicing these types of problems helps strengthen our intuitive understanding of the nature of motion.

Part E. MCAT Review Problems and Solutions

PROBLEMS

1. A net resultant force acting on an object will have which effect?
   (a) The velocity of the object will remain constant.
   (b) The velocity of the object remains constant, but the direction in which the object moves will change.
   (c) The velocity of the object will change.
   (d) None of the above.

2. A box rests on a level table. Which of the following is an action-reaction pair of forces?
   (a) The weight of the box and the upward force of the table on the box.
   (b) The weight of the table and the upward force of the floor on the table.
   (c) The force of the table on the floor and the force of the floor on the table.
   (d) The weight of the box and the force of the floor on the table.

3. Object A has a mass that is twice as great as that of object B. If a force acting on object A is half the value of a force acting on object B, which statement is true.
   (a) The acceleration of A will be twice that of B.
   (b) The acceleration of A will be half that of B.
   (c) The acceleration of A will be equal to that of B.
   (d) The acceleration of A will be one-fourth that of B.

4. What is the force required to impart an acceleration of 10 m/s^2 to an object with a mass of 2.0 kg?
   (a) 0.2 N
   (b) 5 N
   (c) 12 N
   (d) 20 N

5. Newton's second law can be stated as which of the following?
   (a) For every action there is an equal and opposite reaction.
   (b) Force and the acceleration it produces are directly proportional.
   (c) An object at rest tends to remain at rest unless acted upon by a force.
   (d) None of the above.

6. If a car is traveling due south with decreasing speed, then the direction of the car's acceleration is
   (a) due east.
   (b) due west.
   (c) due north.
   (d) due south.

7. A bird flies 2 km due north in 0.25 h and then flies 1 km due south in 0.75 h. What is the bird's average velocity for this trip?

    (a) 3 km/h north

    (b) 3 km/h south

    (c) 1 km/h north

    (d) 1 km/h south

8. If the average speed of a plane is 500 km/h, how long will it take to fly 125 km?

    (a) 4.00 h

    (b) 2.00 h

    (c) 0.50 h

    (d) 0.25 h

9. If the mass of a moving object is doubled, the inertia of the object will be

    (a) half as great as its original value.

    (b) twice as great as its original value.

    (c) four times as great as its original value.

    (d) unchanged from its original value.

10. For an object experiencing zero acceleration, which statement is true?

    (a) The object must be at rest.

    (b) The object may be at rest.

    (c) The object must slow down.

    (d) The object may speed up.

SOLUTIONS

1. *MCAT strategies*

Recognizing that according to Newton's second law a net force on an object will produce an acceleration, and acceleration is a change in velocity with respect to time, we can quickly eliminate answers (a) and (b). Furthermore, answer (b) does not make sense since an object cannot have a constant velocity and be changing direction. So, answer (c) is correct since it states that the velocity of the object will change. No further analysis is necessary.

2. *MCAT strategies*

No answers can be easily eliminated so we proceed with our problem-solving strategy.

Recognize the principle

We recognize this as an application of Newton's third law and proceed by drawing a force diagram.

Sketch the problem

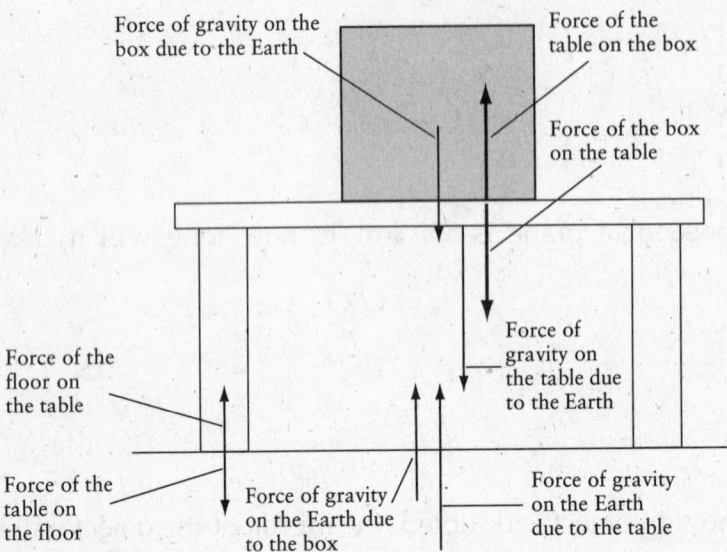

Force of gravity on the box due to the Earth

Force of the table on the box

Force of the box on the table

Force of the floor on the table

Force of gravity on the table due to the Earth

Force of the table on the floor

Force of gravity on the Earth due to the box

Force of gravity on the Earth due to the table

Identify the relationships and Solve

There are actually eight forces involved in the box, table, floor, Earth system as described in the sketch. Remember, the easiest way to determine if two forces form an action-reaction pair is by switching the nouns in the description of a force. Therefore, the reaction to the force of the <u>table</u> on the <u>box</u> is the force of the <u>box</u> on the <u>table</u>. Proceeding in this way we find that the correct answer is (c) the force of the table on the floor and the force of the floor on the table.

What does it mean?

Just because two forces are equal in magnitude and opposite in direction does not mean they form an action-reaction pair. For example, since the table is level and the box is not accelerating, the force of gravity on the box (its weight) and the force of the table on the box (normal force) are equal and opposite in direction but do not form an action-reaction pair.

3. MCAT strategies

No answers can be easily eliminated so we proceed with our problem-solving strategy.

Recognize the principle

This is an application of Newton's second law. We can solve this problem by setting up a ratio of the accelerations of the two objects.

Sketch the problem

None required.

Identify the relationships

From Newton's second law the acceleration of objects A and B are given by the expressions:

$$a_A = \frac{F_A}{m_A} \quad \text{and} \quad a_B = \frac{F_B}{m_B}$$

Taking the ratio of these two accelerations we find

$$\frac{a_A}{a_B} = \frac{F_A m_B}{F_B m_A}$$

Solve

From the question we are given that $m_A = 2m_B$ and $F_B = 2F_A$. Substituting these into our expression for the ratio of the two accelerations we find

$$\frac{a_A}{a_B} = \frac{F_A m_B}{F_B m_A} = \frac{F_A m_B}{(2F_A)(2m_B)} = \frac{1}{4}$$

Therefore the correct answer is (d) the acceleration of A will be one-fourth that of B.

What does it mean?

Without knowing the actual masses or forces involved, we were able to determine how one object would accelerate relative to the other, simply by applying Newton's second law.

4. MCAT strategies

No answers can be easily eliminated so we proceed with our problem-solving strategy.

Recognize the principle

We will need to apply Newton's second law.

Sketch the problem

None required.

Identify the relationships

Newton's second law states that $\vec{a} = \frac{\Sigma \vec{F}}{m}$ or $\Sigma \vec{F} = m\vec{a}$.

Solve

Substituting in for the acceleration and mass we solve for the force:

$$\Sigma \vec{F} = m\vec{a} = (2 \text{ kg})(10 \text{ m/s}^2) = 20 \text{ N}$$

Therefore, the correct answer is (d).

What does it mean?

Since no directions were provided in the question, we assumed that this was a one-dimensional problem and therefore did not need to take into account the vector nature of the force and acceleration.

5. MCAT strategies

This question involves no analysis but simply requires us to remember Newton's laws of motion:

Newton's first law: If the total force acting on an object is zero, the object will move with a constant velocity.

Newton's second law: $\vec{a} = \frac{\Sigma \vec{F}}{m}$. Forces thus cause acceleration.

Newton's third law: For every action (force) there is a reaction (force) of equal magnitude and opposite direction. Therefore, the correct answer is (b).

6. MCAT strategies

Since the motion of the car is in the southern direction and is only changing its speed not its direction, then the acceleration must be in the direction of north or south. Therefore, answers (a) and (b) can be quickly eliminated. Since the car is traveling south and slowing down, the direction of the acceleration must be north. If the car was accelerating in the southern direction then its speed would increase. So, the correct answer is (c) due north.

7. MCAT strategies

Since the bird ends up north of its starting point, we can conclude that its average velocity must be in the northern direction. Therefore, we can quickly eliminate answers (b) and (d) since they are in the southern direction. Without further analysis it is difficult to choose between answers (a) and (c) so we proceed as usual.

Recognize the principle

We must understand the definition of the average velocity of an object moving in one direction:

$$v_{ave} = \frac{\Delta x}{\Delta t}$$

Sketch the problem

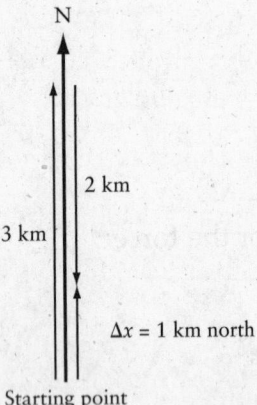

Identify the relationships and Solve

From the diagram we see that the resultant change in position for the bird is $\Delta x = 1$ km north of its starting point. Also, the total time for the bird's flight is $\Delta t = (0.25 + 0.75)$ h $= 1.0$ h. Solving for the average velocity of the bird's flight we find

$$v_{ave} = \frac{\Delta x}{\Delta t} = \frac{1 \text{ km north}}{1 \text{ h}} = 1 \text{ km/h north}$$

Therefore, the correct answer is (c).

What does it mean?

The average velocity is different from the average speed. The average speed for the bird's flight would be the total distance divided by the total time, 3 km/h, tempting us to incorrectly answer (a).

8. MCAT strategies

No answers can be easily eliminated so we proceed with our problem-solving strategy.

Recognize the principle

The average speed of an object is equal to the total distance traveled divided by the time required to travel that distance.

Sketch the problem

None required.

Identify the relationships and Solve

The total distance traveled by the plane is 125 km and the average speed of the plane is 500 km/h. From these two values we can solve for the time required for the plane to travel the 125 km:

$$\text{Time required} = \frac{\text{Distance traveled}}{\text{Average speed}} = \frac{125\ \text{km}}{500\ \text{km/h}} = 0.25\ \text{h}$$

Therefore, the correct answer is (d).

What does it mean?

Once again we need to be careful differentiating between average speed and average velocity. In this case, however, we can assume the plane only traveled in one direction so the average speed and average velocity would be the same.

9. MCAT strategies

We recognize that mass is the quantitative measure of the inertia of an object. Mass and inertia are directly proportional. If we double the mass of an object then the inertia will also double. Therefore the correct answer is (b) twice as great as the original.

10. MCAT strategies

Choices (c) and (d) are eliminated immediately because they involve the speed of the object changing. If an object's speed changes then its velocity changes and that means it is accelerating. Choice (a) is eliminated because although the object may be at rest it is not necessary for the acceleration to be zero. Zero acceleration simply implies that the velocity is not changing. So, the correct answer is (b).

3 Forces and Motion in One Dimension

Part A. Summary of Key Concepts and Problem-Solving Strategies

KEY CONCEPTS

Forces and Motion

A force is a push or a pull exerted by one object on another. Force is a *vector* quantity. All calculations of motion start with Newton's second law. For one-dimensional motion the forces and acceleration are directed along a line, and Newton's second law reads

$$\Sigma F = ma$$

A *free-body diagram* is a very useful first step in the application of Newton's second law.

APPLICATIONS

Motion with constant acceleration

A common and extremely important case is motion with constant acceleration. When a is constant, the position, velocity, and acceleration are related by the following expressions:

$$x = x_0 + v_0 t + \frac{1}{2}at^2$$
$$v = v_0 + at$$
$$v^2 = v_0^2 + 2a(x - x_0)$$

Common types of forces:

Gravity: Near the surface of the Earth, the magnitude of the gravitational force on an object of mass m is given by

$$F_{grav} = mg$$

where $g = 9.8$ m/s^2. The direction of this force is always downward. This force is also called the *weight* of the object. When an object is moving freely under the action of gravity, the motion is called *free fall*, and the acceleration is $a = g$ downward.

Normal force: When two surfaces are in contact, there is a normal force between them. This force is perpendicular to the contact surface, hence the word *normal* meaning *perpendicular*.

Friction: When two surfaces are in contact, a frictional force opposes the direction of relative motion or pending motion of the surfaces. If the surfaces are slipping relative to each other, the opposing force is *kinetic friction* given by

$$F_{friction} = \mu_K N$$

where N is the normal force. If the surfaces are not slipping, the resistive force is *static friction* given by

$$F_{friction} \leq \mu_S N$$

Drag forces: When an object moves through a fluid such as air or water, there is a drag force arising from contact between the fluid molecules and the moving object. We consider drag forces in two regimes:

- For motion through the air, the magnitude of the drag force is described approximately by

$$F_{drag} = \frac{1}{2} \rho A v^2$$

This regime is appropriate for skydivers and cars.

- For motion of cells and bacteria through water, the drag force is given by

$$\vec{F}_{drag} = -Cr\vec{v}$$

PROBLEMS-SOLVING STRATEGIES

In this chapter several new concepts were introduced and discussed in the context of specific examples. Many of these examples involved the application of Newton's second law of motion which relates the net force on an object to the acceleration it experiences. To tackle these problems it is important to organize your thoughts and the given information by drawing a sketch of the situation and a *free-body diagram*. The sketch should include coordinate axes along with other information such as the values of forces and the initial velocity. In addition, you should construct a simplified diagram showing all the forces acting on each object involved in the problem. Free-body diagrams allow us to reduce the problem down to the forces involved and will become increasingly important as we move to two-dimensional problems in the next chapter.

Also in this chapter was the introduction of a new type of problem call *reasoning and relationship problems*. In these problems we are interested in an approximate solution rather than an exact value. In many ways, these problems are more like problems we deal with in the "real" world where not all the required information is available to us and we must either gather that information or approximate values based on our experiences. The reasoning involved in an approximate analysis can sometimes give us clear physical insight into a problem, without the "distraction" of a lot of mathematics. By solving these reasoning and relationship problems we become better problem solvers.

Problem Solving: Constructing Free-Body Diagrams and Applying Newton's Laws

---

Recognize the principle

Begin by recognizing the objects of interest and listing all the forces acting on each of them.

Sketch the problem

- Start with a drawing that shows all the objects of interest in the problem along with all the forces acting on each one.

- Make a separate sketch showing the forces acting on each object; this is the free-body diagram for that object. For clarity, you can represent each object by a simple "dot."

- Forces in a free-body diagram should be represented by arrows, with the direction of the arrow showing the direction of the force. In each free-body diagram show only the forces acting on that particular object.

Identify the relationships

A force in a free-body diagram may be known (i.e., its direction and magnitude may be given), or it may be unknown. Represent unknown quantities by a symbol. You will typically use equations derived from Newton's second law to solve for these unknowns.

Solve

The information contained in a free-body diagram can usually be used directly in writing Newton's second law for an object. A few algebraic steps will then lead to values for the unknown quantities.

What does it mean?

Always *consider what your answer means* and check that it makes sense.

Problem Solving: Dealing with Reasoning and Relationship Problems

Recognize the principle

Determine the key physics ideas that are central to the problem, and that connect the quantity that you want to calculate with the quantities you know. In the examples found in this section of your textbook, this physics involves motion with constant acceleration.

Sketch the problem

Make a drawing that shows all of the given information, and everything else that you know about the problem. For problems in mechanics, your drawing should include all of the forces, velocities, etc.

Identify the relationships

Identify the important physics relationships; for problems concerning motion with constant acceleration, these are the relationships between position, velocity, and acceleration. For many reasoning and relationships problems, values for some of the essential unknown quantities may not be given. You then must use common sense to make reasonable estimates for these quantities. Don't worry or spend time trying to obtain precise values of every quantity. Accuracy to within a factor of 3 or even 10 is usually fine, since the goal is to calculate the quantity of interest to within an order of magnitude (a factor of 10). Don't hesitate to use the Internet, the library, and (especially) your own intuition and experiences.

Solve

Since an exact mathematical solution is not required, cast the problem into one that is easy to solve mathematically. In the examples in this section of your textbook we were able to use the results for motion with constant acceleration.

What does it mean?
Always *consider what your answer means* and check that it makes sense.

Part B. Frequently Asked Questions

1. *What's the difference between mass and weight?*

The simple answer is that mass is how much matter an object has and weight is the force of gravity on that mass. Actually, it's difficult to describe mass in more fundamental terms. Mass is certainly related to the amount of matter, but more importantly it is the quantitative measure of the inertia of an object and inertia is the measure of an object's ability to resist changes in motion. Weight on the other hand is a force. It is the force an object experiences as a result of the gravitational interaction with the Earth.

2. *Why is it easier to keep an object moving at constant velocity than to get it moving?*

This has to do with the difference between static and kinetic friction. As you begin pushing on an object, the force of static friction grows proportionally until the maximum force of static friction is obtained. At this point the object begins to move and the frictional force becomes kinetic friction. Since for a given interface the coefficient of kinetic friction is less than the coefficient of static friction, the force of kinetic friction is always less than the maximum force of static friction. So, it requires more force to get an object to start moving than to keep it moving.

3. *Why is it static friction between a wheel and the road? Shouldn't this be kinetic friction since the wheel is moving?*

This is a good question and requires us to look carefully at the interface between the wheel and the road. As a wheel rotates without slipping, the point of contact between the wheel and the road remains static. That is, there is no relative motion (no slipping) between the point of the wheel in contact with the road and the road. On the other hand, if a wheel is slipping as it rotates there is relative motion between the point of the wheel in contact with the road and the road. When there is no relative motion between two objects (the wheel and the road), the force of friction is static friction. When there is relative motion between the surfaces (slipping), the force of friction is kinetic friction.

4. *Since for a given situation the coefficient of static friction and the normal force on an object are both constant, shouldn't the force of static friction always be at its maximum value, $F_{friction} = \mu_S N$?*

No. Consider a box sitting on a level table. As long as there are no forces applied to the box (other than the force of gravity and the normal force), there will be no frictional force. As you begin to push the box horizontally to try to get it to move, the force of static friction will grow proportionally to your applied force until its maximum value is obtained. Once your applied force exceeds the maximum value for the force of static friction, the box will begin to move and it becomes a situation involving kinetic friction. So, the force of static friction varies depending on the applied force.

5. *When dealing with cables, how do I know which way to draw the tension in my free-body diagram?*

It all depends on which object you are interested in studying. For example, consider the following diagram in which there are two objects connected by a cable and being pulled horizontally on a frictionless surface by a force F.

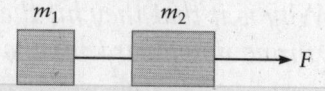

Drawing the free-body diagrams for each object we have the following:

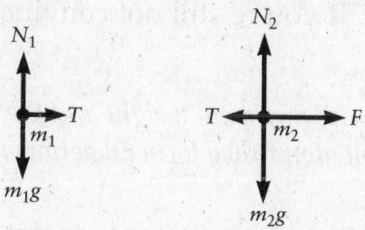

Even though the magnitude of the tension is the same, the direction the tension acts on the two objects is different. Tension can never push on an object, it always pulls so we always draw the tension going away from an object in the direction of the cable.

6. *When an object rests on the floor the normal force between the floor and the object is mg (where m is the mass of the object). Isn't the normal force always equal to mg?*

No, the normal force on an object is not always equal to mg. In this example, $N = mg$ because the table is horizontal and the object is not accelerating. This is not always the case. For example, consider an object sitting on the floor of an elevator which is accelerating upward. Two forces are acting on the object, the force of gravity, mg, and the normal force from the floor on the object. Since the object is accelerating upward, the normal force must be greater than the force of gravity, $N > mg$. Another example when $N \neq mg$ is when the surface is not horizontal. We will discuss this in the next chapter on two-dimensional motion.

7. *I have two balls of the same mass but with different diameters. If I drop the balls from the same height at the same time, should they hit the ground at the same time even if I include air drag?*

While it is true that they will hit the ground at the same time if we ignore the effect of air drag, this is not true when we include air drag. The force of gravity on the two objects is the same since they have the same mass; however, the force of air drag on the two objects is not the same. Drag force is proportional to the cross-sectional area perpendicular to the direction of motion according to the expression $F_{drag} = \frac{1}{2}\rho A v^2$. Since the cross-sectional area is larger for the object with the larger diameter, this object will experience a larger drag force and will therefore take longer to hit the ground.

8. *Suppose I'm standing at the edge of a cliff and throw a ball directly upward at the same speed that I throw a ball directly downward. Why is it that they hit the ground below going the same speed? Shouldn't the ball I throw downward hit the ground going faster?*

Neglecting air drag, when you throw a ball upward with a speed v_0, it will return to your hand going exactly the same speed going downward. If instead of catching the ball you let it continue falling from the edge of a cliff, this is identical to throwing the ball downward with speed v_0 from the edge of the cliff. So, the two balls will hit the ground below going the same speed. If you're still not convinced, try doing the calculation with some actual numbers.

9. *Suppose I'm standing still on the horizontal ground. Since my weight and the normal force on my feet are equal and opposite in direction, don't they form an action-reaction pair according to Newton's third law?*

No. Remember from the FAQs in the last chapter that the easiest way to determine an action-reaction pair is by switching the nouns. For this example, there are two forces acting on you: (1) the force of gravity on <u>you</u> due to the <u>Earth</u> (called your weight) and (2) the normal force on <u>you</u> due to the <u>floor</u>. The corresponding reaction forces for these two forces are (1) the force of gravity on the <u>Earth</u> due to <u>you</u> and (2) the normal force on the <u>floor</u> due to <u>you</u>. Although the two forces acting on you are equal in magnitude and opposite in direction, they do not form an action-reaction pair.

10. *When a ball thrown vertically upward reaches its maximum height it momentarily stops. Does its acceleration and velocity momentarily go to zero?*

This is a great question which can be answered with the following thought experiment: Suppose you toss a ball straight up under the influence of gravity. As we know, the velocity of the ball when it reaches its maximum height becomes zero. Now, suppose that its acceleration also becomes zero. What would happen to the ball? Well, if the ball is stationary and not accelerating, then it should remain stationary. Clearly this does not happen so the acceleration does not become zero at the top of its trajectory. Newton's second law tells us that if there is a net force on an object then it will accelerate. When the ball reaches its maximum height gravity still exists and so the ball must still be accelerating. The gravitational acceleration slows the ball down as it rises and speeds it up as it falls. At the moment it changes direction (at the top of the trajectory) the velocity only momentarily becomes zero. There are many other examples of objects whose velocity becomes zero when the acceleration does not.

Part C. Selection of End-of-Chapter Answers and Solutions

QUESTIONS

Q3.8 (a) Suppose a tire rolls without slipping on a horizontal road. Explain the role friction plays in this motion. What two surfaces are involved in this frictional force? Is it static friction or kinetic friction? (b) Suppose a racecar driver wants to get his car started very quickly and "burns rubber" as he leaves the starting line. Is he exploiting static friction or kinetic friction? (c) Normally, the coefficient of kinetic friction between two surfaces is smaller than the coefficient of static friction for the same two surfaces. Assuming that is true for the case of friction between a car tire and the road, explain why the car will accelerate faster if the driver does not "burn rubber."

Answer

(a) The two surfaces which are involved in the frictional force are the tire and the road. At the point of contact between the tire and the road the tire is not sliding with respect to the road, therefore it is static friction. (b) When the driver "burns rubber" he spins the tires. This means that there is relative motion between the tire and road at the point of contact; therefore it must be kinetic friction. (c) Since the frictional force between the tires and the road is the force responsible for the car's acceleration, by maximizing the frictional force we can maximize the car's acceleration. This can be done by exploiting static friction rather than kinetic friction between the tires and the road, thus not "burning rubber."

Q3.16 Two balls of the same diameter are dropped simultaneously from a very tall bridge. One ball is solid lead, and the other is hollow plastic and has a much smaller mass than the solid lead ball. Use a free-body diagram to explain why the solid lead ball reaches the ground first. Hint: Include the air drag force in your analysis.

Answer

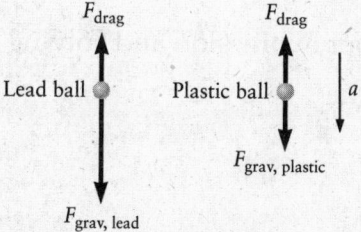

As the balls begin to fall, the drag force on each grows as their velocities increase. Eventually the drag force will increase to the point where it is equal in magnitude to the force of gravity on the plastic ball. This is the situation drawn in the free-body diagram. At that point the plastic ball no longer accelerates and continues to fall at its terminal speed. However, the drag force on the lead ball is still smaller than the force of gravity, so the lead ball continues to accelerate. At some later time the drag force on the lead ball will grow to be equal in magnitude to the force of gravity and the lead ball will stop accelerating and fall at its terminal speed. We see from this analysis that the terminal speed for the lead ball is much higher (and is reached later) than that of the plastic ball. This results in the lead ball reaching the ground first.

PROBLEMS

P3.11 An object with an initial velocity of 12 m/s accelerates uniformly for 25 s. (a) If the final velocity is 45 m/s, what is the acceleration? (b) How far does the object travel during this time?

Solution

Recognize the principle

This is a situation where an object is accelerating at a constant rate so we can apply the expressions describing motion for a constant acceleration. Since this is one-dimensional motion and the velocities are both positive, the object does not change direction. In other words, both the initial and final velocities are in the same direction.

Sketch the problem

$v_0 = 12$ m/s $v = 45$ m/s

Identify the relationships

(a) We are given the initial velocity of the object $v_0 = 12$ m/s, the final velocity $v = 45$ m/s, and the time $t = 25$ s to accelerate. From these we can use the following expression to calculate the acceleration:

$$v = v_0 + at$$

(b) Using the acceleration we calculate in part (a) and the initial and final velocities, we can determine the distance, $x - x_0$, traveled using the following:

$$v^2 = v_0^2 + 2a(x - x_0)$$

Solve

(a) Plugging the known parameters into the proper expression and solving for the acceleration,

$$a = \frac{v - v_0}{t} = \frac{(45 \text{ m/s}) - (12 \text{ m/s})}{25 \text{ s}}$$

$$a = 1.32 \text{ m/s}^2 = \boxed{1.3 \text{ m/s}^2}$$

(b) Plugging the known parameters into the proper expression and solving for the distance $x - x_0$,

$$x - x_0 = \frac{v^2 - v_0^2}{2a} = \frac{(45 \text{ m/s})^2 - (12 \text{ m/s})^2}{2(1.32 \text{ m/s}^2)}$$

$$x - x_0 = 713 \text{ m} \approx \boxed{710 \text{ m}}$$

What does it mean?

Notice in part (b) of this problem we chose to use the expression $v^2 = v_0^2 + 2a(x - x_0)$ to solve for the distance. We could have just as easily used the expression

$x = x_0 + v_0^2 + \frac{1}{2} at^2$ and obtained the same answer. There are often multiple approaches to solving physics problems. Also, because the speed increases, we expect the acceleration to be in the same (positive) direction as the velocity.

P3.23 A ball is thrown upward with a speed of 35 m/s from the edge of a cliff of height $h = 15$ m (similar to Fig. P3.22). (a) What is the speed of the ball when it passes by the cliff on its way down to the ground? (b) What is the speed of the ball when it hits the ground? Ignore air drag. Assume the ball is thrown straight up.

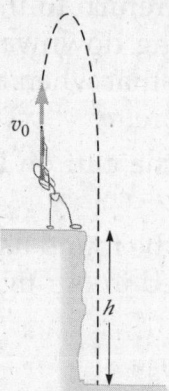

Figure P3.22

Solution

Recognize the principle

This is a situation where an object is accelerating at a constant rate so we can apply the expressions describing motion with a constant acceleration. Furthermore, the object is in free fall so we know the acceleration to be g (downward).

Sketch the problem

Defining up as the positive y direction and choosing our origin at the bottom of the cliff, we have the following sketch of the problem.

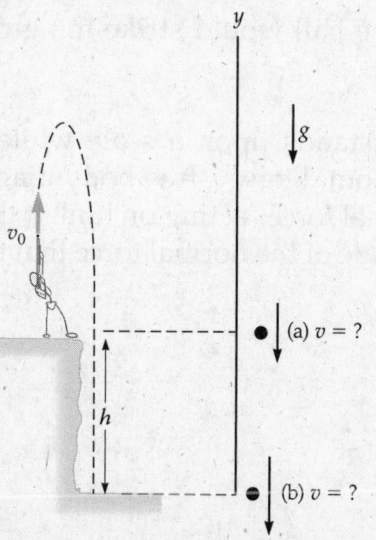

Identify the relationships

Since the object is in free fall it is accelerating at a constant rate. Therefore, we can apply the expressions for motion with constant acceleration,

$$y = y_0 + v_0 t + \frac{1}{2} a t^2$$
$$v = v_0 + at$$
$$v^2 = v_0^2 + 2a(y - y_0)$$

Solve

(a) Since the effects of air drag can be ignored the ball will return to the edge of the cliff with the same speed as it was thrown but heading downward. We can prove this by using the third expression and recognizing that when the ball returns to the same point it was thrown, $y - y_0 = 0$. Therefore, $v^2 = v_0^2$. So, the ball will be traveling at a speed of $\boxed{35 \text{ m/s}}$ when it passes the cliff on its way down.

(b) The velocity of the ball, just before striking the ground, can be found using the third expression as well. With the coordinate system as defined in our figure we have the following information:

$$y_0 = 15 \text{ m}$$
$$y = 0$$
$$a = -g$$
$$v_0 = 35 \text{ m/s}$$

Plugging these values into the third expression and solving for v we find,

$$v = \pm\sqrt{v_0^2 - 2g(y - y_0)} = \sqrt{(35 \text{ m/s})^2 + 2(9.8 \text{ m/s}^2)(15 \text{ m})} = \pm 39 \text{ m/s}$$

Since the ball is traveling downward just before it strikes the ground and we've defined up as our positive direction, we choose the negative solution. Therefore the velocity of the ball just before striking the ground is −39 m/s. The speed of the ball is just the magnitude of the velocity, $\boxed{39 \text{ m/s}}$.

What does it mean?

If we included air drag we would find that the ball would strike the ground at speed much less than 39 m/s.

P3.34   A tall strongman of mass $m = 95$ kg stands upon a scale while at the same time pushing on the ceiling in a small room. Draw a free-body diagram of the strongman (Fig. P3.34) and indicate all normal forces acting on him. If the scale reads 1100 N (about 240 lb), what is the magnitude of the normal force that the ceiling exerts on the strongman?

Figure P3.34

Solution

Recognize the principle

We recognize this as an application of Newton's laws of motion. In particular, since the object (the man) is not accelerating, then the net force on him must be zero.

Sketch the problem

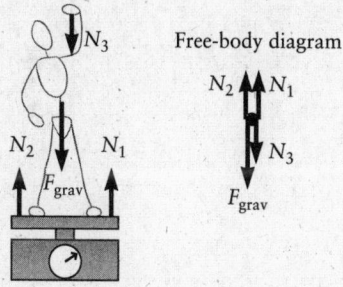

Identify the relationships

After drawing the free-body diagram we can apply Newton's second law,

$$\Sigma F = N_1 + N_2 - F_{grav} - N_3 = 0$$

From Newton's third law, the force on the scale is equal to the sum of the normal forces the scale exerts on the strongman,

$$F_{scale} = N_1 + N_2$$

Also, we know that $F_{grav} = mg$.

Solve

Using the values provided in the question and substituting the expression for F_{scale} in for $N_1 + N_2$ in the first expression, we can solve for the magnitude of the normal force that the ceiling exerts on the strongman.

$$\Sigma F = F_{scale} - mg - N_3 = 0$$

$$N_3 = F_{scale} - mg = 1100 \text{ N} - (95 \text{ kg})(9.8 \text{ m/s}^2) = \boxed{170 \text{ N}}$$

What does it mean?

If you think about it, this answer is rather obvious. If the strongman was not pushing on the ceiling, the scale would simply read his weight, $(95 \text{ kg})(9.8 \text{ m/s}^2) = 930 \text{ N}$. Since he is pushing on the ceiling, the scale reading will increase by an amount equal to his push on the ceiling.

P3.41 The coefficient of kinetic friction between a refrigerator (mass 100 kg) and the floor is 0.20, and the coefficient of static friction is 0.25. If you apply the minimum force needed to get the refrigerator to move, what will the acceleration then be?

Solution

Recognize the principle

This is an application of Newton's second law including the concept of friction.

Sketch the problem

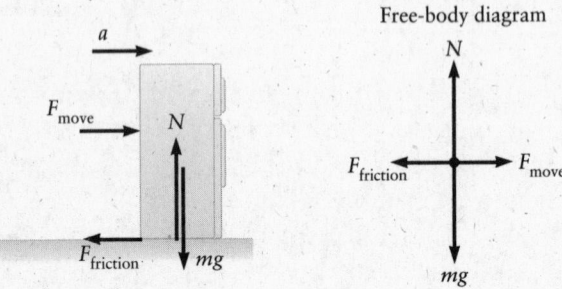

Free-body diagram

Identify the relationships

We apply Newton's second law to both the vertical and horizontal directions. Analysis of the vertical direction allows us to determine the normal force on the refrigerator and analysis of the horizontal direction provides us with information about the refrigerator's acceleration. The minimum force to move the refrigerator will be one that just exceeds the maximum force of static friction. Therefore, F_{move} = $F_{static\ friction}$. After the refrigerator begins to move the frictional force reduces to that of kinetic friction, $F_{kinetic\ friction} = \mu_K N$.

Solve

Applying Newton's second law to the vertical direction we find,

$$\Sigma F = N - mg = 0 \rightarrow N = mg$$

Applying Newton's second law to the horizontal direction we find,

$$\Sigma F = F_{move} - F_{kinetic\ friction} = F_{move} - \mu_K N = ma$$

The maximum force of static friction is,

$$F_{static\ friction} = \mu_S N = \mu_S mg$$

Therefore,

$$F_{move} = F_{static\ friction} = \mu_S mg$$

Combining these expressions we find,

$$a = \frac{F_{move} - \mu_K mg}{m} = \frac{\mu_S mg - \mu_K mg}{m} = g(\mu_S - \mu_K)$$

Inserting the values into the equation and solving for a,

$$a = g(\mu_S - \mu_K) = 9.8 \text{ m/s}^2 (0.25 - 0.20) = \boxed{0.49 \text{ m/s}^2}$$

What does it mean?

Note that the mass of the refrigerator was not necessary to solve for the acceleration. So, you could generalize this to any object you are trying to push along the floor.

P3.53   A ball is thrown directly upward with an initial velocity of 15 m/s. If the ball starts at an initial height of 3.5 m, how long is the ball in the air? Ignore air drag.

Solution

Recognize the principle

This is a situation where an object is accelerating at a constant rate so we can apply the expressions describing motion with a constant acceleration. Furthermore, the object is in free fall so we know the acceleration to be g (downward).

Sketch the problem

Defining up as the positive y direction and choosing the origin of our coordinate system at the point where the ball hits the ground, we have the following sketch:

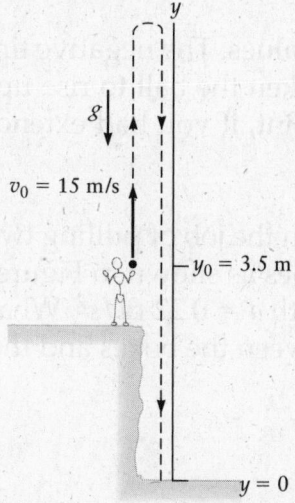

Identify the relationships

Since the ball is in free fall it is accelerating at a constant rate. Therefore, we can apply the expressions for motion with constant acceleration,

$$y = y_0 + v_0 t + \frac{1}{2} a t^2$$

$$v = v_0 + at$$

$$v^2 = v_0^2 + 2a(y - y_0)$$

Solve

Starting with the first expression and setting $y = 0$ and $a = -g$ we can determine the total time that the ball is in the air,

$$0 = y_0 + v_0 t - \frac{1}{2} g t^2$$

Solving for t using the quadratic formula we have,

$$t = \frac{-v_0 \pm \sqrt{v_0^2 - 4\left(-\frac{1}{2} g\right)(y_0)}}{2\left(-\frac{1}{2} g\right)} = \frac{v_0 \mp \sqrt{v_0^2 + 2 g y_0}}{g}$$

Inserting the numerical values into this expression we have,

$$t = \frac{(15 \text{ m/s}) \mp \sqrt{(15 \text{ m/s})^2 + 2(9.8 \text{ m/s}^2)(3.5 \text{ m})}}{(9.8 \text{ m/s}^2)}$$

$$t = +3.3 \text{ s or } t = -0.22 \text{ s}$$

Choosing the positive solution we find that the total time the ball is in the air is
$\boxed{t = 3.3 \text{ s}}$.

What does it mean?

The solution to the quadratic formula produces two values. The negative time also has physical meaning. It is the time it would have taken the ball to rise up to the height of 3.5 m and be traveling at 15 m/s at that point, if you had extended the ball's trajectory "backwards" in time to ground level.

P3.63 You work for a moving company and are given the job of pulling two large boxes of mass $m_1 = 120$ kg and $m_2 = 290$ kg using ropes as shown in Figure P3.63. You pull very hard, and the boxes are accelerating with $a = 0.22$ m/s^2. What is the tension in each rope? Assume there is no friction between the boxes and the floor.

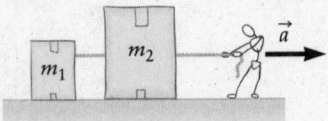

Figure P3.63

Solution

Recognize the principle

This is an application of Newton's second law as well as understanding the concept of the force of tension.

Sketch the problem

We begin by drawing a separate free-body diagram for each object.

Free-body diagrams

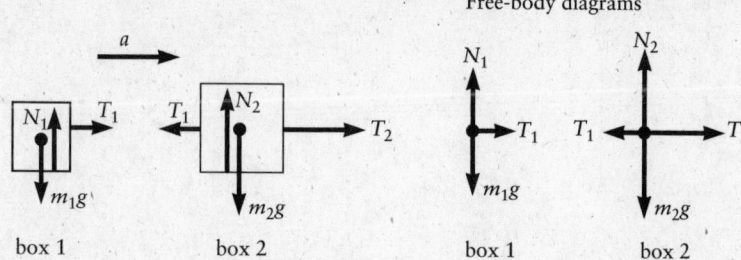

box 1   box 2   box 1   box 2

Identify the relationships and Solve

Since the objects are accelerating to the right, let's define that direction as positive. With this choice for a coordinate system, any force which acts in the direction of the acceleration will be positive and any force in the opposite direction will be negative. Since there is no friction between the boxes and the floor, we will not concern ourselves with the analysis in the vertical direction.

Applying Newton's second law in the horizontal direction to each box and substituting in the given quantities we can solve for the tension in each rope.

Box #1:

$$\Sigma F_1 = T_1 = m_1 a$$

Therefore,

$$T_1 = (120 \text{ kg})(0.22 \text{ m/s}^2) = \boxed{26 \text{ N}}$$

Box #2:

$$\Sigma F_2 = T_2 - T_1 = m_2 a$$

Therefore,

$$T_2 = T_1 + m_2 a = 26 \text{ N} + (290 \text{ kg})(0.22 \text{ m/s}^2) = \boxed{90 \text{ N}}$$

What does it mean?

Note that the tension found for the second rope is equivalent to that of pulling one block of mass $M = m_1 + m_2$, as you would expect.

P3.70 Hang time. LeBron James decides to jump high enough to dunk a basketball. What is the approximate force between the floor and his feet while he is jumping from the floor? *Hint:* Start by estimating the distance he moves while in contact with the floor and the height he must jump for his hands to reach just above the rim.

Solution

Recognize the principle

We recognize this is an application of Newton's second law. Also, once his feet leave the ground, LeBron is in free fall so we can apply the appropriate expressions for motion with constant acceleration.

Sketch the problem

Identify the relationships

The force between his feet and the floor is the normal force. In order to determine the normal force we must first determine the acceleration LeBron undergoes while his feet are still in contact with the ground. To determine this acceleration we must first calculate his vertical launch speed required to just reach the rim. So, our plan of attack is as follows: (1) from his free-fall motion determine his vertical launch speed, (2) using this launch speed determine his vertical acceleration while in contact with the ground, and (3) apply Newton's second law to calculate the normal force on his feet.

To make these calculations we will need to estimate two distances. First, the height, h, LeBron James jumps while in the air. As a rough estimate, which will be correct within an order of magnitude, let's use $h = 1$ m. The second distance to estimate is the distance d he moves while in contact with the ground. Since he won't move much, we can estimate this to be about half a meter, $d = 0.5$ m. One final quantity we will need is his mass. Since he weighs just over 200 lbs, this is about 100 kg.

Solve

For part (1) of our analysis we define up as the positive y direction and use the following expression for motion with constant acceleration:

$$v^2 = v_0^2 + 2a(y - y_0)$$

The distance traveled in this case is just h, so $y - y_0 = h$. Since his velocity at the top of his jump is zero we have $v = 0$. Also, his acceleration while in the air is $a = -g$. Inserting these values into the expression and solving for v_0, we have

$$v^2 = v_0^2 - 2gh$$
$$0 = v_0^2 - 2gh$$
$$v_0 = \sqrt{2gh}$$

For part (2), we start with the same expression,

$$v^2 = v_0^2 + 2a(y - y_0)$$

From part (1) we know that his velocity just before leaving the ground is $\sqrt{2gh}$. His initial velocity (with his knees bent) is zero and he travels a distance d while still in contact with the ground. So, with this information we can calculate his vertical acceleration while in contact with the ground.

$$v^2 = v_0^2 + 2a(y - y_0)$$
$$2gh = 0 + 2ad$$

Solving for a we find,

$$a = \frac{gh}{d}$$

Inserting the values for h and d in this expression we have,

$$a \approx \frac{(10 \text{ m/s}^2)(1 \text{ m})}{0.5 \text{ m}} = 20 \text{ m/s}^2$$

Note we have used $g = 10 \text{ m/s}^2$ since we are only looking for an approximate solution.

Finally we can find the normal force by applying Newton's second law.

$$\Sigma F = ma$$
$$N - mg = ma$$

Therefore,

$$N = m(a + g) \approx (100 \text{ kg})(20 \text{ m/s}^2 + 10 \text{ m/s}^2) = \boxed{3000 \text{ N}}$$

What does it mean?

The acceleration LeBron experiences as he accelerates upward before leaving the floor is about 3 g's. You can experience the same acceleration by jumping off a 1-m high table and bending your knees as you land.

P3.77   You are a secret agent and find that you have been pushed out of an airplane without a parachute. Fortunately you are wearing a large overcoat (as secret agents often do). Thinking quickly, you are able to spread out and hold the overcoat so that you increase your overall area by a factor of 4. If your terminal speed would be 43 m/s without the coat, what is your new terminal speed with the coat? Could you survive impact with the ground? How about over a lake?

Solution

Recognize the principle

We will need to apply the concept of terminal speed.

Sketch the problem

No sketch required for this problem.

Identify the relationships

For an object moving through air the terminal speed is approximated by the expression,

$$v_{\text{term}} = \sqrt{\frac{2mg}{\rho A}}$$

The best way to solve this problem is to consider the ratio of the terminal speed with the coat open to that when the coat is closed. What we know is that the new area from stretching out the coat is 4 times greater than with the coat closed. In equation form we would write this as,

$$A_{\text{new}} = 4A_{\text{old}}$$

$$\frac{v_{\text{term (new)}}}{v_{\text{term (old)}}} = \frac{\sqrt{\dfrac{2mg}{\rho A_{\text{new}}}}}{\sqrt{\dfrac{2mg}{\rho A_{\text{old}}}}} = \sqrt{\frac{A_{\text{old}}}{A_{\text{new}}}} = \sqrt{\frac{A_{\text{old}}}{4A_{\text{old}}}} = \sqrt{\frac{1}{4}} = \frac{1}{2}$$

Solve

Inserting the given values we find,

$$v_{\text{term (new)}} = \frac{1}{2}\, v_{\text{term (old)}} = \frac{1}{2}\,(43 \text{ m/s}) = \boxed{22 \text{ m/s or } 49 \text{ mi/h}}$$

What does it mean?

The agent has a good chance of survival if he lands in water. A 15-m Olympic platform diver will impact the water at 17 m/s which is not too far from our 22 m/s. Survival for a landing on land would be as problematic as jumping 15 m into an empty pool.

P3.79 The force exerted on a bacterium by its flagellum is 4×10^{-13} N. Find the velocity of the bacterium in water. Assume a size $r = 1\ \mu$m.

Solution

Recognize the principle

Here we must apply the concepts of drag force and terminal velocity.

Sketch the problem

We begin by drawing a sketch of the situation and a free-body diagram.

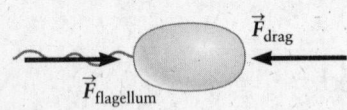

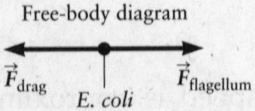

Free-body diagram

Identify the relationships

Since we know that the bacterium is not accelerating (moving at a constant velocity), the net force on the bacterium is zero. Application of Newton's second law will allow us to calculate the velocity of the bacterium. Here we approximate the bacterium as spherical in shape and use the expression for the drag force, $F_{\text{drag}} = -Crv$.

Solve

Applying Newton's second law,

$$\Sigma F = F_{\text{flagellum}} + F_{\text{drag}} = ma$$

$$F_{\text{flagellum}} - Crv = 0$$

Solving for v,

$$v = \frac{F_{\text{flagellum}}}{Cr}$$

Inserting the values to find v,

$$v \approx \frac{(4 \times 10^{-13}\,\text{N})}{(0.02\,\text{N}\cdot\text{s/m}^2)(1 \times 10^{-6}\,\text{m})}$$

$$\boxed{v \approx 2 \times 10^{-5}\,\text{m/s}}\ \text{or}\ 20\ \mu\text{m/s}$$

What does it mean?

At this speed it would take the bacterium over an hour to swim across a Petri dish 10 cm across.

P3.97 High dive. The cliff-divers of Acapulco are famous for diving from steep cliffs that overlook the ocean into places where the water is very shallow. (a) Suppose a cliff-diver jumps from a cliff that is 25 m above the water. What is the speed of the diver just before he enters the water? (b) If the water is 4.0 m deep, what is the acceleration of the diver after he enters the water? Assume this acceleration is constant and it begins at the moment his hands enter the water.

Solution

Recognize the principle

As the cliff-diver falls he is in free fall so we can apply the equations of motion with constant acceleration. Once he enters the water his acceleration is also constant so the same expressions can be used.

Sketch the problem

Defining up as the positive y direction and choosing the origin at the surface of the water we have the following figure:

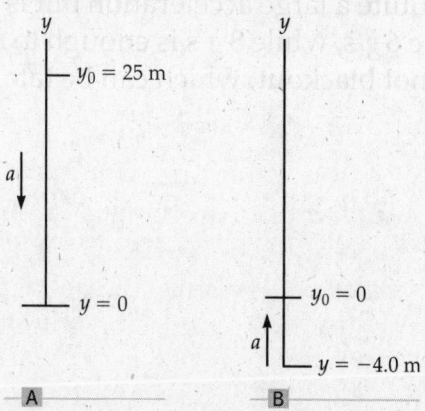

Identify the relationships

For both the fall through the air and the landing in the water the acceleration is constant so we can use the following expression,

$$v^2 = v_0^2 + 2a(y - y_0)$$

Solve

(a) Using our choice for the coordinate system, the initial velocity of the diver is $v_0 = 0$, and the acceleration $a = -g$. The initial height is $y_0 = 25$ m and the final height is the surface of the water at $y = 0$ m. Inserting these values into the expression we find,

$$v^2 = 0 + 2(-g)(0 - y_0)$$

$$v^2 = 2gy_0$$

$$v = \pm\sqrt{2(9.8 \text{ m/s}^2)(25 \text{ m})} = \pm 22 \text{ m/s}$$

Since the diver is moving on the $-y$ direction, we choose the negative solution. So, the diver's velocity just before he enters the water is -22 m/s. However, we were asked to find the speed of the diver which is just the magnitude of the velocity, so the correct answer is $\boxed{22 \text{ m/s}}$.

(b) We assume the diver uses the full 4.0 m of water to slow down. The diver's initial velocity was found in part (a), $v_0 = -22$ m/s, and has a final velocity of $v = 0$. Since the initial position, $y_0 = 0$ m is at the surface of the water, the final position will be $y = -4.0$ m. Again, plugging these values into the expression and solving for a we find,

$$0 = v_0^2 + 2a(y - 0)$$

$$a = \frac{-v_0^2}{2y}$$

$$a = \frac{-(-22 \text{ m/s})^2}{2(-4.0 \text{ m})}$$

$$\boxed{a = 61 \text{ m/s}^2}$$

What does it mean?

Notice that the acceleration is positive. This makes sense since the diver is traveling downward but is slowing down. This is quite a large acceleration but is it dangerous? The value of 61 m/s^2 is slightly above 6 g's, while 8 g's is enough to make a person blackout. So the diver will probably not blackout, which can be fatal when occurring in water.

Part D. Additional Worked Examples and Capstone Problems

The following worked examples provide you with practice drawing free-body diagrams, calculating forces, and applying Newton's second law. The first capstone problem provides you practice in graphing and helps develop your intuitive understanding of the relationship between force, acceleration, velocity, and position. Finally, the second capstone problem combines several key concepts in this chapter. Although these five problems do not incorporate all the material discussed in this chapter, they do highlight several of the key concepts. If you can successfully solve these problems then you should feel confident in your understanding of these key concepts, so use these problems as a test of your understanding of the chapter material.

WE 3.1 Tension in a Cable with Mass

Consider again the spider and the bug in Figure 3.20 in your textbook only this time let the lower piece of silk have a mass m_{silk}. If the bug has a mass m_{bug} and the acceleration is a upward, what is the value of the tension at the top end of the silk "cable" that connects them?

Solution

Recognize the principle

We have already seen that the tension increases as we move upward along a cable whose mass is nonzero. Using a free-body diagram, we can apply Newton's second law to both the silk and the bug and then solve for the tension at the top end of the piece of silk.

Sketch the problem

We begin by drawing a sketch of the problem and the free-body diagram for the lower strand of silk and the bug.

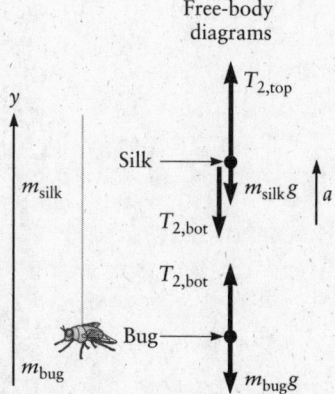

Identify the relationships

Using the forces in the free-body diagram, we write Newton's second law for the bug:

$$m_{bug}a = \Sigma F = F_{grav,bug} + T_{2,bot} = -m_{bug}g + T_{2,bot}$$

Rearranging gives

$$T_{2,bot} = m_{bug}(g + a)$$

We can also apply Newton's second law to the silk:

$$m_{silk}a = \Sigma F = -m_{silk}g + T_{2,top} - T_{2,bot}$$

Solve

Solving for the tension at the top end of the silk, we find

$$T_{2,top} = m_{silk}(g + a) + T_{2,bot}$$

Inserting the result for $T_{2,bot}$ from the first expression then gives

$$T_{2,top} = m_{silk}(g + a) + m_{bug}(g + a) = \boxed{(m_{silk} + m_{bug})(g + a)}$$

What does it mean?

As expected, we find that the tension at the top of the silk is greater than the tension at the bottom ($T_{2,top} > T_{2,bot}$).

WE 3.2 Please Open the Parachute

Estimate the terminal speed of a skydiver who has not yet opened her parachute. Take the mass of the skydiver to be $m = 60$ kg.

Solution

Recognize the principle

To find the drag force, we must know the skydiver's mass m and frontal area A. The mass is given but the area is not, so we need to estimate the frontal area of a human body (the skydiver) falling through the air.

Sketch the problem

We begin by drawing a sketch of the problem and the free-body diagram.

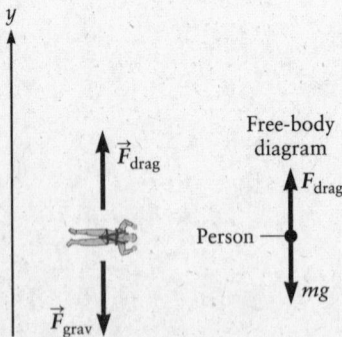

Identify the relationships

For a person of height 2 m and "width" 0.25 m, the frontal area is approximately the area of a rectangle with this height and width, so we estimate $A = 0.5$ m^2.

Solve

The density of air is $\rho = 1.3$ kg/m^3. Inserting this value along with $m = 60$ kg and our estimate for A into our result for the terminal speed gives

$$v_{\text{term}} = \sqrt{\frac{2mg}{\rho A}} = \sqrt{\frac{2(60 \text{ kg})(9.8 \text{ m/s}^2)}{(1.3 \text{ kg/m}^3)(0.5 \text{ m}^2)}}$$

$$v_{\text{term}} \approx \boxed{40 \text{ m/s}}$$

What does it mean?

This is about 100 mi/h and is much larger than the value of 3.0 m/s we found for the terminal speed with an open parachute. That is why parachutes are important!

WE 3.3 Drag Force on a Bacterium

Calculate the drag force on an *E. coli* moving at a speed of 40 μm/s (= 40 × 10^{-6} m/s) in water, where $C = 0.02 \text{ N} \cdot \text{s/m}^2$. Assume the *E. coli* is spherical with a radius $r = 1$ μm (= 1×10^{-6} m). Compare this force to the weight of the *E. coli* ($m = 4 \times 10^{-15}$ kg).

Solution

Recognize the principle

The drag force is given by the expression:

$$F_{\text{drag}} = -Crv$$

Hence, to evaluate F_{drag} we need to know v, C, and r, all of which are given.

Sketch the problem

Once again we begin by drawing a sketch of the problem and the free-body diagram.

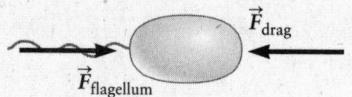

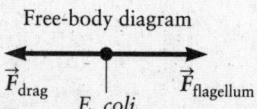

Free-body diagram

Identify the relationships and Solve

Inserting the values of v, C, and r given above into our expression for drag force, we find for the magnitude of the force,

$$F_{\text{drag}} = Crv = (0.02 \text{ N} \cdot \text{s/m}^2)(1 \times 10^{-6} \text{ m})(40 \times 10^{-6} \text{ m/s})$$

$$F_{\text{drag}} = \boxed{8 \times 10^{-13} \text{ N}}$$

The weight of an *E. coli* is

$$mg = (4 \times 10^{-15} \text{ kg})(9.8 \text{ m/s}^2) = \boxed{4 \times 10^{-14} \text{ N}}$$

so the drag force is about 20 times larger than its weight.

What does it mean?

For an object such as a bacterium, a cell, or a virus particle moving through a liquid similar to water, the drag force is quite substantial and is usually much larger than the object's weight.

CP 3.1 Describing One-Dimensional Motion: Going from Words to Graphs

(a) Plot the position, velocity, and acceleration versus time graphs for an object that is shot straight up in the air (neglect air drag). (b) Including air drag, plot the position, velocity, and acceleration versus time graphs for an object which is dropped from a tall building. (c) Plot the position, velocity, and acceleration versus time graphs for an object which starts from rest but has a force applied to it in the direction of its motion which increases linearly with time. No other forces act on this object, not even gravity.

Solution

Recognize the principle

In the first situation we are neglecting air drag so the only force we need to consider is the force of gravity. The next situation involves air drag in addition to the force of gravity. By thinking about the relationship between the speed of the object and the drag force it experiences we can predict what the position, velocity, and acceleration versus time graphs will look like. In the third situation we'll need to think about how force and acceleration are related (Newton's second law) and predict the velocity and position versus time graphs from the acceleration versus time graph.

Sketch the problem

The free-body diagram for each situation is as follows:

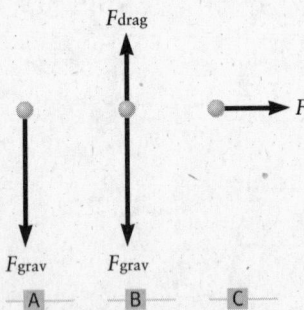

Identify the relationships

(a) Let's begin by defining up as the positive x-direction and assume that the object starts at the origin. The only force on the object is the force of gravity directed in the negative direction. Therefore, the object will have an acceleration of $-g$ after it is launched. Since the acceleration is negative and constant, the velocity of the object as it rises will steadily decrease to zero when the object is at its highest point. After turning around the object will fall as its velocity steadily increases in the negative direction. Since the acceleration is constant throughout the motion, the slope of the velocity graph will also remain constant. As a result of this motion the position versus time graph will form a parabolic curve.

(b) Again, let's define up as the positive *x*-direction. As the object begins to fall under the force of gravity, its speed will increase. As the object falls, the drag force points upward and continues to grow until it equals the force of gravity on the object. At that point the two forces cancel and the acceleration becomes zero. The object continues to fall at a constant speed, the terminal speed.

(c) Newton's second law tells us that the acceleration of an object is proportional to the net force on the object. For this situation the force grows linearly with time so the acceleration must also grow linearly with time. Since the slope of the velocity versus time curve is the acceleration, then the velocity versus time curve must continuously increase in slope, representing an increasing acceleration. The position versus time graph will also continuously increase in slope at an even greater rate.

Solve

Putting this all together we come up with the following graphs.

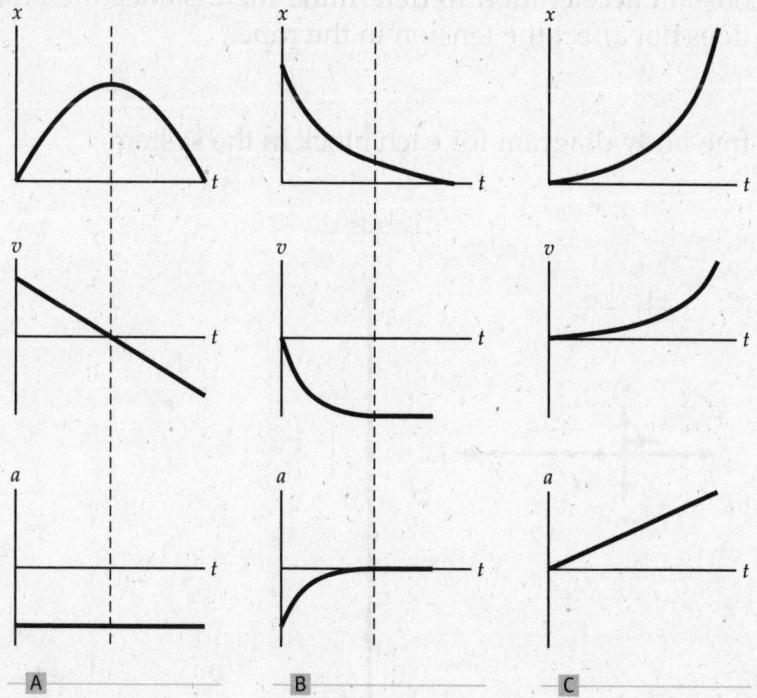

What does it mean?

Just by understanding a few relationships, we can analyze various situations and reason our way through predicting the motion of an object. We'll have more practice with this in Chapter 4.

CP 3.2 Cables, Pulleys, and Friction: Drawing Free-Body Diagrams and Applying Newton's Second Law

In the figure shown three blocks are connected by ropes. The coefficient of kinetic friction between the blocks and the table is 0.1. The pulley is massless and friction-less. (a) Once in motion, determine the tension in the ropes and the acceleration of the system of blocks. (b) Assuming the blocks start from rest, how far will they move in 2 s?

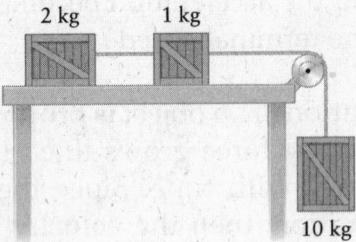

Solution

Recognize the principle

This is an application of Newton's second law including the force of friction. Once we determine the acceleration of the system of blocks we can apply the expressions for motion with constant acceleration to determine the distance the blocks move in 2 s. The pulley does not affect the tension in the rope.

Sketch the problem

We begin by drawing a free-body diagram for each block in the system.

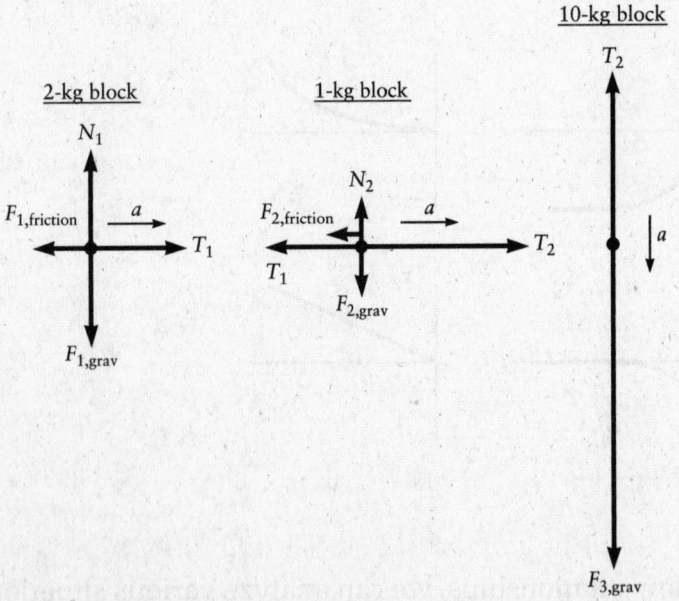

Identify the relationships and Solve

(a) Now that we've drawn free-body diagrams for each block we can apply Newton's second law to each block in both the vertical and horizontal directions. Since we know the direction of the acceleration (this is obvious from the figure), we can use that direction to define a positive direction. With that in mind, any force in the direction of the acceleration will be positive and forces in the opposite direction will be negative.

2-kg block
Vertical direction

$$\Sigma F = N_1 - F_{1,grav} = N_1 - m_1 g = m_1 a$$

Since $a = 0$ in this direction, this reduces to,

$$N_1 = m_1 g$$

Horizontal direction

$$\Sigma F = T_1 - F_{1,friction} = T_1 - \mu_K N_1 = m_1 a$$

Substituting N_1 into this expression and solving for T_1 we find,

$$T_1 = m_1 a + \mu_K m_1 g \ (\#1)$$

1-kg block
Vertical direction

$$\Sigma F = N_2 - F_{2,grav} = N_2 - m_2 g = m_2 a$$

Since $a = 0$ in this direction, this reduces to,

$$N_2 = m_2 g$$

Horizontal direction

$$\Sigma F = T_2 - T_1 - F_{2,friction} = T_2 - T_1 - \mu_K N_2 = m_2 a$$

Substituting N_2 into this expression we find,

$$T_2 - T_1 - \mu_K m_2 g = m_2 a \ (\#2)$$

10-kg block
Vertical direction

$$\Sigma F = F_{3,grav} - T_2 = m_3 g - T_2 = m_3 a$$

Solving for T_2 we find,

$$T_2 = m_3 (g - a) \ (\#3)$$

We now have three equations relating our three unknowns, T_1, T_2, and a. To solve the equations we can substitute Equations #1 and #3 into #2 and solve for the acceleration.

$$m_3 (g - a) - (m_1 a + \mu_K m_1 g) - \mu_K m_2 g = m_2 a$$

$$a = \frac{(m_3 - \mu_K m_1 - \mu_K m_2)g}{m_1 + m_2 + m_3}$$

Plugging in the values we find,

$$a = \frac{[10 \text{ kg} - (0.1)(2 \text{ kg}) - (0.1)(1 \text{ kg})](9.8 \text{ m/s}^2)}{13 \text{ kg}} = \boxed{7.3 \text{ m/s}^2}$$

Substituting the acceleration into Equations #1 and #3 we can solve for the tensions in the ropes.

$$T_1 = (2\text{ kg})(7.3\text{ m/s}^2) + (0.1)(2\text{ kg})(9.8\text{ m/s}^2) = \boxed{17\text{ N}}$$

$$T_2 = 10\text{ kg}(9.8\text{ m/s}^2 - 7.3\text{ m/s}^2) = \boxed{25\text{ N}}$$

(b) Now that we know the acceleration of the system of blocks, we can apply our expressions for motion with constant acceleration to determine how far the blocks move in 2 s. The initial velocity of the blocks is $v_0 = 0$, the acceleration is $a = 7.3\text{ m/s}^2$, and the time is $t = 2$ s. Plugging these values into the following expression we can solve for the distance.

$$x - x_0 = v_0 t + \frac{1}{2}at^2 = 0 + \frac{1}{2}(7.3\text{ m/s}^2)(2\text{ s})^2 = \boxed{15\text{ m}}$$

What does it mean?

The value for the acceleration seems reasonable since it is less than g which it should be! Another approach would have been to treat the three blocks as one 13-kg block and applied Newton's second law. The end results would have been the same.

Part E. MCAT Review Problems and Solutions

PROBLEMS

1. A rock is dropped from a height of 19.6 m above the ground. How long does it take the rock to hit the ground?

   (a) 2 s

   (b) 4 s

   (c) 4.9 s

   (d) 9.8 s

2. A spacecraft exploring a distant planet releases from rest a probe to explore the planet's surface. The probe falls freely a distance of 40 m during the first 4.0 s after its release. What is the acceleration due to gravity on this planet?

   (a) 4.0 m/s^2

   (b) 5.0 m/s^2

   (c) 10 m/s^2

   (d) 16 m/s^2

3. Which property is constant for a body in free fall?

   (a) acceleration

   (b) displacement

   (c) velocity

   (d) speed

4. Which graph represents the relationship between vertical speed (v) and time (t) for an object falling freely near the surface of the Earth?

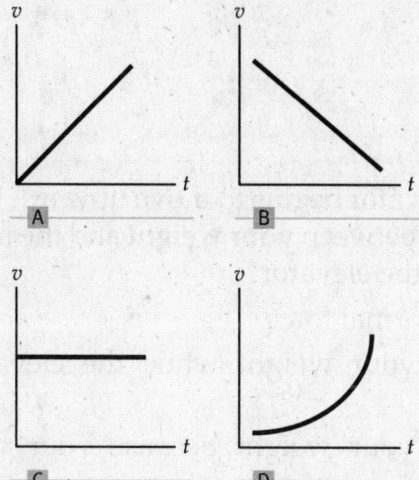

(a) A

(b) B

(c) C

(d) D

5. A ball is dropped from the roof of a very tall building. What is its velocity after falling for 5.0 s?

(a) 1.96 m/s downward

(b) 9.80 m/s downward

(c) 49.0 m/s downward

(d) 98.0 m/s downward

6. What is the weight of a 2.0-kg object on or near the surface of the Earth?

(a) 4.9 N

(b) 16 lbs

(c) 19.6 N

(d) 64 kg · m/s²

7. An object accelerates at 2.5 m/s² when acted upon by a net force of 5.0 N. The mass of the object is

(a) 0.5 kg

(b) 2.0 kg

(c) 12.5 kg

(d) 25 kg

8. On a large asteroid, the force of gravity on a 10-kg object is 20 N. What is the acceleration due to gravity on the asteroid?

   (a) 0.5 m/s^2

   (b) 2.0 m/s^2

   (c) 9.8 m/s^2

   (d) 98 m/s^2

9. You are standing in an elevator as the elevator begins to move upward. Which statement best describes the relationship between your weight and the normal force between your feet and the floor of the elevator?

   (a) Your weight is always equal to the normal force.

   (b) The normal force is greater than your weight while the elevator is accelerating upward.

   (c) The normal force is greater than your weight because your weight decreases.

   (d) The normal force is less than your weight while the elevator is accelerating upward.

10. Suppose you push horizontally on a box that rests on the floor. Which graph best represents the relationship between the frictional force between the box and the floor and your pushing force on the box as you get the box to move?

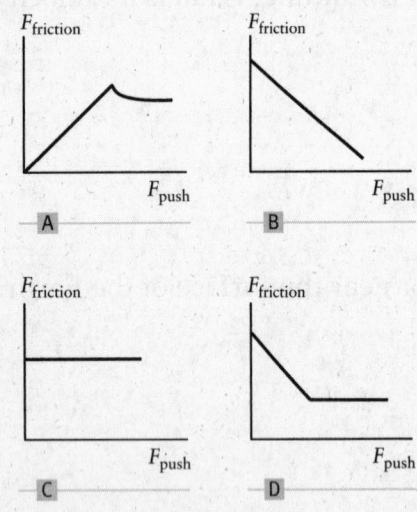

   (a) A

   (b) B

   (c) C

   (d) D

SOLUTIONS

1. *MCAT strategies*

No answers can be easily eliminated so we proceed with our problem-solving strategy.

Recognize the principle

We recognize this as a free-fall problem and so we are able to apply the expressions for motion with constant acceleration.

Sketch the problem

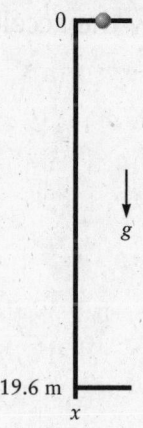

Identify the relationships

By defining the coordinate system as shown in the diagram, we have the following information:

$$x_0 = 0$$
$$x = 19.6 \text{ m}$$
$$v_0 = 0$$
$$a = 9.8 \text{ m/s}^2$$

Also, we can use the following expression to find t:

$$x = x_0 + v_0 t + \frac{1}{2}at^2$$

Solve

Plugging the values into our expression we find,

$$19.6 = 0 + 0t + \frac{1}{2}(9.8 \text{ m/s}^2)t^2$$

Solving for t,

$$t = \sqrt{\frac{2(19.6 \text{ m})}{9.8 \text{ m/s}^2}} = \sqrt{4.0 \text{ s}^2} = 2.0 \text{ s}$$

Therefore, the correct answer is (a).

What does it mean?

By choosing the origin of our coordinate system at the initial position of the rock and defining down as the positive *x*-axis, we simplified the expression we needed to solve.

2. MCAT strategies

No answers can be easily eliminated so we proceed with our problem-solving strategy.

Recognize the principle

We recognize this as a free-fall problem and so we are able to apply the expressions for motion with constant acceleration. This time we do not know the acceleration due to gravity.

Sketch the problem

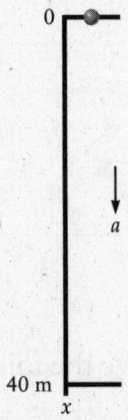

Identify the relationships

By defining the coordinate system as shown in the sketch, we have the following information:

$$x_0 = 0$$
$$x = 40 \text{ m}$$
$$v_0 = 0$$
$$t = 4.0 \text{ s}$$

Also, we can use the following expression to find the acceleration:

$$x = x_0 + v_0 t + \frac{1}{2} a t^2$$

Solve

Plugging the values into our expression we find,

$$40 \text{ m} = 0 + 0 \,(4.0 \text{ s}) + \frac{1}{2} a \,(4.0 \text{ s})^2$$

Solving for a,

$$a = \frac{2(40\,\text{m})}{(4.0\,\text{s})^2} = \frac{80\,\text{m}}{16\,\text{s}^2} = 5.0 \text{ m/s}^2$$

Therefore, the correct answer is (b).

What does it mean?

Again by choosing the origin of our coordinate system at the initial position of the probe and defining down as the positive x-axis, we simplified the expression we needed to solve.

3. MCAT strategies

By definition, a free-falling object experiences a constant acceleration due to gravity. Therefore the correct answer is (a) acceleration. No further analysis is required.

4. MCAT strategies

A free-falling object experiences a constant acceleration due to gravity. Therefore, the speed of the object, released from rest, should increase uniformly (linearly) with time as it falls. Answer (b) is eliminated since it represents an object whose speed decreases linearly with time. Answer (c) can be eliminated since it represents an object moving at constant speed, therefore zero acceleration. Finally, answer (d) can be eliminated since it is a curve thus representing an object whose acceleration changes with time. The only answer which correctly describes a free-falling object is (a). No further analysis is required.

5. MCAT strategies

No answers can be easily eliminated so we proceed with our problem-solving strategy.

Recognize the principle

We recognize this as a free-fall problem and so we are able to apply the expressions for motion with constant acceleration. This time we are asked to find the velocity of the object.

Sketch the problem

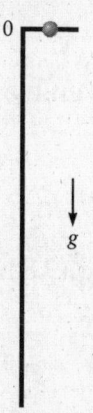

Identify the relationships

By defining the coordinate system as shown in the sketch, we have the following information:

$$x_0 = 0$$
$$v_0 = 0$$
$$a = 9.8 \text{ m/s}^2$$
$$t = 5.0 \text{ s}$$

Also, we can use the following expression to find the acceleration:

$$v = v_0 + at$$

Solve

Plugging the values into our expression and solving we find,

$$v = 0 + (9.8 \text{ m/s}^2)(5.0 \text{ s}) = 49 \text{ m/s}$$

Therefore, the correct answer is (c).

What does it mean?

Again by choosing the origin of our coordinate system at the initial position of the ball and defining down as positive, we simplified the expression we needed to solve. Our answer is positive since we defined down as positive.

6. MCAT strategies

Although the answers use different units, all are units of force. Therefore, no answers can be easily eliminated so we proceed with our problem-solving strategy.

Recognize the principle

This is an application of Newton's second law. Specifically, the weight of an object is the force of gravity on that object. Also, the acceleration of an object as a result of the force of gravity is $g = 9.8 \text{ m/s}^2$ downward as long as no other forces are present.

Sketch the problem

None required.

Identify the relationships

The weight of an object is the force of gravity on the object and is given by,

$$F_{grav} = mg$$

Solve

Using $m = 2$ kg, we find the weight of the object to be $F_{grav} = mg = (2 \text{ kg})(9.8 \text{ m/s}^2) = 19.6$ N. Therefore, the correct answer is (c).

What does it mean?

It's easy to remember that an apple has a mass of about 1 kg, therefore a weight of approximately 10 N.

7. MCAT strategies

No answers can be easily eliminated so we proceed with our problem-solving strategy.

Recognize the principle

This is a straightforward application of Newton's second law. We are given the net force and the acceleration of an object and are asked to find its mass.

Sketch the problem

None required.

Identify the relationships

From the statement of the problem we have,

$$F_{net} = \Sigma F = 5.0 \text{ N}$$
$$a = 2.5 \text{ m/s}^2$$

Also, Newton's second law states that

$$\Sigma F = ma$$

Solve

Plugging in the given values and solving for the mass we find,

$$a = \frac{\Sigma F}{a} = \frac{5.0 \text{ N}}{2.5 \text{ m/s}^2} = 2.0 \text{ kg}$$

So the correct answer is (b).

What does it mean?

Checking the units for our final answer, we note that we have used the correct expression for Newton's second law.

8. MCAT strategies

Unless we are dealing with an asteroid which is larger than Earth (not very likely), then the acceleration due to the asteroid's gravity should be less than 9.8 m/s². Therefore, answers (c) and (d) can be eliminated.

Recognize the principle

Again this is a straightforward application of Newton's second law. We are given the force of gravity on an object and its mass and asked for the acceleration due to gravity.

Sketch the problem

None required.

Identify the relationships

From the statement of the problem we have,

$$F_{grav} = 20 \text{ N}$$
$$m = 10 \text{ kg}$$

Also, Newton's second law states that

$$\Sigma F = ma$$

Solve

Plugging in the given values and solving for the mass we find,

$$a = \frac{\Sigma F}{m} = \frac{F_{grav}}{m} = \frac{20 \text{ N}}{10 \text{ kg}} = 2.0 \text{ m/s}^2$$

So the correct answer is (b).

What does it mean?

Again, checking the units of our final answer we can be sure that we have used the correct expression.

9. MCAT strategies

No answers can be easily eliminated so we proceed with our problem-solving strategy.

Recognize the principle

This is an application of Newton's second law. Since we are not given numerical values, we will need to think about the relative size of the forces involved in order to answer the question.

Sketch the problem

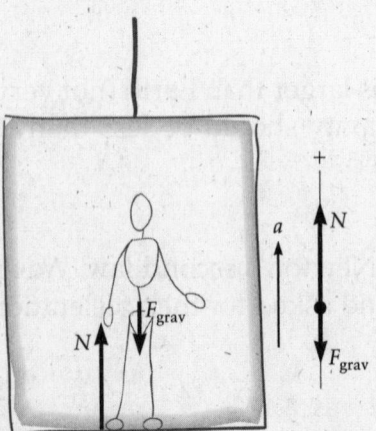

Identify the relationships

By defining the coordinate system as in the sketch, choosing the positive direction as the direction of the acceleration, and applying Newton's second law we obtain the following expression,

$$\Sigma F = ma$$

$$N - F_{\text{grav}} = ma$$

Therefore,

$$N = F_{\text{grav}} + ma$$

Solve

Since, according to our coordinate system, your acceleration in the elevator is positive, then the normal force must be greater than the force of gravity on you (your weight). So, the correct answer is (b).

What does it mean?

If the elevator was accelerating downward then a would be negative and the normal force would be less than the force of gravity. This is consistent with the fact that you feel less force on your feet as an elevator begins to drop.

10. MCAT strategies

No answers can be easily eliminated so we proceed with our problem-solving strategy.

Recognize the principle

We need to understand the two types of frictional force that exist, the force of static friction and the force of kinetic friction.

Sketch the problem

None required.

Identify the relationships

If the surfaces are slipping relative to each other, the opposing force is kinetic friction given by

$$F_{\text{friction}} = \mu_K N$$

where N is the normal force. If the surfaces are not slipping, the resistive force is static friction given by

$$F_{\text{friction}} \leq \mu_S N$$

Solve

As you begin to push horizontally on the crate the force of static friction will grow proportionally to your force so that it matches your force. This must be the case since the object remains stationary and therefore the net force on the crate must be zero. Once the maximum force of static friction is attained the crate will begin to move. At this point the force of friction becomes kinetic friction and remains constant. Since the coefficient of kinetic friction is less than the coefficient of static friction, the force of kinetic friction is less than the maximum force of static friction. Putting this all together we see that the correct answer is (a).

What does it mean?

Since the maximum force of static friction is greater than the force of kinetic friction, it takes a greater force to get the crate to start moving than to keep it moving.

4 Forces and Motion in Two and Three Dimensions

CONTENTS

Part A. Summary of Key Concepts and Problem-Solving Strategies

KEY CONCEPTS

Translational equilibrium

When the total force on an object is zero, Newton's second law tells us that the acceleration is also zero. This means that all of the vector components of $\vec{F}_{\text{total}} = \Sigma \vec{F}$ and $\vec{a}$ are zero. Such an object is in *translational equilibrium*.

Analyzing motion in two and three dimensions

We use Newton's second law to calculate the acceleration of an object.

$$\Sigma \vec{F} = m\vec{a}$$

The acceleration can then be used to find the velocity and displacement.

Inertial and noninertial reference frames

A *reference frame* is an observer's choice of coordinate system, including an origin, for making measurements. An *inertial reference frame* is one that moves with a constant velocity, while a *noninertial* frame is one that is accelerating. Newton's laws are only obeyed in inertial reference frames. They are not obeyed in noninertial frames.

APPLICATIONS

Projectile motion

In many cases the force of air drag is small. The force on a projectile near the Earth's surface, such as a baseball, is then just the force of gravity, and the acceleration has a constant magnitude g directed downward. The displacement and velocity for such a projectile are described by the relations for motion with constant acceleration.

For projectile motion, the vertical and horizontal motions are independent. Two objects can have very different horizontal velocities, but they still fall at the same rate.

The trajectory for simple projectile motion starting from the origin is:

- Symmetric in time. The time spent traveling to the point of maximum height is equal to the time spent falling back down to the initial height.

- Symmetric in space. The trajectory has a parabolic shape.

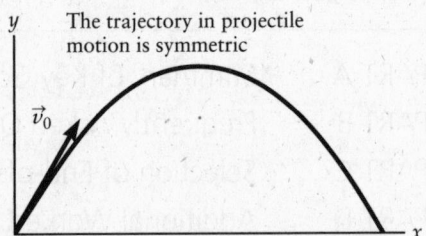

PROBLEM-SOLVING STRATEGIES

Two main problem-solving strategies were introduced in the chapter. The first is how to tackle problems in statics and the second involves the application of Newton's second law.

When the total force on an object is zero, Newton's second law tells us that the acceleration is also zero. This means that all of the vector components of $\Sigma \vec{F} = 0$ and $\vec{a}$ are zero. Such an object is in *translational equilibrium*. Examples of situations involving translational equilibrium are hanging objects, stationary blocks on inclines, and objects moving at constant velocity. The forces involved may include gravity, tension, friction, normal forces, and air drag.

When an object is not in translational equilibrium the net force on the object is not zero and so the object must be accelerating. We still use Newton's second law, but now the sum of the forces results in acceleration of the mass, $\Sigma \vec{F} = m\vec{a}$. If the mass and forces are known, the acceleration can be determined and then used to find the velocity and displacement of the object. If the acceleration is known, we can often determine some unknown quantity such as a force or the mass of the object. Examples of motion in two and three directions include projectile motion, sliding blocks, and rolling balls. The forces may include gravity, tension, normal forces, and air drag.

Problem Solving: Plan of Attack for Problems in Statics

Recognize the principle

Any time the net force on an object is zero the object is considered to be translational equilibrium.

Sketch the problem

A quick sketch of the problem will help you visualize the physical situation. The sketch should include a coordinate system.

Identify the relationships

- Determine all the forces acting on the object of interest and construct a free-body diagram showing all of the forces on the object.
- Express all of the forces on the object in terms of their components along x, y, and perhaps z.
- Apply the conditions $\Sigma F_x = 0$, $\Sigma F_y = 0$, and $\Sigma F_z = 0$

Solve

Solve these equations for the unknown quantities. As always, the number of equations must be greater than or equal the number of unknown quantities in order to solve for the unknowns.

What does it mean?

Always *consider what your answer means*, and check that it makes sense.

Problem Solving: Applying Newton's Second Law

Recognize the principle

Any time the net force on an object is not zero the object is accelerating and we can analyze its motion using Newton's second law.

Sketch the problem

Your picture should define a coordinate system and contain all of the forces in the problem. It is usually a good idea to also show all of the given information.

Identify the relationships

- Find all of the forces acting on the object of interest (the object whose motion you wish to describe), and construct a free-body diagram.
- Express all of the forces in terms of their components along x, y, and perhaps z.
- Apply Newton's second law in component form,
$$\Sigma F_x = ma_x, \Sigma F_y = ma_y, \text{and } \Sigma F_z = ma_z$$
- If the acceleration is constant along x, y, or z, you can apply the kinematic equations from Chapter 3 for the motion along that direction.

Solve

Solve the equations resulting from Newton's second law for the unknown quantities in terms of the known quantities. The number of equations must equal the number of unknown quantities.

What does it mean?

Always *consider what your answer means*, and check that it makes sense.

Part B. Frequently Asked Questions

1. *How do I decide where to place the origin of my coordinate system?*

For the most part that's really up to you. Certainly when you draw a free-body diagram the origin of the coordinate system represents the object of interest. However, for projectile motion you can define the origin wherever you'd like. It is often convenient to place the origin at the point where the object begins its motion, and sometimes the mathematics is a little easier by choosing the end point of the motion as your origin. As you tackle more problems you'll develop an intuition as to where it makes sense to place the origin, but the results will not depend on your choice.

2. *Do we always have up as the positive y-axis and horizontal as the x-axis?*

No. Although this is a common way to define the coordinate system for projectiles, it is not the only way. You could define down as the positive y-axis or up as the positive x-axis, or even use z as the horizontal coordinate. Our choice of the coordinate system does not affect the outcome of the problem. The choice of coordinate system can affect the complexity of the mathematics required to solve the problem. For instance, for a standard projectile problem if you chose the x-axis to be 20° above the horizontal and the y-axis to be 20° shifted from vertical, you would complicate the problem with unnecessary sines and cosines.

3. *If the object is not accelerating, how can it be moving?*

If an object is not accelerating we can conclude that the net force on the object is zero. Since it is not accelerating its velocity must remain constant. While it is true that a velocity of zero is a constant velocity, constant velocity is NOT necessarily zero. Consider an object falling under the influence of gravity and air drag. As the object's speed increases, the drag force increases until it is equal in magnitude to the force of gravity but opposite in direction. At that point the net force on the object is zero and the object stops accelerating. Does it stop in mid air? No, it simply continues to fall but at a constant velocity. We call this its terminal velocity. There are many other examples of objects moving without accelerating.

4. *At the top of a trajectory, the vertical component of the velocity becomes zero. Doesn't this mean that the vertical acceleration becomes zero as well?*

This is a great question which can be answered with the following thought experiment: Suppose you toss a ball straight up under the influence of gravity. As we know, the velocity of the ball when it reaches its maximum height becomes zero. Now, suppose that its acceleration also becomes zero. What would happen to the ball? Well, if the ball is stationary and not accelerating, then it should remain stationary. Clearly this does not happen so the acceleration does not become zero at the top of its trajectory. Newton's second law tells us that if there is a net force on an object then it will accelerate. At the top of a projectile's trajectory, gravity still exists and so the ball must still be accelerating. The gravitational acceleration slows the ball down as it rises and speeds it up as it falls. At the moment it changes direction (at the top of the trajectory) the vertical component of its velocity only momentarily becomes zero. There are many other examples of objects whose velocity becomes zero when the acceleration does not.

5. *When analyzing the forces on an object on an incline, I always have trouble determining the components of the force of gravity on the object. Is the component which is along the incline always mgsin(θ)?*

No. It depends on where you define the angle θ and how you define your coordinate system. Consider the following diagram of a block on an incline:

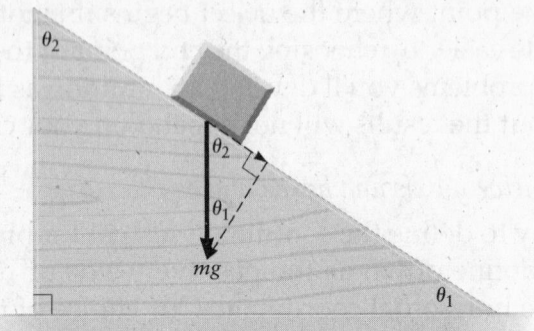

Comparing the triangle formed by the incline and the triangle formed by the gravitational force *mg* and its components, we can identify the two angles θ_1 and θ_2 in the component triangle. The component of the force of gravity on the object which is directed downward along the incline (parallel to the incline) is given by

$$mg \sin \theta_1 = mg \cos \theta_2$$

Similarly, the component of the force of gravity on the object which is directed into the incline (perpendicular to the incline) is given by

$$mg \cos \theta_1 = mg \sin \theta_2$$

Depending on which angle is given in the problem and how you define your system, the component you're looking for may be the sine, cosine, or even the tangent of an angle. It's best to analyze every situation rather that remember it as always being one way.

6. *When is the normal force on an object important to consider?*

Most often the normal force on an object is important for determining the frictional force on that object. Both the static and kinetic friction forces depend on the coefficients of friction and the normal (or contact) force between the object and a surface. Normal forces cannot make an object move, but they can prevent an object from moving. For example, an object sitting on a horizontal table is being acted upon by the force of gravity (its weight) and the normal force from the table. These two forces cancel each other out so the object remains at rest. If the normal force were eliminated by removing the table, the object would accelerate under the influence of gravity.

7. *When dealing with blocks on inclines, how do I know which way to draw the force of friction?*

That all depends on which way the block is moving or would move if there was no friction. Remember, the force of friction is always in a direction opposing the relative motion or pending motion of the two surfaces which are in contact. So, if a block just sitting or sliding on an inclined surface with no forces other than friction, the force of gravity, and the normal force acting on the block, then the direction of the frictional force will be up the incline since it's sliding (or tends to slide) down the incline. If the block is being pulled up the incline, then the force of friction will be directed down the incline.

8.   *Can the acceleration of a projectile be greater than 9.8 m/s²?*

If the projectile is in free fall then the only force acting on the projectile is the force of gravity and so it will accelerate at 9.8 m/s² downward. If the projectile has other forces acting on it like the air drag on a windy day or the thrust of its rockets, then it could certainly have an acceleration less than or greater than 9.8 m/s².

9.   *When asked for the final velocity of a projectile, how do I find the direction?*

We can determine the direction of the final velocity (or the velocity at any point) by first finding the components of the velocity in both the horizontal and vertical directions. If the only force acting on the projectile is the force of gravity, then this is quite easy since we can use the kinematic equations for motion with constant acceleration we derived in Chapter 3. If other forces are involved then we need to apply Newton's second law as we've done in a few worked examples. Once we find the components of the velocity, then we simply use the Pythagorean theorem and some trigonometry to determine the speed and direction. The following example will help illustrate this:

> Suppose an object is moving with a speed in the horizontal direction of 10 m/s and a downward speed of 5 m/s. As showing in the diagram, the speed of the object is given by using the Pythagorean theorem and the angle θ is found using $\tan\theta$.

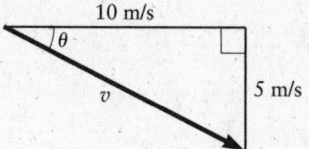

$$v^2 = (5 \text{ m/s})^2 + (10 \text{ m/s})^2 \rightarrow v = \sqrt{25 \text{ m}^2/\text{s}^2 + 100 \text{ m}^2/\text{s}^2} = \boxed{11 \text{ m/s}}$$

$$\tan\theta = \left(\frac{5 \text{ m/s}}{10 \text{ m/s}}\right) \rightarrow \theta = \tan^{-1}\left(\frac{5 \text{ m/s}}{10 \text{ m/s}}\right) = \boxed{27°}$$

> So, we would say that the velocity of the object is "11 m/s, 27° below the horizontal." Using our convention that angles are measured counterclockwise from the horizontal axis we would state this as "11 m/s, −27°" or "11 m/s, 333°."

10.   *As a projectile travels in a parabolic path, is its acceleration −9.8 m/s² on the way up and +9.8 m/s² on the way down?*

No. The sign (+/−) of the acceleration of an object is determined by our choice of coordinate system, not by the direction the object is traveling. For example, suppose we choose up as a positive direction. With this choice of coordinate system, the object will have an acceleration of −9.8 m/s² on its way up and on its way down. However, if we choose down as a positive direction for our coordinate system, then the acceleration of the object will be +9.8 m/s² throughout its trajectory. Although the effect of this downward acceleration is to slow the object down on its way up, and speed it up on its way down, the direction of the acceleration never changes.

Part C. Selection of End-of-Chapter Answers and Solutions

QUESTIONS

Q4.7   Consider again the wedge in Question 6, but now assume a block is placed onto it as shown in Figure Q4.7. There is again no friction between the wedge and the table, and there is also no friction between the block and the wedge. Explain why in this case the wedge *will* be accelerated to the left. How does this situation differ from that in Question 6? *Hint*: Compare the free-body diagrams for the wedge in the two cases.

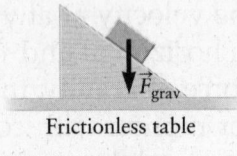

Frictionless table

Figure Q4.7

Answer

We begin by drawing the free-body diagram for the wedge and comparing this with the free-body diagram from Q4.6. In Q4.6 there are three forces acting on the wedge: the force of gravity, the normal force of the table on the wedge, and the applied force, F. In this situation, there are also three forces acting on the wedge: the force of gravity, the normal force of the table on the wedge, and the normal force of the block on the wedge.

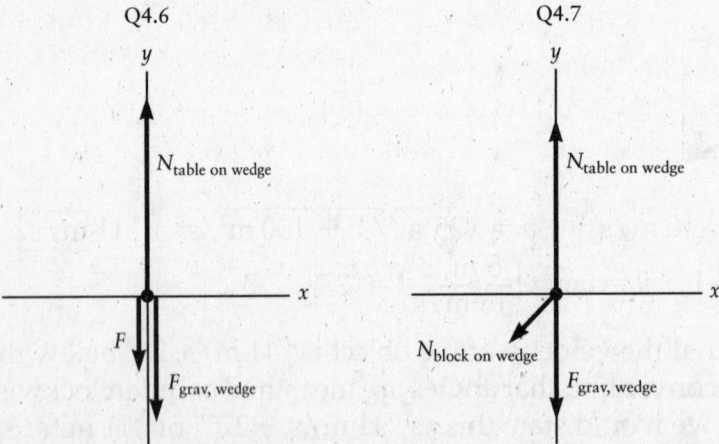

We see that in the y direction there is still no net force so on the wedge does not accelerate in that direction. However, we now have a net force in the $-x$ direction resulting in the acceleration of the wedge in that direction.

Q4.20   Two balls are thrown into the air with the same initial speed, directed at the same initial angle with respect to the horizontal. The mass of ball 1 is 5 times the mass of ball 2, and the force of air drag is negligible.

(a) Which ball has the larger acceleration as it moves through the air?
(b) Which ball lands first?

(c) Which ball reaches the greatest height?

(d) For which ball is the force of gravity larger at the top of the trajectory?

Answer

(a) Since air drag is negligible, both balls are considered to be in free fall and accelerate at the same rate due to gravity, 9.8 m/s² downward. The horizontal acceleration for the balls is zero.

(b) Again, since we can neglect air drag we can apply the expressions for kinematic motion for objects moving with constant acceleration (see Chapter 3). Both balls are launched at the same angle with the same speed and therefore follow the same trajectory, landing at the same point at the same time. This is clear by looking at the kinematic equations and observing that they don't depend on the mass of the object.

(c) By the same reasoning as part (b), both balls will reach the same maximum height.

(d) Since the force of gravity on an object is proportional to its mass ($F_{grav} = mg$), ball 1 will experience 5 times the force of gravity as ball 2.

PROBLEMS

P4.3   Several forces act on a particle as shown in Figure P4.3. If the particle is in translational equilibrium, what are the values of F_3 (the magnitude of force 3) and θ_3 (the angle that force 3 makes with the x axis)?

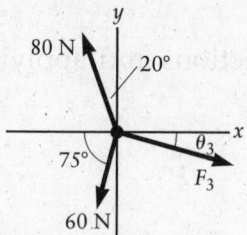

Figure P4.3

Solution

Recognize the principle

We will need to apply Newton's second law. Since the particle is in translational equilibrium as stated in the problem, then the net force on the particle must be zero.

Sketch the problem

We begin by drawing the free-body diagram for the particle and identify the x and y components of all the forces acting on the particle.

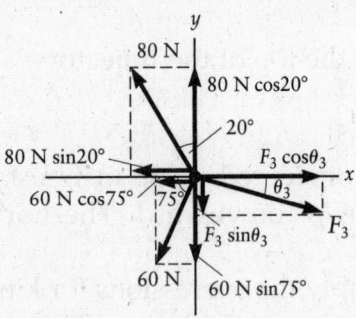

Identify the relationships

Now we can write the components of the all the forces and apply Newton's second law.

For 80-N force,

$$F_{80Nx} = -(\sin 20°)80 \text{ N}$$
$$F_{80Ny} = (\cos 20°)80 \text{ N}$$

For 60-N force,

$$F_{60Nx} = -(\cos 75°)60 \text{ N}$$
$$F_{60Ny} = -(\sin 75°)60 \text{ N}$$

For F_3,

$$F_{3x} = F_3 \cos \theta_3$$
$$-F_{3y} = F_3 \sin \theta_3$$

Applying Newton's second law in both the x and y directions and applying the conditions for static equilibrium, we have,

$$\Sigma F_x = -(\sin 20°)80 \text{ N} - (\cos 75°)60 \text{ N} + F_{3x} = 0$$
$$\Sigma F_y = (\cos 20°)80 \text{ N} - (\sin 75°)60 \text{ N} - F_{3y} = 0$$

Solve

Solving for components of F_3 and computing its magnitude and angle from these components:

$$F_{3x} = (\sin 20°)80 \text{ N} + (\cos 75°)60 \text{ N} = 42.9 \text{ N}$$
$$F_{3y} = (\cos 20°)80 \text{ N} - (\sin 75°)60 \text{ N} = 17.2 \text{ N}$$

$$F_3 = \sqrt{F_{3x}^2 + F_{3y}^2} = \sqrt{(42.9 \text{ N})^2 + (17.2 \text{ N})^2} = \boxed{46 \text{ N}}$$

$$\theta_3 = \tan^{-1}\left(\frac{-F_{3y}}{F_{3x}}\right) = \boxed{-22°}$$

What does it mean?

From the free-body diagram we can visually confirm that if the three vectors are added together (tip-to-tail method) they will sum to zero (they will form a triangle) as required by the equilibrium condition.

P4.22 An airplane is flying horizontally with a constant velocity of 200 m/s at an altitude of 5000 m when it releases a package. (a) How long does it take the package to reach the ground? (b) What is the distance between the airplane and the package when the package hits the ground? (c) What is the distance between the airplane and the package's landing spot, as measured along the x direction, when the package is released? Ignore air drag.

Solution

Recognize the principle

This is a problem involving projectile motion.

Sketch the problem

We begin by drawing a sketch of the problem, defining a coordinate system and including given information.

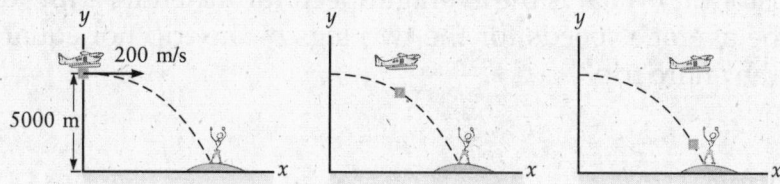

Identify the relationships

This is projectile problem. Since we can neglect air drag, the acceleration in the x direction is zero and the acceleration in the y direction is $-g$. Both of these accelerations remain constant so we can apply the expressions for motion with constant acceleration we derived in Chapter 3. Since the airplane is flying horizontally, there will be no initial y component to the velocity. So, using this information and the quantities given in the problem, the expressions for both the x and y directions are,

$$x = x_0 + v_{0,x}t + \frac{1}{2}a_x t^2 \rightarrow x - x_0 = (200 \text{ m/s})\, t$$

$$y = y_0 + v_{0,y}t + \frac{1}{2}a_y t^2 \rightarrow 0 = 5000 \text{ m} + \frac{1}{2}(-g)t^2$$

Solve

(a) We can solve for the time it takes the package to reach the ground using the expression for the y direction,

$$0 = 5000 \text{ m} - \frac{1}{2}gt^2$$

$$t = \sqrt{\frac{2(5000 \text{ m})}{g}} = \sqrt{\frac{2(5000 \text{ m})}{9.8 \text{ m/s}^2}} = \boxed{32 \text{ s}}$$

(b) Since the airplane and the package have the same horizontal speed (200 m/s), then when the package hits the ground the airplane will be directly overhead. At that point the distance between the airplane and the package is simply the altitude of the airplane which is $\boxed{5000 \text{ m}}$.

(c) Since we determined the time of flight for the package in part (a), we can simply use this time to determine the horizontal distance the package traveled in this time using the expression for the x direction. This is also the horizontal distance the airplane travels in the 32 s.

$$x - x_0 = (200 \text{ m/s})t = (200 \text{ m/s})(32 \text{ s}) = \boxed{6400 \text{ m}}$$

What does it mean?

Knowing the altitude and velocity of the airplane, it is possible to calculate how far ahead of a target to drop the payload. Of course, we neglected the effects of air drag which would have complicated the solution; however, in many situations this is a reasonable approximation.

P4.40 An airplane flies from Boston to San Francisco (a distance of 5000 km) in the morning and then immediately returns to Boston. The airplane's speed relative to the air is 250 m/s. The wind is blowing at 50 m/s from west to east, so it is "in the face" of the airplane on the way to San Francisco and it is a tailwind on the way back. (a) What is the average speed of the airplane relative to the ground on the way to San Francisco? (b) What is the average speed relative to the ground on the way back to Boston? (c) What is the average speed for the entire trip? (d) Why is the average of the average speeds for the two legs of the trip not equal to the average speed for the entire trip?

Solution

Recognize the principle

We can apply the equations and concepts of relative velocities and kinematics in one dimension.

Sketch the problem

The following sketch helps illustrate the problem.

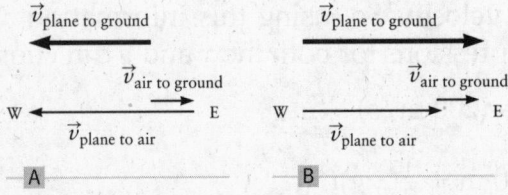

Identify the relationships

The velocity of the plane relative to the ground can be found by the vector addition of the velocity of the airplane with respect to the air and the velocity of the air with respect to the ground. Average speed is the total distance divided by the total time for the journey.

Solve

(a) On the way to San Francisco, the airplane moves with a speed of 250 m/s in the west direction relative to the air, however the air is moving in the opposite direction with a speed of 50 m/s relative to the ground. So, relative to the ground, the airplane is moving with an average speed of 250 m/s − 50 m/s = $\boxed{200 \text{ m/s.}}$

(b) On the way back to Boston, the airplane moves with a speed of 250 m/s in the east direction relative to the air, and the air is moving in the same direction with a speed of 50 m/s relative to the ground. So, relative to the ground, the airplane is moving with an average speed of 250 m/s + 50 m/s = $\boxed{300 \text{ m/s.}}$

(c) The time it takes the airplane to fly to San Francisco is the distance divided by the average speed,

$$t_{\text{San Francisco}} = \frac{5{,}000{,}000 \text{ m}}{200 \text{ m/s}}$$

$$t_{\text{San Francisco}} = 25{,}000 \text{ s}$$

The time it takes the airplane to fly back to Boston is the distance divided by that journey's average speed,

$$t_{\text{Boston}} = \frac{5{,}000{,}000 \text{ m}}{300 \text{ m/s}}$$

$$t_{\text{Boston}} = 17{,}000 \text{ s}$$

So, the average speed for the entire trip is then the total distance divided by the total time,

$$v_{\text{total}} = \frac{10{,}000{,}000 \text{ m}}{25{,}000 \text{ s} + 17{,}000 \text{ s}}$$

$$\boxed{v_{\text{total}} = 240 \text{ m/s}}$$

(d) Since the times for the two legs of the trip are different, the airplane spends more time traveling at the lower speed. This results in the average speed for the entire trip being a bit less than the average of the speeds in (a) and (b).

What does it mean?

Note that even though the headwind and the tailwind are the same, the overall average speed for the entire round trip is less that the air speed of the plane. One might expect a tailwind and headwind of equal magnitude would have canceled out. This is not the case here!

P4.53 A skier is traveling at a speed of 40 m/s when she reaches the base of a frictionless ski hill. This hill makes an angle of 10° with the horizontal. She then coasts up the hill as far as possible. What height (measured vertically above the base of the hill) does she reach?

Solution

Recognize the principle

We can apply Newton's second law and the equations for motion under a constant acceleration.

Sketch the problem

We begin by drawing a free-body diagram of the situation and defining a coordinate system. For this situation it seems reasonable to define the *x* axis up the slope and the *y* axis perpendicular to the slope.

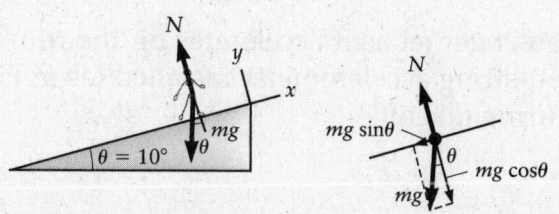

Identify the relationships

Using the free-body diagram we can apply Newton's second law for both the x and y directions,

$$\Sigma F_x = -mg\sin\theta = ma_x$$
$$\Sigma F_y = N - mg\cos\theta = ma_y = 0$$

We can find the distance traveled up the slope using one of our kinematic expression for motion with constant acceleration,

$$v_x^2 = v_{0x}^2 + 2a_x(x - x_0)$$

Solve

From the equation applying Newton's second law along the x direction we have,

$$a_x = -g\sin\theta$$

Plugging this acceleration into our kinematic expression and knowing that her final speed is zero, we can calculate the distance up the incline she traveled,

$$v_x^2 = v_{0x}^2 + 2a_x(x - x_0)$$
$$0 = v_{0x}^2 + 2a_x(x - 0)$$
$$x = \frac{-v_{0x}^2}{2a_x}$$

Inserting the value for the acceleration up the incline,

$$x = \frac{-v_{0x}^2}{2(-g\sin\theta)}$$

Inserting the values,

$$x = \frac{(40\text{ m/s})^2}{2(9.8\text{ m/s}^2)\sin 10°}$$
$$x = 470\text{ m}$$

To find the vertical height, h, we need to multiply by $\sin\theta = \sin 10°$,

$$h = (470\text{ m})(\sin 10°)$$

$$\boxed{h = 82\text{ m}}$$

What does it mean?

By choosing a coordinate system with one axis along the direction of motion, we simplified the calculations.

P4.58 Consider a commercial passenger jet as it accelerates on the runway during takeoff. This jet has a rock-on-a-string accelerometer installed (as in Fig. 4.30). Estimate the angle of the string during takeoff.

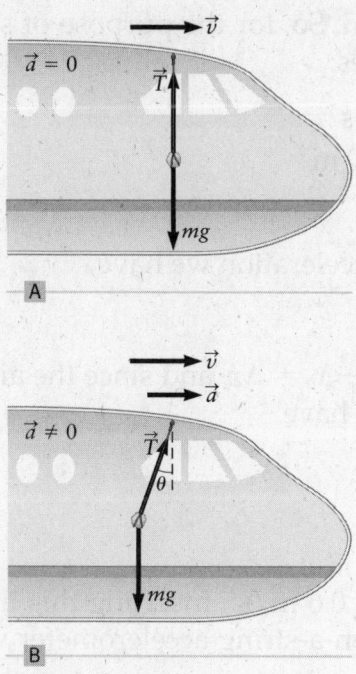

Figure 4.30

Solution

Recognize the principle

We can apply the kinematic expression in one dimension and Newton's second law.

Sketch the problem

The following sketch helps to illustrate the problem.

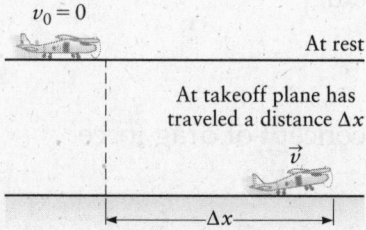

Identify the relationships

We assume the airplane starts from rest and has a constant acceleration until the plane leaves the ground. Using the expression, $v^2 = v_0^2 + 2a\Delta x$, we can determine the acceleration. We will need to find or estimate the distance traveled by the airplane (the takeoff distance), and the takeoff velocity.

For most airplanes one can look up approximate values for takeoff velocity and takeoff distance. These values depend on a variety of factors including air temperature, wind speed, and airplane weight. For a Boeing 747 takeoff velocities can typically range from 180 km/h to 250 km/h (50 m/s to 70 m/s), and takeoff distances can range from about 2000 m to 4000 m. So, for the purpose of solving this problem we will make the following estimates:

Estimate the takeoff velocity, v to be 60 m/s

Estimate the takeoff distance, Δx to be 3000 m

Solve

Using our expression for motion with constant acceleration we have,

$$v^2 = v_0^2 + 2a(x - x_0)$$

The distance traveled down the runway is just $x - x_0 = \Delta x$, and since the airplane starts from rest, $v_0 = 0$. Inserting these values, we have

$$v^2 = 0 + 2a(\Delta x)$$

$$a = \frac{v^2}{2\Delta x}$$

Using $v = 60$ m/s and $\Delta x = 3000$ m we find $a = 0.6$ m/s². Inserting this into the expression derived in the textbook for the rock-on-a-string accelerometer we can solve for the angle,

$$a = g\tan\theta$$

$$\theta = \tan^{-1}\left(\frac{a}{g}\right) = \tan^{-1}(0.6 \text{ m/s}^2/9.8 \text{ m/s}^2)$$

$$\boxed{\theta \approx 4°}$$

What does it mean?
The acceleration is significantly less than g, which seems reasonable.

P4.59   Estimate the terminal velocity for a golf ball.

Solution

Recognize the principle
We need to apply Newton's second law and the concept of drag force.

Sketch the problem
We begin by drawing a sketch of the problem and a free-body diagram.

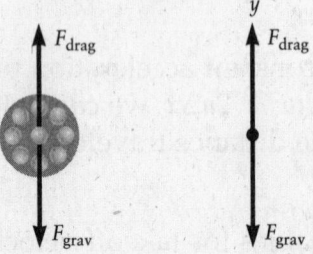

Identify the relationships

At terminal velocity the acceleration of the golf ball will be zero. Since the motion will be along y, we can apply Newton's second law along that direction and solve for the terminal velocity.

$$\Sigma F_y = F_{\text{drag}} - F_{\text{grav}} = ma_y = 0$$

Using our expression for the drag force and $F_{\text{grav}} = mg$, we have,

$$\frac{1}{2}\rho A v_{\text{term}}^2 - mg = 0$$

$$v_{\text{term}} = \sqrt{\frac{2mg}{\rho A}}$$

To determine the terminal velocity we need the mass of a typical golf ball and its radius.

Estimate the mass, m, to be 50 g.
Estimate the radius, r, to be 2 cm.

Solve

Inserting these values and using $A = \pi r^2$,

$$v_{\text{term}} = \sqrt{\frac{2(0.05 \text{ kg})(9.8 \text{ m/s}^2)}{(1.3 \text{ kg/m}^3)\pi(0.02 \text{ m})^2}}$$

$$\boxed{v_{\text{term}} \approx 20 \text{ m/s}}$$

What does it mean?

This is about 45 mi/h. If dropped from a tall building this would be the maximum speed attained by the ball.

P4.64 Consider the system of blocks in Figure P4.64, with $m_2 = 5.0$ kg and $\theta = 35°$. If the coefficient of static friction between block 1 and the inclined plane is $\mu_S = 0.25$, what is the largest mass m_1 for which the blocks will remain at rest?

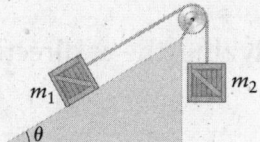

Figure P4.64

Solution

Recognize the principle

We will need to apply Newton's second law, the expression for maximum static friction, and the condition for translational equilibrium.

Sketch the problem

We begin by drawing a sketch of the problem and a free-body diagram for each of the blocks.

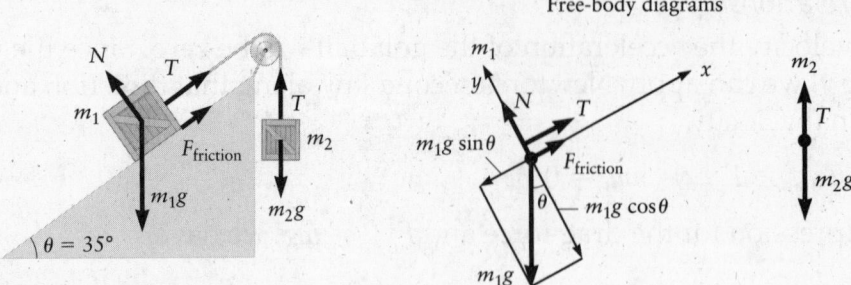

Free-body diagrams

Identify the relationships

From the free-body diagram for m_2, we can apply Newton's second law and set $a = 0$, indicating that the system remains at rest.

$$\Sigma F = T - m_2 g = ma = 0$$

Similarly we can apply Newton's second law for m_1 in both the x and y directions with the x-axis aligned with the incline.

$$\Sigma F_x = T + F_{\text{fric}} - m_1 g \sin \theta = m_1 a_x = 0$$

$$\Sigma F_y = N - m_1 g \cos \theta = m_1 a_y = 0$$

Also, we have the expression for the maximum static friction, $F_{\text{fric}} = \mu_S N$.

Solve

Solving the expression involving m_2 we have,

$$\Sigma F = T - m_2 g = ma = 0$$

$$T = m_2 g$$

From our analysis of m_1 we can solve for the normal force and use this to determine the force of static friction,

$$N = m_1 g \cos \theta$$

$$F_{\text{fric}} = \mu_S m_1 g \cos \theta$$

Inserting this into the expression we obtained by analyzing the x direction we get,

$$T + \mu_S m_1 g \cos \theta - m_1 g \sin \theta = 0$$

Solving for tension, T,

$$T = -\mu_S m_1 g \cos \theta + m_1 g \sin \theta$$

Combining this with our previous expression for T we have,

$$-\mu_S m_1 g \cos \theta + m_1 g \sin \theta = m_2 g$$

Solving for m_1, we have

$$m_1 = \frac{m_2}{-\mu_S \cos \theta + \sin \theta}$$

and substituting in the numerical values we find

$$m_1 = \frac{m_2}{-\mu_S \cos\theta + \sin\theta} = \frac{5.0 \text{ kg}}{-0.25\cos 35° + \sin 35°} = \boxed{14 \text{ kg}}$$

What does it mean?

By carefully defining our coordinate system we were able to simplify the mathematics necessary to solve the problem.

P4.79 Two blocks of mass m_1 and m_2 are sliding down an inclined plane (Fig. P4.79). (a) If the plane is frictionless, what is the magnitude of the contact force between the two masses? (b) If the coefficients of kinetic friction between m_1 and the plane is $\mu_1 = 0.25$ and between m_2 and the plane is $\mu_2 = 0.25$, what is the contact force between the two masses? (c) If $\mu_1 = 0.15$ and $\mu_2 = 0.25$, what is the contact force?

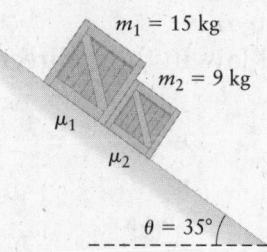

Figure P4.79

Solution

Recognize the principle

We will need to apply Newton's second law to each block.

Sketch the problem

We begin by drawing a sketch of the problem and a free-body diagram for each of the blocks. Here we have included the force of friction which can be neglected for part (a) of the problem.

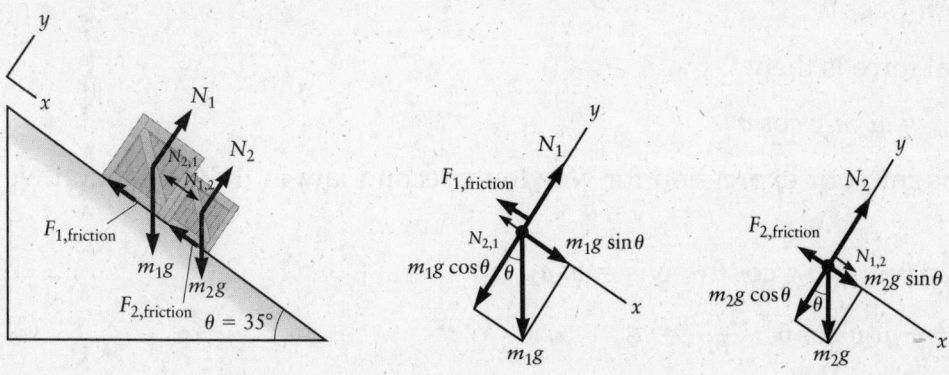

Identify the relationships

In part (b) there is friction, as seen from the diagram. Applying Newton's second law to m_1 for each direction we find,

$$\Sigma F_{1,x} = m_1 g \sin \theta - F_{1,\text{friction}} - N_{2,1} = m_1 a_{1,x}$$

$$\Sigma F_{1,y} = N_1 - m_1 g \cos \theta = m_1 a_{1,y}$$

Our expression for friction is $F_{1,\text{friction}} = \mu_1 N_1$ and since there is no motion in the y direction, $a_{1,y} = 0$. Inserting this into our expression for Newton's second law in the y direction we have,

$$N_1 - m_1 g \cos \theta = 0$$

$$N_1 = m_1 g \cos \theta$$

The frictional force is then,

$$F_{1,\text{friction}} = \mu_1 m_1 g \cos \theta$$

Inserting this into our expression for Newton's second law in the x direction we have,

$$m_1 g \sin \theta - \mu_1 m_1 g \cos \theta - N_{2,1} = m_1 a_{1,x}$$

$$g \sin \theta - \mu_1 g \cos \theta - \frac{N_{2,1}}{m_1} = a_{1,x} \tag{1}$$

Now we turn our attention to m_2, once again applying Newton's second law. Summing forces for each component on m_2,

$$\Sigma F_{2,x} = m_2 g \sin \theta - F_{2,\text{friction}} + N_{1,2} = m_2 a_{2,x}$$

$$\Sigma F_{2,y} = N_2 - m_2 g \cos \theta = m_2 a_{2,y}$$

Our expression for friction is $F_{2,\text{friction}} = \mu_2 N_2$ and since there is no motion in the y direction, $a_{2,y} = 0$. Inserting this into our expression for Newton's second law in the y direction we have,

$$N_2 - m_2 g \cos \theta = 0$$

$$N_2 = m_2 g \cos \theta$$

The frictional force is then,

$$F_{2,\text{friction}} = \mu_2 m_2 g \cos \theta$$

Inserting this into our expression for Newton's second law in the x direction we have,

$$m_2 g \sin \theta - \mu_2 m_2 g \cos \theta + N_{1,2} = m_2 a_{2,x}$$

$$g \sin \theta - \mu_2 g \cos \theta + \frac{N_{1,2}}{m_2} = a_{2,x} \tag{2}$$

From Newton's third law, $N_{1,2} = N_{2,1}$. If this normal force is nonzero, then the accelerations must be equal, $a_{1,x} = a_{2,x}$.

Inserting Equations (1) and (2) from above into this,

$$a_{1,x} = a_{2,x}$$

$$g \sin \theta - \mu_1 g \cos \theta - \frac{N_{1,2}}{m_1} = g \sin \theta - \mu_2 g \cos \theta + \frac{N_{1,2}}{m_2} \tag{3}$$

Solve

(a) We can solve expression (3) for the situation where the plane is frictionless by simply setting $\mu_1 = \mu_2 = 0$.

$$g \sin \theta - \frac{N_{1,2}}{m_1} = g \sin \theta + \frac{N_{1,2}}{m_2} \rightarrow \frac{N_{1,2}}{m_1} = -\frac{N_{1,2}}{m_2}$$

Therefore, $N_{1,2} = \boxed{0}$.

(b) Similarly, for this situation $\mu_1 = \mu_2$ and expression (3) gives,

$$g \sin \theta - \mu_1 g \cos \theta - \frac{N_{1,2}}{m_1} = g \sin \theta - \mu_1 g \cos \theta + \frac{N_{1,2}}{m_2} \rightarrow \frac{N_{1,2}}{m_1} = -\frac{N_{1,2}}{m_2}$$

Therefore, $N_{1,2} = \boxed{0}$.

(c) Now $\mu_1 = 0.15$ and $\mu_2 = 0.25$. Substituting these into expression (3) and solving for $N_{1,2}$ we find,

$$-\mu_1 g \cos \theta - \frac{N_{1,2}}{m_1} = -\mu_2 g \cos \theta + \frac{N_{1,2}}{m_2}$$

$$\frac{N_{1,2}}{m_1} + \frac{N_{1,2}}{m_2} = \mu_2 g \cos \theta - \mu_1 g \cos \theta$$

$$N_{1,2}\left(\frac{1}{m_2} + \frac{1}{m_1}\right) = (\mu_2 - \mu_1) g \cos \theta$$

$$N_{1,2} = \frac{(\mu_2 - \mu_1) g \cos \theta}{\frac{1}{m_2} + \frac{1}{m_1}}$$

Inserting the values,

$$N_{1,2} = \frac{(0.25 - 0.15)(9.8 \text{ m/s}^2) \cos 35°}{\frac{1}{15 \text{ kg}} + \frac{1}{9.0 \text{ kg}}}$$

$$\boxed{N_{1,2} = 4.5 \text{ N}}$$

What does it mean?

There is only a normal force between the two masses when $\mu_2 > \mu_1$, as is the case in part (c).

P4.84 A golf ball is hit with an initial velocity of magnitude 60 m/s at an angle of 65° with respect to the horizontal (x) direction. At the same time, a second golf ball is hit with an initial speed v_0 at an angle 35° with respect to x. If the two balls land at the same time, what is v_0?

Solution

Recognize the principle

Once again this is an example of projectile motion.

Sketch the problem

We begin by sketching the problem and defining a coordinate system.

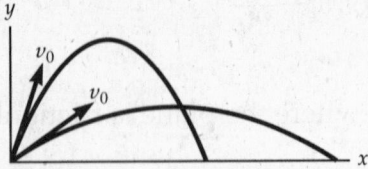

Identify the relationships

Using the coordinate system defined in the sketch we see that both balls have an initial height of $y = 0$. Furthermore, they both land at the same height, also $y = 0$. Since air drag can be neglected, the acceleration of both balls in the y direction is $-g$ and zero in the x direction. Also, the two golf balls have the same time of flight. We can apply the expression of motion with constant acceleration for the y direction for both of the golf balls and obtain the following two expressions:

$$y_1 = y_{1,0} + v_{1,0,y}t + \frac{1}{2}a_y t^2 \rightarrow 0 = 0 + 60 \text{ m/s} \sin 65° - \frac{1}{2}gt^2$$

$$y_2 = y_{2,0} + v_{2,0,y}t + \frac{1}{2}a_y t^2 \rightarrow 0 = 0 + v_{2,0} \sin 35° - \frac{1}{2}gt^2$$

Solve

Setting these two expressions equal to each other and solving for $v_{2,0}$ we find, Since the two hit golf balls have the same time of flight,

$$60 \text{ m/s} \sin 65° - \frac{1}{2}gt^2 = v_{2,0} \sin 35° - \frac{1}{2}gt^2$$

$$60 \text{ m/s} \sin 65° = v_{2,0} \sin 35°$$

$$v_{2,0} = \frac{60 \text{ m/s} \sin 65°}{\sin 35°} = \boxed{95 \text{ m/s}}$$

What does it mean?

Can you think of another way we could have solved this problem? Well, since this is a symmetric projectile problem (i.e., the balls land at the same height as they are launched), we could have used the expression $t_{\text{lands}} = \frac{2v_0 \sin \theta}{g}$ and obtained the same answer. Note also that if two objects are launched at different initial angles and have the same time of flight, then the initial velocity for the lower trajectory must have a larger initial velocity.

Part D. Additional Worked Examples and Capstone Problems

The following worked examples provide you with practice drawing free-body diagrams, projectile motion, and applying Newton's second law. The first capstone problem provides you practice in graphing and helps develop your intuitive understanding of the relationship between force, acceleration, and velocity. Finally, the second capstone problem gives you experience in interpreting graphs.

Although these five problems do not incorporate all the material discussed in this chapter, they do highlight several of the key concepts. If you can successfully solve these problems then you should feel confident in your understanding of these key concepts, so use these problems as a test of your understanding of the chapter material.

WE 4.1 Aiming a Rifle

Consider again the problem of firing a bullet at a target some distance L away. At what angle *above* the target must the rifle be aimed so as to hit the bull's-eye? Assume $L = 150$ m and the bullet's initial speed is $v_0 = 400$ m/s.

Solution

Recognize the principle

To hit the target, we want the bullet to have traveled a distance $x = L$ at the moment it has returned to its initial height. The range should therefore be L, and we can apply the range formula.

Sketch the problem

The following sketch shows the problem. If the firing angle is θ, the initial components of the velocity are $v_{0x} = v_0 \cos \theta$ and $v_{0y} = v_0 \sin \theta$.

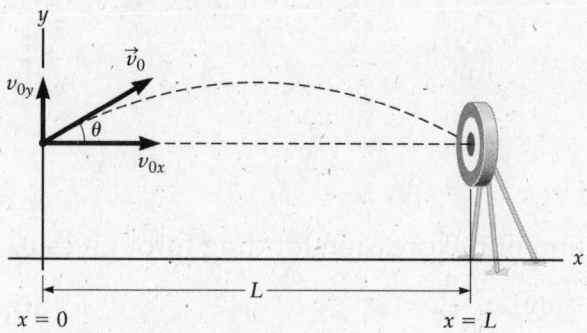

Identify the relationships

Since the bullet lands at the same height which it is fired, we can use the expression for the range of a projectile,

$$x_{\text{lands}} = L = \frac{v_0^2 \sin (2\theta)}{g}$$

Solve

We can rearrange this expression to find θ:

$$\sin (2\theta) = \frac{gL}{v_0^2}$$

$$\theta = \frac{1}{2} \sin^{-1} \left(\frac{gL}{v_0^2} \right) = \frac{1}{2} \sin^{-1} \left[\frac{(9.8 \text{ m/s}^2)(150 \text{ m})}{(400 \text{ m/s})^2} \right] = \boxed{0.26°}$$

What does it mean?

The result for θ is very small, so in this case we must aim only very slightly above the x axis to hit the target.

WE 4.2 Terminal Velocity for a Falling Baseball

Recall from Chapter 3 that an object falling in the presence of air drag reaches a velocity at which the force of gravity is just balanced by the drag force. Calculate the terminal velocity for a baseball weighing 1.4 N. The radius of a baseball is $r = 3.6$ cm.

Solution

Recognize the principle

When the ball travels at its terminal velocity, the acceleration is zero. Applying Newton's second law for motion along the vertical (y) direction, we have

$$\Sigma F_y = F_{\text{drag}} + F_{\text{grav}} = ma_y = 0$$

Sketch the problem

We begin by sketching the problem and drawing the free-body diagram.

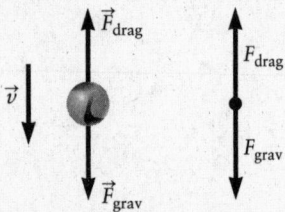

Identify the relationships

Substituting into the above equation our expression for drag force gives us

$$F_{\text{drag}} = \frac{1}{2} \rho A v_{\text{term}}^2 = -F_{\text{grav}} = mg$$

$$\frac{1}{2} \rho A v_{\text{term}}^2 = mg$$

$$v_{\text{term}} = \sqrt{\frac{2mg}{\rho A}}$$

Solve

Inserting the given values of m, ρ, and the area $A = \pi r^2$ into our expression for v_{term} leads to

$$v_{\text{term}} = \sqrt{\frac{2(1.4 \text{ N})}{(1.3 \text{ kg/m}^3)\pi(0.036 \text{ m})^2}} = \boxed{23 \text{ m/s}}$$

What does it mean?

This terminal velocity is about 50 mi/h, a speed easily achieved for either a thrown or batted ball. Hence, we again conclude that air drag is important for this projectile.

WE 4.3 Effect of Wind on a Soccer Ball

Suppose you are playing soccer on a rather windy day. You kick the soccer ball so that it travels on a trajectory like that shown in Figure B below and observe that the ball spends 1.5 s in the air. The wind is blowing horizontally in a direction perpendicular to your kick (along the z axis in Fig. A), with a speed of 10 m/s (about 20 mi/h). Calculate the approximate distance the ball is deflected horizontally by the wind. That is, find the z coordinate of the ball's landing point. A soccer ball has a mass of approximately 0.43 kg and a radius of 11 cm.

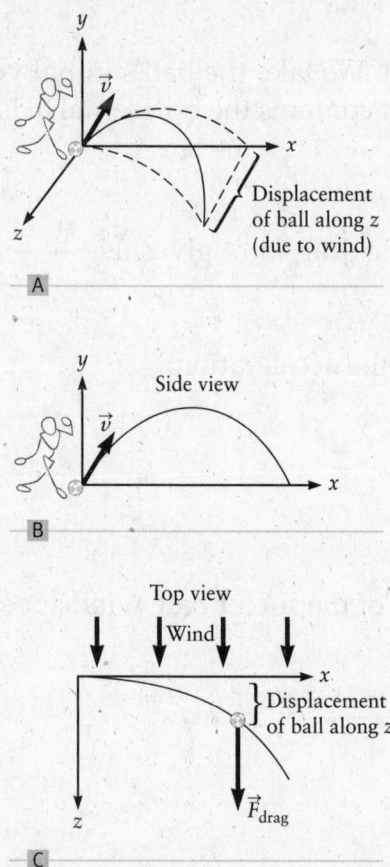

Recognize the principle

The wind is blowing horizontally along z, leading to a drag force on the ball directed along z (Fig. C). We want to find how far the ball moves along the +z direction as a result of this force. Because we are after an approximate calculation, we assume the drag force is constant and compute the acceleration along z using our expression for F_{drag}. We then calculate the displacement along z using our relations for motion with constant acceleration.

The ball spends a time $t = 1.5$ s in the air, so we need to estimate how far the ball travels along z during this time. The magnitude of the drag force is

$$F_{drag} = \frac{1}{2}\rho A v_{rel}^2$$

where $\vec{v}_{rel}$ is the velocity of the ball relative to the air. The importance of the relative velocity can be understood by viewing the problem in the reference frame of the ball. There is then a wind speed v_{rel}, and the drag force is given by our expression above, directed opposite to $\vec{v}_{rel}$. In an exact treatment of this problem, we would have to consider that the magnitude and direction of $\vec{v}_{rel}$ varies with time as the ball is deflected by the wind. Because we are interested in an approximate value of the deflection along z, for the purposes of calculating the drag force we approximate $\vec{v}_{rel}$ by the initial velocity of the ball along z relative to the wind. With this approximation, we have $\vec{v}_{rel} = -\vec{v}_{wind}$.

Sketch the problem

Figure A shows the trajectory of the soccer ball. We take the ball's initial velocity to be in the x–y plane; that is, it is in a plane that contains the horizontal axis x and the vertical axis y (Fig. B).

Identify the relationships

Inserting $\vec{v}_{rel} = -\vec{v}_{wind}$ into our expression for the drag force gives us

$$F_{drag} = \frac{1}{2}\rho A v_{wind}^2$$

Next we use Newton's second law to calculate the acceleration:

$$F_{drag} = \frac{1}{2}\rho A v_{wind}^2 = ma_z$$

$$a_z = \frac{\rho A v_{wind}^2}{2m}$$

Solve

Using the given values for the mass and radius of the soccer ball, wind speed, and known air density, we get

$$a_z = \frac{\rho A v_{wind}^2}{2m} = \frac{(1.3\ \text{kg/m}^3)\pi(0.11\ \text{m})^2\ (10\ \text{m/s})^2}{2(0.43\ \text{kg})} = 5.7\ \text{m/s}^2$$

which leads to a deflection along z of

$$z = \frac{1}{2}a_z t^2 = \frac{1}{2}(5.7\ \text{m/s}^2)(1.5\ \text{s})^2 = \boxed{6.4\ \text{m}}$$

What does it mean?

The deflection of the ball along z is quite substantial and is roughly equal to the width of a soccer goal. This deflection is why playing soccer on a windy day can be quite challenging.

CP 4.1 Describing Motion: Going from Words to Graphs

For each of the following situations, sketch both the horizontal and vertical components of the net force, acceleration, and velocity versus time:

(a) An object launched horizontally off a cliff, neglecting air drag.
(b) An object launched at 45° above flat, level ground. Neglect air drag.
(c) An object launched vertically upward. Include air drag.
(d) An object launched horizontally off a cliff. Include air drag.

Solution

Recognize the principle

These are all examples of projectile motion. In the first two situations we are neglecting air drag so the only force we need to consider is the force of gravity. The next two situations involve air drag in addition to the force of gravity. Based on the situation we can determine how the force (both horizontally and vertically) will vary with time. Applying Newton's second law will then provide us with the acceleration, since acceleration is proportional to the force. The velocity as a function of time can be determined by understanding the relationship between velocity and acceleration.

Sketch the problem

The initial free-body diagram for each situation is as follows:

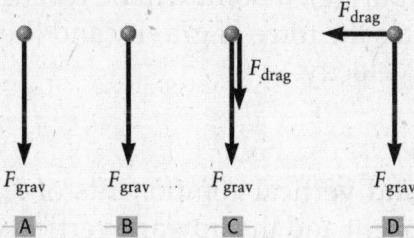

Identify the relationships

Let's tackle the relationships one variable at a time. For all four situations the force of gravity on the object will remain constant and in the same direction (straight down). For parts (c) and (d), the magnitude of the drag force depends on the square of the speed of the object and the direction is always opposite to the direction of the object's velocity (see Chapter 3). The other relationship we'll need is Newton's second law which tells us that there is a linear relationship between the net force on an object and its acceleration.

Putting this all together we can analyze each of the situations.

(a) The only force on the object is the force of gravity directed vertically downward. Therefore, the object will only accelerate vertically downward after it is launched. As a result, the horizontal component of its velocity will remain constant and the vertical component of the velocity will steadily increase downward.

(b) Again, the only force on the object is the force of gravity directed vertically downward. Therefore, the object will only accelerate vertically downward after it is launched. However, this time the object is given a horizontal and vertical component to its initial velocity which are equal due to the 45° launch angle. The horizontal component of its velocity will remain constant and the vertical component of the velocity will steadily decrease until the object reaches its maximum height. At this point the object stops rising and begins to fall. The vertical speed will then steadily increase until the object reaches the ground.

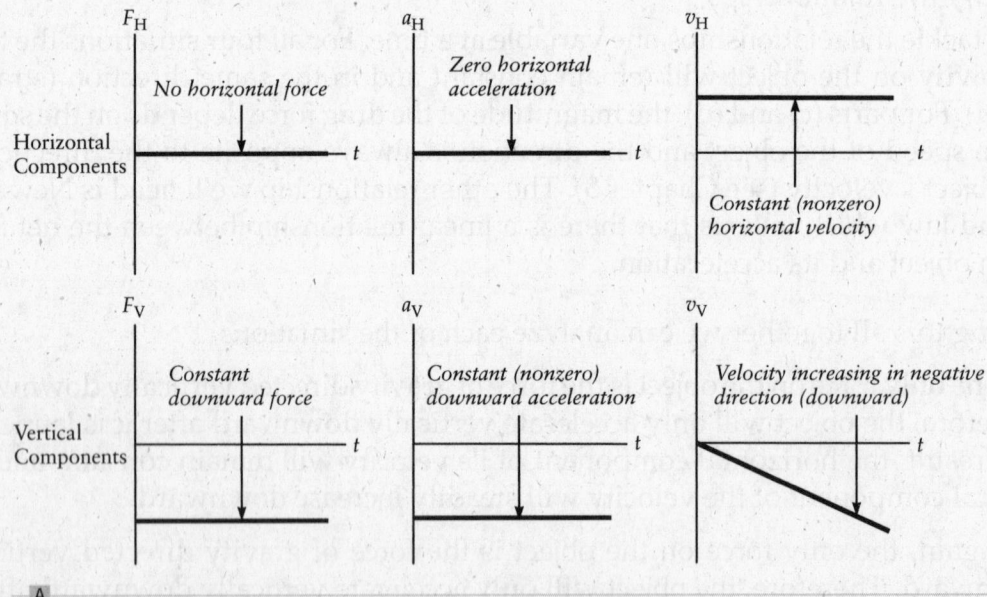

(c) The drag force will steadily decrease as the object rises and loses speed. When the object reaches its maximum height, it will momentarily stop and the drag force will become zero. As the object falls, the drag force points upward and continues to grow until it equals the force of gravity on the object. At that point the two forces cancel and the acceleration becomes zero. The object continues to fall at a constant velocity, the terminal velocity.

(d) After the object is launched the direction of its velocity begins to change, therefore the direction of the drag force changes. Horizontally, there will be only once force acting on the object: the horizontal component of the drag force. This will slow the object down until it finally stops moving horizontally and the horizontal component of the drag force becomes zero, provided the object remains in the air long enough for this to happen. Vertically the object accelerate downward after launch due to the force of gravity and the vertical component of the drag force (directed upward) will grow. As in part (c), a point will be reached when the vertical component of the drag force equals the force of gravity and then the object stops accelerating, falling at its terminal velocity.

Solve

We can sketch the required horizontal and vertical components of F_{net}, a, and v. Choose the time axis as increasing to the right and the upward vertical direction as positive for force, acceleration, and velocity.

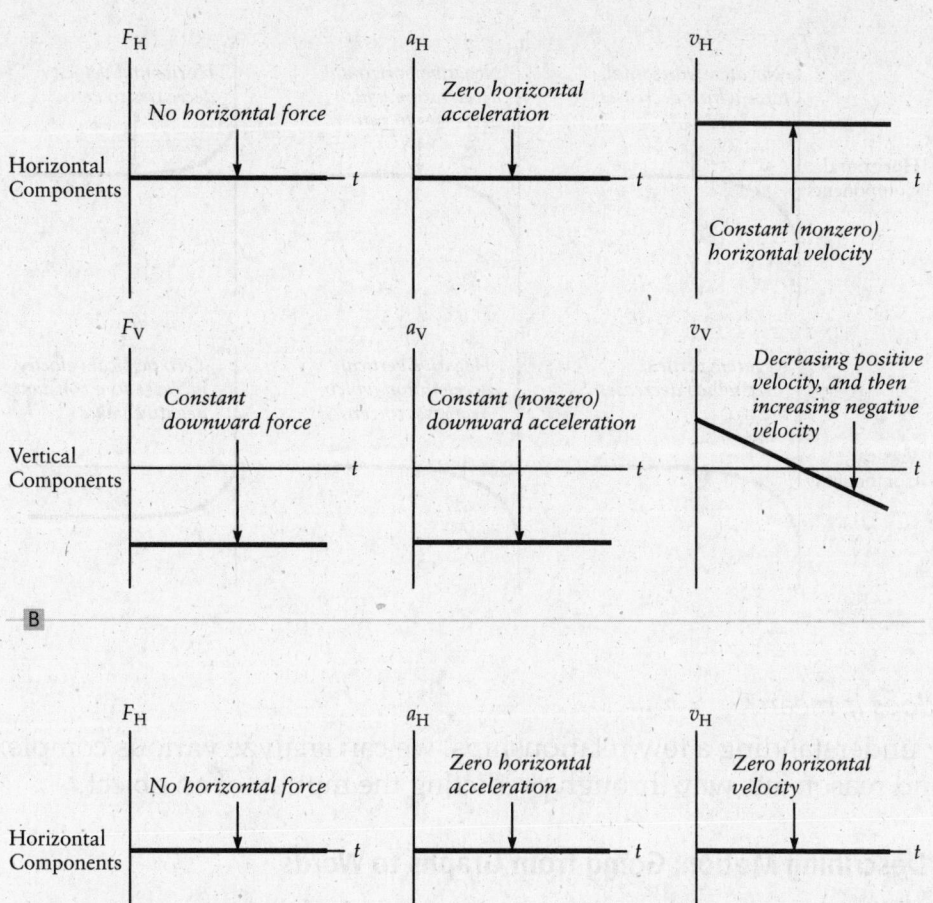

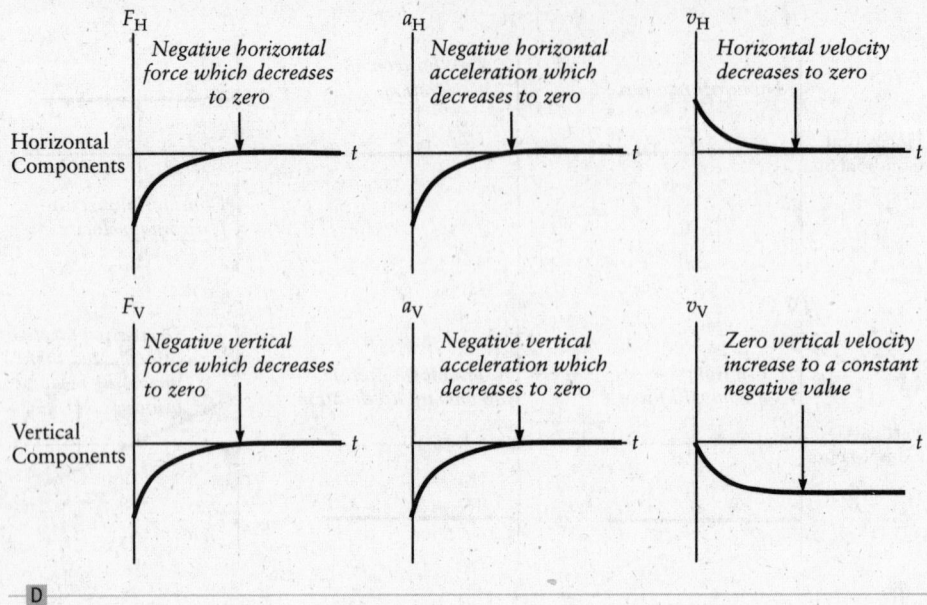

What does it mean?

Just by understanding a few relationships, we can analyze various complex situations and reason our way through predicting the motion of an object.

CP 4.2 Describing Motion: Going from Graphs to Words

For the following sets of data, describe the motion of the object and give an example of a physical situation they might depict:

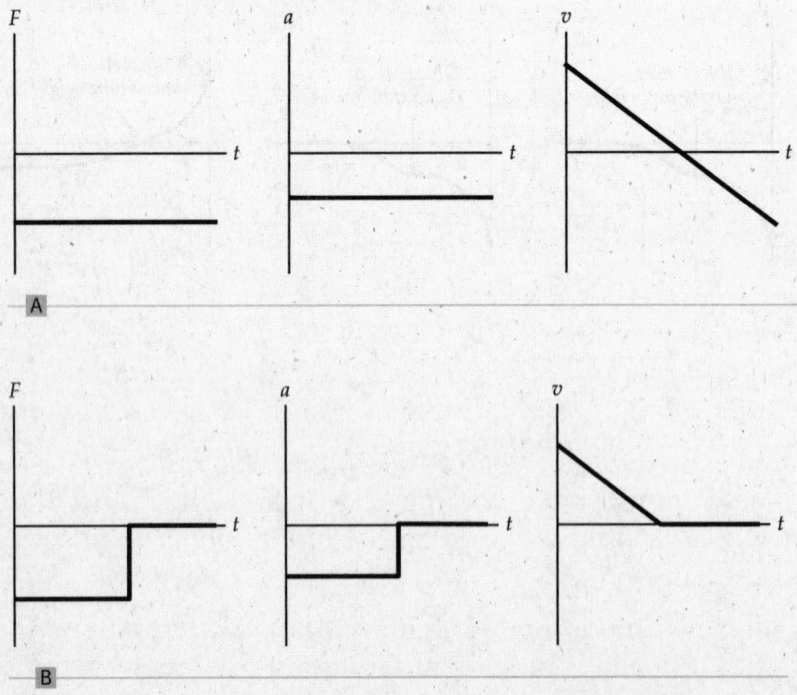

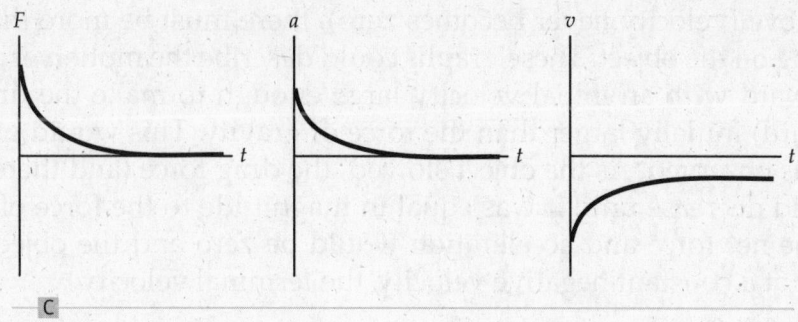

Solution

Recognize the principle

Since the components of force, acceleration, and velocity are not provided, we can assume the motion in each case is one-dimensional. Newton's second law certainly applies and is illustrated by the direct relationship between the force and acceleration in each case. For part (a) the object described moves in both the positive and negative directions.

Sketch the problem

The sketch is provided in the problem.

Identify the relationships

(a) The force and therefore the acceleration are both constant and negative. The velocity begins in the positive direction and steadily decreases to zero before steadily increasing in the negative direction.

(b) For the first period of time the force and acceleration are constant and negative. The force and acceleration then become zero. The velocity starts out positive and steadily decreases to zero where it remains.

(c) The force and acceleration start out positive and decrease at a decreasing rate to zero. The velocity begins negative and decreases at a decreasing rate to a constant negative value.

Solve

(a) This object moves in a positive direction while slowing down at a constant rate (constant acceleration). It momentarily comes to rest ($v = 0$) before moving in the negative direction and steadily increases its speed. These graphs could describe the vertical motion of a projectile neglecting air drag.

(b) This object moves in the positive direction while slowing down at a constant rate (constant acceleration) before coming to rest and remaining at rest. This data could describe the motion of your car when you apply your brakes while approaching a stop sign.

(c) This object begins its motion moving in a negative direction and slows at a decreasing rate until it finally is moving at a constant negative velocity. Since the force is not constant and varies with time (or perhaps velocity), there is probably a drag force involved. Since the net force goes to zero after some time but the object continues to move (velocity never becomes zero), there must be more than just a drag force acting on the object. These graphs could describe the motion of an object thrown downward with an initial velocity large enough to make the drag force (directed upward) initially larger than the force of gravity. This would give a net positive force as observed. As the object slowed, the drag force (and therefore the net force) would decrease until it was equal in magnitude to the force of gravity. At this time the net force and acceleration would be zero and the object would continue to fall at a constant negative velocity, the terminal velocity.

What does it mean?

Once again, by understanding a few relationships, we can analyze various sets of data and reason our way to a plausible explanation of the data.

Part E. MCAT Review Problems

PROBLEMS

1. If two forces act on an object at the same time, the resultant force will be greatest when the angle between the forces is

    (a) 0°
    (b) 45°
    (c) 90°
    (d) 180°

2. Suppose a force acting on a box can have one of the four orientations, A, B, C, or D, indicated below.

    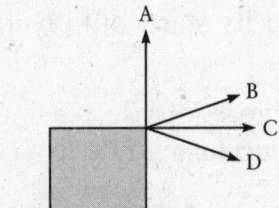

    In which orientation will the force have the smallest vertical component?

    (a) A
    (b) B
    (c) C
    (d) D

3. A missile is fired at an angle with the horizontal. The air resistance is negligible. The initial horizontal velocity component is twice that of the initial vertical velocity component. The trajectory of the missile is best described as

    (a) semicircular.
    (b) translational.
    (c) parabolic.
    (d) hyperbolic.

4. Which trajectory best describes a ball rolling down a curved ramp that ends at point P?

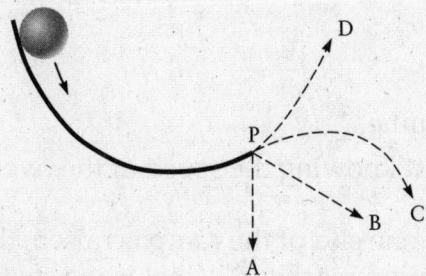

   (a) A
   (b) B
   (c) C
   (d) D

5. An object hangs from two ropes with tensions T_1 and T_2 as shown in the figure. For the object to be in translational equilibrium, the following must be true:

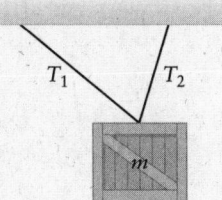

   (a) $T_1 > T_2$
   (b) $T_1 < T_2$
   (c) $T_1 + T_2 = mg$
   (d) $T_1 = T_2$

6. A ball is thrown horizontally with a speed of 6.0 m/s. What is its speed after 3.0 s of flight?
   (a) 30.0 m/s
   (b) 16.0 m/s
   (c) 18.0 m/s
   (d) 5.0 m/s

7. A box of mass m is placed on an incline with angle of inclination θ. The box does not slide. The magnitude of the frictional force on the box is
   (a) $\mu_S mg \sin\theta$
   (b) $mg \sin\theta$
   (c) $mg \cos\theta$
   (d) mg

8. A boat moves through the water in a river at a speed of 8 m/s relative to the water. The boat makes a trip downstream and then makes a return trip upstream to the original starting place. Which trip takes longer?

   (a) The downstream trip takes longer.

   (b) The upstream trip takes longer.

   (c) Both trips take the same amount of time.

   (d) The answer cannot be figured without knowing the speed of the river flow.

9. A box of mass m is placed on a ramp. As one end of the ramp is raised, the box begins to slide when the angle of inclination reaches θ. What is the coefficient of static friction between the box and the ramp?

   (a) mg

   (b) $\cos \theta$

   (c) $\sin \theta$

   (d) $\tan \theta$

10. In the configuration shown, two masses are connected by a light string over a frictionless, massless pulley. Once released the masses move at a constant speed. Determine the coefficient of kinetic friction between the 10-kg block and the table.

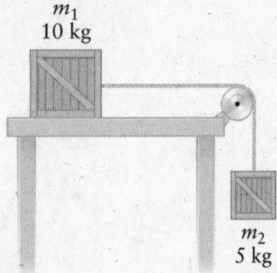

   (a) 0.1

   (b) 0.5

   (c) 1.0

   (d) 2.0

SOLUTIONS

1. MCAT strategies

This problem is simply testing our understanding of vector addition. The resultant of two forces (vectors) is greatest when the two forces are parallel and least when they are antiparallel. So, the correct answer is (a).

2. MCAT strategies

Again, this problem is testing our understanding of vectors and vector components. When a force (vector) is directed horizontally it has no vertical component. So, the correct answer is (c).

3. *MCAT strategies*

There is information given in the problem which is not necessary to the solution and is designed to confuse us. The key point in the problem is that air drag is negligible. Therefore, the missile, like any projectile, will simply follow a parabolic trajectory. So, the correct answer is (c).

4. *MCAT strategies*

Once the ball leaves the track it is only under the influence of gravity and will follow a path similar to a thrown ball. So, the correct answer is (c).

5. *MCAT strategies*

Though we could probably eliminate a couple of the answers quite easily, let's sketch the free-body diagram for this situation so that we can see why answer (b) is correct.

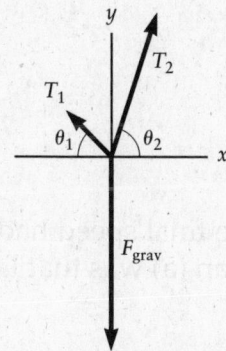

Since the object is in translational equilibrium, the net force in both the x and y directions must be zero. So, we can conclude the following:

$$T_{2,x} = T_{1,x} \rightarrow T_2 \cos \theta_2 = T_1 \cos \theta_1$$

$$T_{1,y} + T_{2,y} = mg \rightarrow T_1 \sin \theta_1 + T_2 \sin \theta_2 = mg$$

But since $\theta_2 > \theta_1$ then $T_2 > T_1$. So, the correct answer is (b).

6. *MCAT strategies*

Since the ball is under the influence of gravity it will accelerate downwards. So, its speed in the vertical direction will increase while its horizontal speed remains constant. The vector sum of these two will certainly be greater than 6 m/s so answer (d) can be quickly eliminated. Let's proceed with our problem-solving strategy to determine the correct answer.

Recognize the principle

This is a projectile problem where the only acceleration is in the downward direction at 9.8 m/s². We can approximate this to 10 m/s² to simplify the calculations.

Sketch the problem

No sketch is required.

Identify the relationships

In the horizontal direction the speed will remain constant since there is no force (and therefore no acceleration) in that direction. If we call this the x direction then $v_x = 6$ m/s. In the vertical direction the ball has no initial speed, but accelerates downward at approximately 10 m/s². If we call the downward direction the positive y direction then we can apply the following expression to calculate the vertical speed at time t,

$$v_y = v_{0y} + a_y t \rightarrow v_y = -10t$$

Solve

Plugging in $t = 3$ s to this expression we find,

$$v_y = -10(3) = -30 \text{ m/s}$$

So the speed after 3 s is,

$$v = \sqrt{v_x^2 + v_y^2} = \sqrt{6^2 + 30^2} \approx 30 \text{ m/s}$$

and the correct answer is (a).

What does it mean?

Note that the last step was not necessary since we knew the final speed had to be at least as large as the y component and no answer other than (a) was that large.

7. MCAT strategies

Since the block is stationary, and therefore not accelerating, Newton's second law tells us that the net force on the block must be zero. So, if we evaluate the forces along the incline we see that the force of friction, which is directed up the incline, must be equal in magnitude to the component of the force of gravity on the block directed down the incline. This component is given by the expression $mg \sin \theta$, so the correct answer is (b).

8. MCAT strategies

To determine the velocity of the boat relative to the shore we need the vector sum of the velocity of the boat relative to the water and the velocity of the water relative to the shore. Clearly the greater velocity results when the boat and river are moving in the same direction. So, the downstream trip will take less time than the upstream trip. The correct answer is (b).

9. MCAT strategies

Since the coefficient of friction has no units, answer (a) can be quickly eliminated. No other answer can be easily eliminated so we proceed as usual.

Recognize the principle

This problem involved the application of Newton's second law and the ability to take components of vectors.

Sketch the problem

We begin by drawing the free-body diagram for the mass, choosing our axis carefully.

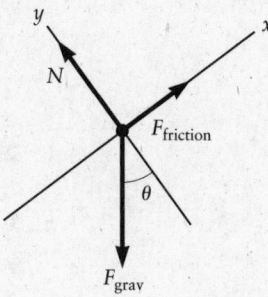

Identify the relationships

We can now apply Newton's second law in both the x and y directions, noting that the accelerations in both directions are zero.

$$\Sigma F_x = F_{\text{friction}} - mg \sin \theta = 0$$

$$\Sigma F_y = N - mg \cos \theta = 0$$

We also know that the frictional force is related to the normal force according to the following expression,

$$F_{\text{friction}} = \mu_S N$$

Solve

Combining these three expressions and solving for the coefficient of friction we have,

$$N = mg \cos \theta \rightarrow F_{\text{friction}} = \mu_S \, mg \cos \theta \rightarrow mg \sin \theta = \mu_S \, mg \cos \theta$$

$$\mu_S = \frac{\sin \theta}{\cos \theta} = \tan \theta$$

So the correct answer is (d).

What does it mean?

This answer is independent of the mass of the object!

10. MCAT strategies

No answer can be quickly eliminated so we proceed with our problem-solving strategy.

Recognize the principle

We will need to apply Newton's second law to both masses, recognizing that since the masses move at a constant velocity then their accelerations are zero.

Sketch the problem

We begin by drawing the free-body diagram for both masses.

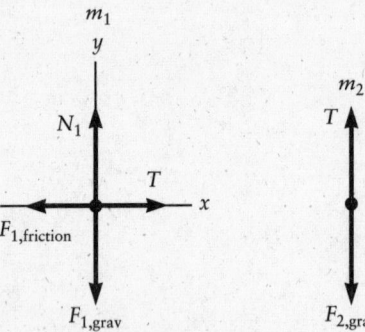

Identify the relationships

Applying Newton's second law:

For m_1:

$$\Sigma F_x = -F_{\text{friction}} + T = 0$$

$$\Sigma F_y = N - m_1 g = 0$$

And combining these with our expression for the frictional force,

$$T = F_{\text{friction}} = \mu_K N = \mu_K (m_1 g)$$

For m_2:

$$\Sigma F = m_2 g - T = 0 \rightarrow T = m_2 g$$

Solve

Setting the two expressions for tension equal to each other we find,

$$\mu_K(m_1 g) = m_2 g \rightarrow \mu_K = \frac{m_2}{m_1} = \frac{5 \text{ kg}}{10 \text{ kg}} = 0.5$$

So the correct answer is (b).

What does it mean?

Since the coefficient of kinetic friction between most surfaces is between 0 and 1, this answer seems reasonable.

5 Circular Motion and Gravitation

Part A. Summary of Key Concepts and Problem-Solving Strategies

KEY CONCEPTS

Circular motion and centripetal acceleration

An object moving in a circle of radius r at a constant speed has an acceleration

$$a_c = \frac{v^2}{r}$$

directed toward the center of the circle. This acceleration arises from the fact that the direction of the velocity is changing. The symbol a_c is called the *centripetal acceleration*. According to Newton's second law, this acceleration must be caused by a total force of magnitude $\Sigma F = ma_c$, so

$$\Sigma F = \frac{mv^2}{r}$$

This force is directed toward the center of the circle. This net force can be due to several possible forces. For example, a rock traveling around a circular path on the end of a string has a centripetal acceleration. The force responsible for this acceleration is the tension in the string. Another example is the gravitational force of the Sun on the Earth. This force keeps the Earth in a nearly circular orbit around the Sun; therefore the Earth has a centripetal acceleration. There are many other examples of forces which cause a centripetal acceleration.

133

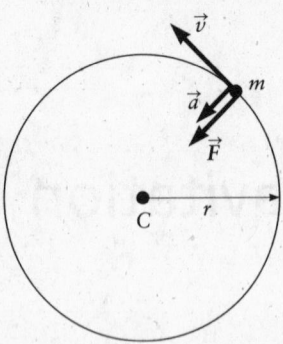

Newton's law of gravitation

There is a gravitational force

$$F_{\text{grav}} = \frac{Gm_1m_2}{r^2}$$

between any two objects which have mass. This force is always attractive. Newton's law of gravitation is an example of an *inverse square law*. This $1/r^2$ dependence suggests a force line model of gravity and tells us something about the geometry of the universe. Gravitation is also an example of "action at a distance." Other forces, including electric forces, exhibit this property.

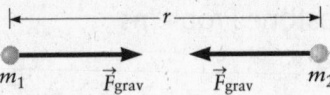

APPLICATIONS

Kepler's laws

Kepler deduced three laws of planetary motion: (1) planetary orbits are elliptical, (2) a planetary orbit sweeps out equal areas in equal times, and (3) the square of the orbital period is proportional to the cube of the average orbital radius,

$$T^2 = \left(\frac{4\pi^2}{GM}\right)r^3$$

These laws apply to planets orbiting a sun and also to satellites and moons orbiting a planet.

Kepler's laws, and hence the motions of the planets, moons, comets, and so forth, are all explained by Newton's law of gravitation, together with Newton's laws of motion. The circular motion of the Moon and the free fall of an apple look different, but they are due to the same force.

PROBLEM-SOLVING STRATEGIES

In this chapter the end-of-chapter questions and problems focus on three main areas. The first area is circular motion. Specifically, when an object moves in a circular path, there is a force directed toward the center of this circular path resulting in a centripetal acceleration. The second area involves applying Newton's law of gravity. The following two schematics will help remind you of the thought process you should use while tackling problems from this chapter. By combing our understanding of circular motion with Newton's law of gravity and his laws of motion, we can derive Kepler's laws of planetary motion. Kepler's laws are the third main area of focus in this chapter.

Problem Solving: Analyzing Circular Motion

Recognize the principle

An object can move in a circular path only if there is a net force directed toward the center of the circle. This force results in an acceleration $a_c = v^2/r$ known as a *centripetal acceleration*.

Sketch the problem

Make a drawing that shows the path followed by the object of interest. This drawing should identify the circular part of the path, the radius of this circle, and the center of the circle.

Identify the relationships

- Find all the forces acting on the object; as in our applications of Newton's laws in Chapters 3 and 4, a free-body diagram is often very useful.

- Using your drawing and free-body diagram, find the components of the forces that are directed *toward the center* of the circle, and the components perpendicular to this direction.

- Apply Newton's second law $\Sigma F = ma$ for motion toward the center of the circle and (if necessary) in the perpendicular direction. The total force directed toward the center of the circle results in the centripetal acceleration. Therefore, applying Newton's second law we find

$$\Sigma F_{\text{center}} = ma_c = \frac{mv^2}{r}$$

Solve

Solve for the quantities of interest. For example, we can solve for the centripetal acceleration from the following:

$$a_c = \frac{v^2}{r}$$

What does it mean?

Always *consider what your answer means* and check that it makes sense.

Problem Solving: Applying Newton's Law of Gravity

Recognize the principle

When applying Newton's law of gravity it is important to measure the distance between two objects from the center of one object to the center of the other object. This distance is the r that is in the denominator of Newton's law of gravity.

Sketch the problem

Sketch the situation, being sure to draw in the distance of separation from the center of one object to the center of the other object.

Identify the relationships

The key relationship is the expression,

$$F_{grav} = \frac{Gm_1m_2}{r^2}$$

In this expression, m_1 and m_2 are the masses of the two interacting objects, r is the distance of separation (center-to-center), and $G = 6.67 \times 10^{-11}$ Nm^2/kg^2. If more than two objects are involved, you'll need to apply this expression several times and do a vector sum of the forces on an object to find the net gravitational force on that object.

Solve

Solve for the unknown quantities such as the net force, the distance of separation, or an unknown mass. Problems may involve the application of Newton's other laws, as we've done in previous chapters.

What does it mean?

Always *consider what your answer means* and check that it makes sense.

Part B. Frequently Asked Questions

1. *Doesn't acceleration imply that the object is speeding up or slowing down? How can an object's speed be constant but it still be accelerating?*

Acceleration means that the object's velocity is changing. Remember, velocity is a vector having both a magnitude (speed) and direction. If either the speed or the direction the object is traveling change, then the velocity of the object changes. It the case of an object moving around in a circular path at a constant speed, the direction of the velocity is continually changing so the object is accelerating. We call this type of acceleration a *centripetal acceleration* since it is directed toward the center of the circular path.

2. *I've heard the term "centripetal force" used before. Is this a force we should include in our analysis?*

Not really. Sometimes people use this term to describe the force causing a centripetal acceleration; however, it's important to note that *centripetal force* is not a force in and of itself. If you do use the term *centripetal force* you should keep in mind that it is always due to a force we are already familiar with. For example, the *centripetal force* responsible for the circular path a rock follows on the end of a string is actually the *tension* in the string. The *centripetal force* responsible for the Moon going around the Earth is the *force of gravity*. It's better to avoid using the term altogether and to look for the net force directed toward the center of the circular path. This net force (whatever it's due to) is causing a centripetal acceleration.

3. *As I round a curve in my car, I feel a force pushing me toward the outside of the curve. What is this force?*

Actually, it's not a force at all. The feeling of being pushed to the outside of the curve comes from the fact that your car is turning and therefore accelerating. You are in a non-inertial frame of reference and therefore cannot apply Newton's laws of motion unless you invent a fictitious force to describe what you are feeling. From the point of view of a stationary observer, there is a force due to your seat belt or car door which turns you with the car. This force is real and if it did not exist (if you were not wearing a seat belt and the car door was missing!) then you would simply continue in a straight line as your car turned away from you. You would obey Newton's first law of motion.

4. *I've heard the term "centrifugal force" used before. Is this a force we should include in our analysis?*

No, this is not a real force. As discussed in the previous FAQ, when viewing things from a non-inertial frame of reference, we can invent fictitious forces to describe what we observe. Some refer to these forces as centrifugal forces as in the case of the car in the previous FAQ or the cell in a centrifuge as discussed in the textbook. These are not real forces.

5. *Suppose I'm swirling a ball around in a circle on the end of a string. If the string breaks, which way will the ball travel?*

Applying Newton's first law will help us answer this question. According to this law an object will continue moving in the direction it's moving unless acted upon by a net force. So, once the string breaks, there is no longer a force of tension acting on the ball and so it will continue moving in whatever direction it was traveling at the instant the string broke. The direction of the velocity of the ball at an instant in time is tangent to the circle so the ball will travel tangent to the circle at the instant the string breaks. The following diagram will help illustrate what happens.

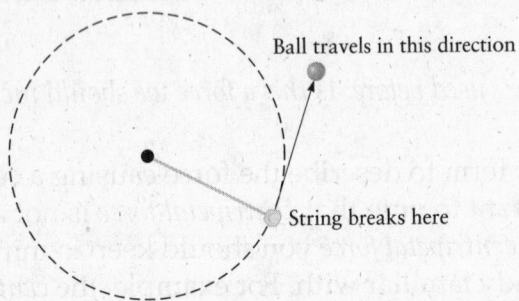

6. *Why is it that the ropes of a swing can support my weight when I'm just sitting on the swing, but cannot support my weight when I'm swinging?*

This is due to the fact that you are not accelerating in the first situation but you are accelerating in the second situation. Let's draw a free-body diagram for each of these situations and apply Newton's second law.

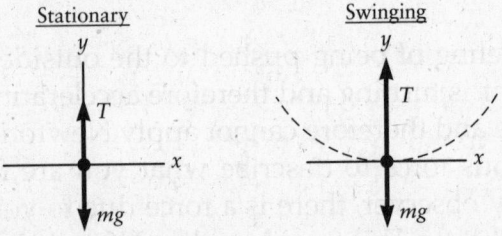

Applying Newton's second law for the situation where you are stationary we see that, since there is no acceleration, the tension in the ropes must equal the force of gravity on you. For the situation where you are swinging and looking at the bottom of the swing, you are accelerating in the y-direction. This is a centripetal acceleration since you are traveling in a circular path. Applying Newton's second law we see that the tension in the ropes must be larger than the force of gravity on you since you are accelerating. So, the tension in the ropes must be greater in the situation when you are swinging than when you are stationary.

7. *What is the purpose of a "G-suit"?*

A G-suit is a garment worn by pilots to prevent blood from pooling in their legs and abdomen when they are subject to large accelerations. One example of this is when a pilot pulls out of a steep dive. The centripetal acceleration they undergo can be so large as to pool the blood in their legs thus depriving their brains of blood. Loss of blood to the brain can cause blackouts or total loss of consciousness. A G-suit usually consists of a pair of pants worn under the flight suit and is fitted with bladders which inflate, thus squeezing the legs and lower abdomen preventing blood from pooling.

8. *The term "weightless" has always confused me. Are astronauts on the space shuttle really weightless?*

No. In fact, astronauts on the space shuttle typically orbit the Earth at between 200 km and 600 km above the surface of the Earth. At this altitude the acceleration due to gravity is between 9.2 m/s^2 and 8.2 m/s^2. This is not that much different than on the surface of the Earth! So, an astronaut's weight is similar to on the Earth. The term weightless really refers to how they feel. Since the space shuttle, along with the astronauts, is in free fall as it orbits the Earth, the astronauts feel weightless. The astronauts still have the force of gravity acting upon them (their weight). The next FAQ will help explain the concept of free fall while in orbit.

9. *How can an object in orbit around the Earth still be in free fall? Doesn't free fall imply that the object is falling toward the Earth?*

This is a good question. The term free fall means that the object is only under the influence of gravity. In fact, an object in orbit is falling toward the Earth, but the Earth is curving away from the object at the same rate. To help explain this let's consider the following thought experiment. Suppose you had a very powerful cannon on the top of a very tall mountain as illustrated in the figure. Now suppose that you fired the cannon ball with different velocities. For low velocities the ball would travel much like we typically observe, landing some distance from the cannon. However, as we increase the launch velocity, there comes a point when the rate at which the ball falls toward the Earth is matched by the rate at which the Earth falls away (curves away) from the ball. When this happens, the ball is in orbit around the Earth since it never lands.

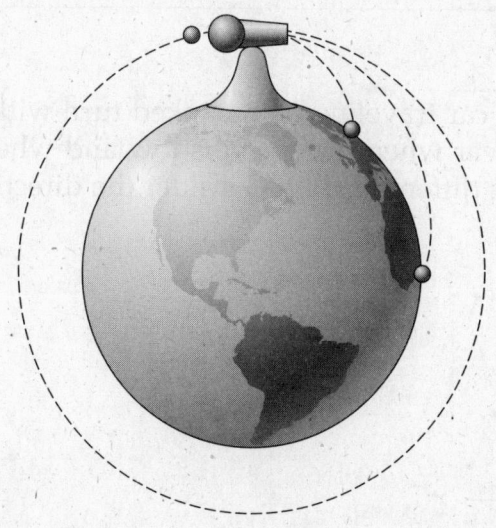

10. *Newton's law of gravity shows that the force between two masses is inversely proportional to the square of the distance between the two objects measured from the center of one object to the center of the other object. Based on this law, if I dug a hole to the center of the Earth, r would become zero at the center; therefore the force would become infinite. Is this true?*

No. Remember, in order to apply Newton's law of gravity in this way we treat objects such as the Earth as a point mass. In order to determine the force of gravity due to non-point masses we need to use some mathematics which is beyond the scope of this course. We can get a sense of how this works by treating the Earth as a series of point masses. Consider the following figure in which we are at the center of the Earth and we are modeling the Earth as many point masses. If we calculated the force of gravity on us due to each of these point masses using Newton's law of gravity and added these forces as vectors, we would find that the net sum was zero, not infinite.

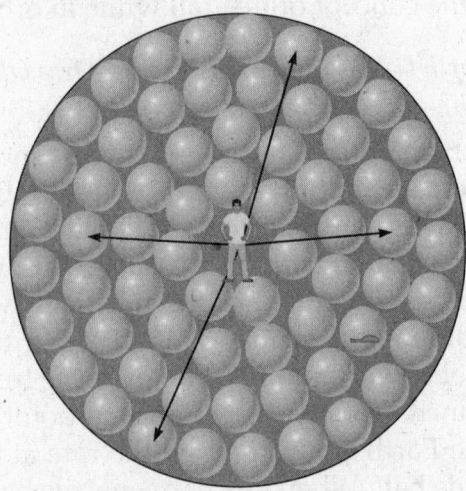

Part C. Selection of End-of-Chapter Answers and Solutions

QUESTIONS

Q5.2   In Example 5.3, we considered a car traveling on a banked turn with friction. Draw free-body diagrams for the car when the speed is low and when the speed is high, and explain why they are different. *Hint*: Consider the direction of the frictional force in the two cases.

Answer

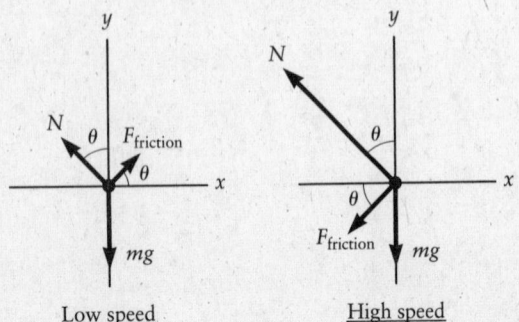

As the car rounds the turn it experiences a centripetal acceleration. From the free-body diagram, the direction of this acceleration is in the $-x$ direction. For low speeds the car "wants" to slip down the bank so the force of static friction is directed up the bank. For high speeds the car "wants" to slip up the bank so the force of static friction is directed down the bank. The net force in the y direction is zero, and the net force in the $-x$ direction results in the centripetal acceleration. The normal force for these two situations is quite different.

Q5.17 The difference in the gravitational force is only about 10% less on an object that is in a low Earth orbit than it is for the same object on the ground. Why is it that an astronaut in orbit experiences weightlessness?

Answer

Weightlessness occurs when an object is freely falling due to gravity. The astronaut experiences weightlessness because the astronaut is freely falling toward the Earth, but has a tangential velocity large enough that she "misses" the Earth. Her velocity allows her to continue moving around and simultaneously falling toward the Earth as the Earth falls away from her. See FAQ#9 for a more detailed explanation.

PROBLEMS

P5.11 A compact disc spins at 2.5 revolutions per second. An ant is walking on the CD and finds that it just begins to slide off the CD when it reaches a point 3.0 cm from the CD's center. (a) What is the coefficient of friction between the ant and the CD? (b) Is this the coefficient of static friction or kinetic friction?

Solution

Recognize the principle

The ant will stay on the CD so long the net force directed toward the center of the CD is sufficient to provide the centripetal acceleration to move in a circular path. In this situation, the force is the force of friction.

Sketch the problem

A free-body diagram helps us set up this problem. Choosing our x direction toward the center of the CD we have,

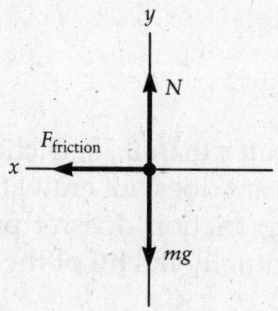

Identify the relationships

(a) The ant travels a distance of one circumference ($2\pi r$) for each rotation. We can therefore find the speed using:

$$v = \frac{2\pi r \text{ m}}{1 \text{ rev}} \times \frac{2.5 \text{ rev}}{1 \text{ s}}$$

The friction in this problem must be static friction, since the ant is not moving relative to the CD. The friction force is given by $F_{\text{friction}} = \mu_s N = \mu_s mg$ where μ_s is the coefficient of static friction. From Newton's second law we have $F = ma$ where the force is the frictional force and the acceleration is the centripetal acceleration v^2/r. Therefore, in order to stay on the CD at that distance, the frictional force must be at least equal to mv^2/r. That is:

$$F_{\text{friction}} = \frac{mv^2}{r}$$

(b) The coefficient of friction must be for $\boxed{\text{static friction}}$, since the ant is not moving relative to the CD.

Solve

(a) If the ant just begins to slide off the CD at a distance of 3.0 cm, its speed is:

$$v = \frac{2\pi (0.03) \text{ m}}{1 \text{ rev}} \times \frac{2.5 \text{ rev}}{1 \text{ s}}$$
$$v = 0.47 \text{ m/s}$$

We know that the force of friction must provide the centripetal acceleration:

$$F_{\text{friction}} = \mu_s mg = \frac{mv^2}{r}$$

The mass can be canceled from the right two terms, and we can solve for the coefficient of static friction:

$$\mu_s \cancel{m}g = \frac{\cancel{m}v^2}{r}$$
$$\mu_s = \frac{v^2}{gr}$$

Inserting the given values yields,

$$\mu_s = \frac{(0.47 \text{ m/s})^2}{(9.8 \text{ m/s}^2)(0.030 \text{ m})}$$

$$\boxed{\mu_s = 0.75}$$

What does it mean?

As long as the CD has a coefficient of static friction greater than 0.75, friction can keep the ant moving in a circle on the CD as long as the ant does not crawl further from the center of the CD. If the coefficient is smaller, friction doesn't provide enough force and so the ant will begin to slide and eventually fall off of the CD.

P5.21 A car of mass 1000 kg is traveling over the top of a hill as shown in Figure P5.21. (a) If the hill has a radius of curvature of 40 m and the car is traveling at 15 m/s, what is the normal force between the hill and the car at the top of the hill? (b) If the driver increases her speed sufficiently, the car will leave the ground at the top of the hill. What is the speed required to make that happen?

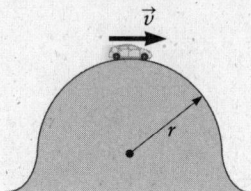

Figure P5.21

Solution

Recognize the principle

When the car is in contact with the hill, it is moving in uniform circular motion. Since it is following a circular path, the car has centripetal acceleration which is a result of a net force directed toward the center of the circular path. This net force is a combination of the force of gravity and the normal force of the road on the car.

Sketch the problem

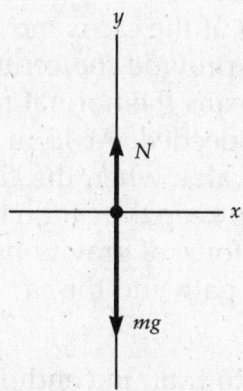

Identify the relationships

(a) Because the car is moving in a circle, we know that the sum of the forces in the y direction is what results in the centripetal acceleration. In this case the centripetal acceleration is in the $-y$ direction when the car is at the top of the hill. Applying Newton's second law in the y direction we find,

$$\Sigma F_y = N - mg = -\frac{mv^2}{r}$$

(b) When the car does not need to be supported by the ground, the normal force N is zero. This means the car must be accelerated in a circle entirely by the gravitational force.

Solve

(a) Solving for the normal force, N, gives:

$$N = mg - \frac{mv^2}{r} = m\left(g - \frac{v^2}{r}\right)$$

Inserting values for g, mass, radius, and speed:

$$N = (1000 \text{ kg})\left((9.8 \text{ m/s}^2) - \frac{(15 \text{ m/s})^2}{(40 \text{ m})}\right)$$

$$\boxed{N = 4200 \text{ N}}$$

(b) Setting the normal force to zero and solving for v gives:

$$0 - mg = -\frac{mv^2}{r}$$

$$v = \sqrt{rg}$$

Inserting values,

$$v = \sqrt{(40 \text{ m})(9.8 \text{ m/s}^2)}$$

$$\boxed{v = 20 \text{ m/s}}$$

What does it mean?

When stationary at the top of the hill, the car's normal force is exactly equal and opposite the gravitational force (9.8×1000 kg = 9800 N). If the car is moving in a circle, however, a net downward force is necessary to provide the centripetal acceleration. Since the force of gravity is constant, this means the normal force is reduced. The faster the car goes, the less normal force is needed. At 15 m/s, the normal force is reduced to 4200 N, less than half of the value when the car was standing still. At 20 m/s, no normal force is needed at all to keep the car on the circular hill. If the car went faster than 20 m/s, the constant force of gravity by itself would not be large enough to keep the car on the circular path and the car would "jump" off of the hill.

P5.34 NASA has built centrifuges to enable astronauts to train in conditions in which the acceleration is very large. The device in Figure P5.34 shows one of these "human centrifuges." If the device has a radius of 8.0 m and attains accelerations as large as $5.0 \times g$, what is the rotation rate?

Figure P5.34

Solution

Recognize the principle

Since the person is moving in a circular path, that person is accelerating. This acceleration is centripetal acceleration.

Sketch the problem

No sketch needed.

Identify the relationships

We can use the expression for centripetal acceleration, $a_c = v^2/r$, and recognize that the centripetal acceleration is equal to $5g$, and the radius is 8 m. The rotation rate in rev/s can then be found by dividing the speed, v, by the circumference $2\pi r$.

Solve

We can solve for v from the expression for the centripetal acceleration, setting $a_c = 5g$,

$$a_c = \frac{v^2}{r} = 5g$$

$$v = \sqrt{5gr}$$

Then the rotation rate is,

$$\text{rotation rate} = \frac{v}{2\pi r} = \frac{\sqrt{5gr}}{2\pi r}$$

Inserting values,

$$\text{rotation rate} = \frac{\sqrt{5(9.8 \text{ m/s}^2)(8 \text{ m})}}{2\pi(8 \text{ m})}$$

$$= \boxed{0.39 \text{ rev/s}}$$

What does it mean?

By placing a person on the end of an 8-m boom and rotating her at a rate of once every 2.5 s, she will experience an acceleration 5 times the acceleration of gravity.

P5.43   When a spacecraft travels from the Earth to the Moon, both the Earth and the Moon exert a gravitational force on the spacecraft. Eventually, the spacecraft reaches a point where the Moon's gravitational attraction overcomes the Earth's gravity. How far from the Earth must the spacecraft be for the gravitational forces from the Moon and the Earth to just cancel?

Solution

Recognize the principle

We can use Newton's law of gravitation to find the gravitational force between the spacecraft and the Earth and the spacecraft and the Moon. For the gravitational force of the Earth and the Moon to just cancel, the forces on the spacecraft must be equal in magnitude, or $F_{\text{Moon}} = F_{\text{Earth}}$. If the distance from the Earth to the spacecraft is $r_{\text{spacecraft}}$, then the distance from the Moon to the spacecraft is the Moon's orbital radius, r_{Moon}, minus $r_{\text{spacecraft}}$, or $r_{\text{Moon}} - r_{\text{spacecraft}}$. Remember when applying Newton's law of gravity, r is the distance measured from the center of one object to the center of the other object.

Sketch the problem

We begin by drawing a sketch of the situation, being sure to label the distances between the objects involved.

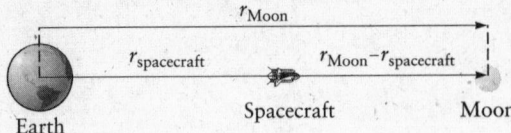

Identify the relationships

Using Newton's law of gravity for F_{Earth} and F_{Moon} and equating the two we find,

$$F_{Earth} = F_{Moon}$$

$$\frac{GM_{Earth}m_{spacecraft}}{(r_{spacecraft})^2} = \frac{GM_{Moon}m_{spacecraft}}{(r_{Moon} - r_{spacecraft})^2}$$

Canceling like terms and solving for $r_{spacecraft}$,

$$\frac{\cancel{G}M_{Earth}\cancel{m}_{spacecraft}}{(r_{spacecraft})^2} = \frac{\cancel{G}M_{Moon}\cancel{m}_{spacecraft}}{(r_{Moon} - r_{spacecraft})^2}$$

$$\frac{(r_{Moon} - r_{spacecraft})^2}{(r_{spacecraft})^2} = \frac{M_{Moon}}{M_{Earth}}$$

$$\frac{r_{Moon} - r_{spacecraft}}{r_{spacecraft}} = \sqrt{\frac{M_{Moon}}{M_{Earth}}}$$

$$r_{Moon} - r_{spacecraft} = (r_{spacecraft})\sqrt{\frac{M_{Moon}}{M_{Earth}}}$$

$$r_{Moon} = r_{spacecraft}\left(1 + \sqrt{\frac{M_{Moon}}{M_{Earth}}}\right)$$

$$r_{spacecraft} = \frac{r_{Moon}}{\left(1 + \sqrt{\frac{M_{Moon}}{M_{Earth}}}\right)}$$

Solve

Inserting the values,

$$r_{spacecraft} = \frac{(3.85 \times 10^8 \text{ m})}{\left(1 + \sqrt{\frac{(0.0735 \times 10^{24} \text{ kg})}{(5.98 \times 10^{24} \text{ kg})}}\right)}$$

$$r_{spacecraft} = \boxed{3.5 \times 10^8 \text{ m}} \text{ from the Earth}$$

What does it mean?

The result shows that the distance where the Earth and the Moon have roughly equal pulls on a spacecraft is 90% of the distance to the Moon. This reflects the relatively smaller size of the Moon compared to the Earth.

P5.51 A newly discovered asteroid is found to have a circular orbit, with a radius equal to 27 times the radius of the Earth's orbit. How long does this asteroid take to complete one orbit around the Sun?

Solution

Recognize the principle

We assume a circular orbit, and would expect that a larger orbit would mean a larger period for this asteroid. We can use Kepler's third law to calculate the period of the orbit from the radius.

Sketch the problem

No sketch needed.

Identify the relationships

Kepler's third law can be written as:

$$T^2_{asteroid} = \left(\frac{4\pi^2}{GM_{Sun}}\right) r^3_{asteroid}$$

Solving for the period, with $r_{asteroid} = 27 r_{Earth}$, gives

$$T_{asteroid} = \sqrt{\left(\frac{4\pi^2}{GM_{Sun}}\right)(27 r_{Earth})^3}$$

Solve

Inserting values for the mass of the Sun and radius of the Earth:

$$T_{asteroid} = \sqrt{\left(\frac{4\pi^2}{(6.67 \times 10^{-11} \text{ N} \cdot \text{m}^2/\text{kg}^2)(1.99 \times 10^{30} \text{ kg})}\right)(27 \times 1.50 \times 10^{11} \text{ m})^3}$$

$$= \boxed{4.4 \times 10^9 \text{ s}}$$

What does it mean?

As expected, the period of this asteroid is much larger than Earth's period. Since there are 3.1×10^7 s in 1 year, this is about 140 years!

P5.59 During an eclipse, the Sun, Earth, and Moon are arranged in a line as shown in Figure P5.59. There are two types of eclipses: (a) a lunar eclipse, when the Earth is between the Sun and the Moon, and (b) a solar eclipse, when the Moon is between the Sun and the Earth. Calculate the percentage change in your weight when going from one type of eclipse to the other.

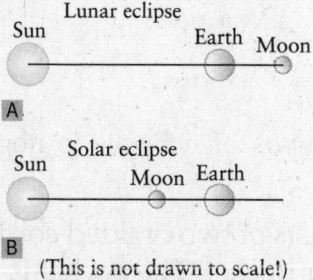

(This is not drawn to scale!)

Figure P5.59

Solution

Recognize the principle

We can calculate the gravitational force between any two objects using Newton's law of gravity. The total force of gravity on you from the Moon and the Sun is the superposition (vector sum) of those forces.

Sketch the problem

No sketch needed.

Identify the relationships

For the situation where the Moon and the Sun are on the same side (above your head), the force from both oppose your weight:

$$F_{same} = \frac{GM_{Sun}m}{r_{ES}^2} + \frac{GM_{Moon}m}{r_{EM}^2} \text{ directed toward the Sun}$$

where r_{ES} and r_{EM} are the distances from the Earth to the Sun and the Earth to the Moon, respectively. For the situation where the Moon and the Sun are on opposite sides, the total force exerted by those objects is,

$$F_{opposite} = \frac{GM_{Sun}m}{r_{ES}^2} - \frac{GM_{Moon}m}{r_{EM}^2} \text{ directed toward the Sun}$$

The percentage change can be found by subtracting these two forces, dividing by your weight, mg, and multiplying by 100%.

Solve

The percentage change is then,

$$\text{percentage change} = \frac{(F_{same} - F_{opposite})}{mg} \times 100\%$$

$$= \frac{\dfrac{2GM_{Moon}m}{r_{EM}^2}}{mg} \times 100\%$$

$$= \frac{2GM_{Moon}}{r_{EM}^2 g} \times 100\%$$

Inserting values,

$$\text{percentage change} = \frac{2(6.67 \times 10^{-11}\text{ N} \cdot \text{m}^2/\text{kg}^2)(0.0735 \times 10^{24}\text{ kg})}{(3.85 \times 10^8\text{ m})^2(9.8\text{ m/s}^2)} \times 100\%$$

$$= \boxed{6.7 \times 10^{-4}\ \%}$$

What does it mean?

The difference in your weight during these two types of eclipses is negligibly small.

P5.64 An ancient and deadly weapon, a *sling* consists of two braided cords, each about half an arm's length long, attached to a leather pocket. The pocket is loaded with a projectile made of lead, carved rock, or clay and made to swing in a vertical circle as shown in Figure P5.64. The projectile is released by letting go of one end of the cord. (a) If a Roman soldier can swing the sling at a rate of 7.5 rotations per second, what is the maximum range of his 100-g projectile? (Ignore air drag.) (b) What is the maximum tension in each cord during the rotation?

Figure P5.64

Solution

Recognize the principle

From our understanding of projectile motion (see Chapter 3), we know that the projectile will have maximum range if it is released such that the initial velocity vector makes an angle of 45° with the horizontal. We can determine the speed of the projectile when it is released from the information on the rate at which the soldier can rotate the sling. The problem then becomes a projectile motion problem. The maximum tension in the cords of the sling occur when the mass is at the bottom of the circular loop. Drawing a free-body diagram will help us with this analysis.

Sketch the problem

The first figure illustrates how the sling should be released so that $\theta = 45°$ and the maximum range is obtained. The second figure is a free-body diagram of the mass when it is at the bottom of the circular path.

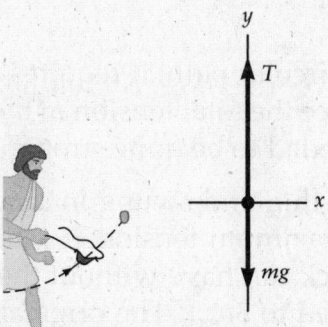

Identify the relationships

First we can determine the speed of the mass while it is in the sling from the rate of rotation.

$$v = (2\pi r)\,(7.5 \text{ rev/s})$$

(a) Assuming the projectile lands at the same height as it was released the range of the projectile is given by,

$$\text{range} = \frac{v_0^2 \sin(2\theta)}{g}$$

(b) Applying Newton's second law to the mass at the bottom of the circular path and recognizing that the acceleration is centripetal acceleration we find,

$$\Sigma F = ma$$
$$T - mg = m\frac{v^2}{r}$$

Solve

We can estimate the radius of the circular path to be around 1 m. Substituting this into our expression for the speed we find,

$$v = (2\pi(1 \text{ m}))(7.5 \text{ rev/s}) = 47 \text{ m/s}$$

(a) Plugging in our value for speed and $\theta = 45°$ we find the maximum range of the projectile to be

$$\text{range} = \frac{v_0^2 \sin(2\theta)}{g} = \frac{(47 \text{ m/s})^2 \sin(90)}{9.8 \text{ m/s}^2} \approx \boxed{225 \text{ m}}$$

(b) Substitution in the numerical vales and solving for tension we get,

$$T = (0.100 \text{ kg})\left[(9.8 \text{ m/s}^2) + \frac{(47 \text{ m/s})^2}{(1.0 \text{ m})}\right]$$
$$\approx 220 \text{ N}$$

Since this tension is divided equally between the two cords, the tension in one cord is

$$T_{\text{cord}} = \frac{T}{2} \approx \boxed{110 \text{ N} \approx 25 \text{ lbs}}$$

What does it mean?

Since the projectile leaves the sling tangent to the circular path, it requires a great deal of skill to accurately use this weapon. Also, since the total tension in the cords must be supported by the soldier's arm, he also needed to be quite strong.

P5.73   A rock of mass m is tied to a string of length L and swung in a horizontal circle of radius r. The string can withstand a maximum tension T_{max} before it breaks. (a) What is the maximum speed v_{max} the rock can have without the string breaking? (b) The speed of the rock is now increased to $3v_{\text{max}}$. The original single string is then replaced by N pieces that are all identical to the original string. What is the minimum value of N required so that the strings do not break? Ignore the force of gravity on the rock.

Solution

Recognize the principle

Since the rock is moving in a horizontal circle, the tension must provide a force equal to its mass times a centripetal acceleration.

Sketch the problem

We begin by drawing a sketch of the situation and then the free-body diagram for the rock.

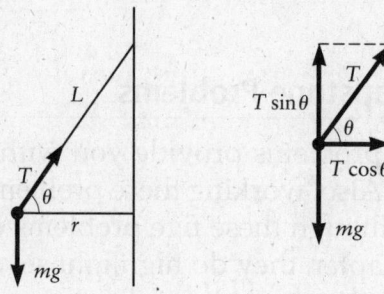

Identify the relationships

The horizontal force directed toward the center of the circular path can be seen from the free-body diagram as $T \cos\theta$. This force must equal the force required to hold the rock in a circular motion, namely $\frac{mv^2}{r}$. The vertical components of the forces are $+ T \sin\theta$ and $-mg$. Using geometry, $\cos\theta = \frac{r}{L}$ and $\sin\theta = \frac{\sqrt{L^2 - r^2}}{L}$. From the free-body diagram, we can sum the forces along each axis.

$$\Sigma F_x = T \frac{r}{L} = ma_x = m \frac{v^2}{r}$$

$$\Sigma F_y = T \frac{\sqrt{L^2 - r^2}}{L} - mg = ma_y = 0$$

Along y, the acceleration is zero since the object is not accelerating (or moving) in the vertical direction.

Solve

(a) Simplifying the two above equations,

$$T \frac{r}{L} = \frac{mv^2}{r}$$

$$T \frac{\sqrt{L^2 - r^2}}{L} - mg = 0$$

Solving for the speed in the first equation,

$$T \frac{r}{L} = \frac{mv^2}{r}$$
$$v^2 = \frac{Tr^2}{mL}$$
$$v = \sqrt{\frac{Tr^2}{mL}}$$

For a maximum tension T_{max}, there is a corresponding maximum speed,

$$\boxed{v_{max} = \sqrt{\frac{T_{max} r^2}{mL}}}$$

(b) If the speed is increased by a factor of 3 and all other factors are held constant, the tension must increase by a factor of 9 as can be seen from the above equation.

What does it mean?

If the original string is replaced by N identical strings with the same maximum tension, then in order to hold 9 times more tension, that string must be replaced by 9 strings, so $\boxed{N = 9}$.

Part D. Additional Worked Examples and Capstone Problems

The following worked examples and capstone problems provide you with additional practice with circular motion and gravity. Also, working these problems will give you greater insight into these concepts. Although these five problems do not incorporate all the material discussed in this chapter, they do highlight several of the key concepts. If you can successfully solve these problems then you should have confidence in your understanding of these key concepts, so use these problems as a test of your understanding of the chapter material.

WE 5.1 Centripetal Acceleration and Force

Consider again the bug in Example 5.2 of your textbook. If the coefficient of static friction between the bug and the CD is $\mu_S = 0.80$, will the bug be able to stay on the CD while the disc spins?

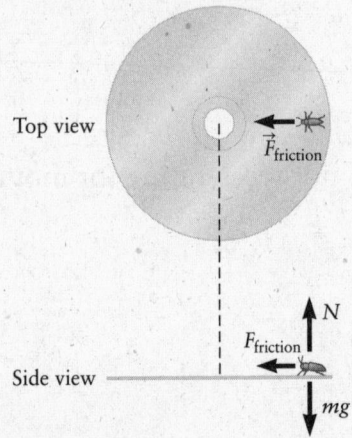

Solution

Recognize the principle

The force that enables the bug to move in a circle is due to static friction, and static friction can be as large as $F_{\text{friction,max}} = \mu_S N$. We can calculate this maximum frictional force and then compare it with the force required to make the bug move in a circle, which was found in Example 5.2. If $F_{\text{friction,max}}$ is smaller than the required force found in Example 5.2, the bug will slide off the CD.

Sketch the problem

The figure describes the problem.

Identify the relationships and Solve

The normal force between the CD and the bug is $N = mg$, so

$$F_{\text{friction,max}} = \mu_S N = \mu_S mg = (0.80)(0.0050 \text{ kg})(9.8 \text{ m/s}^2) = \boxed{0.039 \text{ N}}$$

This value is smaller than the required force calculated in Example 5.2, so the bug will slide off the CD.

What does it mean?

The force $\sum F = mv^2/r$ is the total force required to make an object undergo uniform circular motion. The source of this force depends on the situation. In this example, it is due to friction, whereas for the rock twirling in a circle on a string, it is due to tension in the string.

WE 5.2 People and Gravity

Find the approximate gravitational force between two people standing next to each other.

Solution

Recognize the principle

We can treat the two people as spheres and calculate the gravitational force by assuming that all the mass of each person is at the center of the corresponding sphere. Although most people are not spheres, we can estimate the gravitational force by making this approximation.

Sketch the problem

The figure shows the situation where we approximate the people as point masses m_1 and m_2.

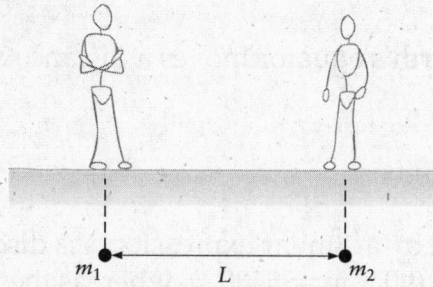

Identify the relationships

To calculate the gravitational force, we need to estimate the mass of each person (m_1 and m_2) and their separation L. The people are assumed to be close together, so we take $L = 20$ cm $= 0.20$ m. We take each mass to be 80 kg, which is typical for a person of average size.

Solve

Using our expression for the gravitational force between two masses, we get

$$F_{\text{grav}} = \frac{Gm_1m_2}{L^2} = \frac{(6.67 \times 10^{-11} \text{ N} \cdot \text{m}^2/\text{kg}^2)(80 \text{ kg})(80 \text{ kg})}{(0.20 \text{ m})^2} = \boxed{1.1 \times 10^{-5} \text{ N}}$$

What does it mean?

This force is a very small (about a tenth the weight of a mosquito) and again shows that the gravitational attraction between terrestrial objects is generally very weak. The force of gravity is most noticeable when one or both of the objects involved have a very large mass, such as the Moon or the Earth.

WE 5.3 Orbital Speeds and the Earth's Rotation

Calculate the speed v_R of a point on the Earth's equator due to the Earth's rotation. Compare this speed to the speed of a satellite in low Earth orbit.

Solution

Recognize the principle

In both cases, the speed can be found from the period and radius of the "orbit" together with the relation $T = 2\pi r/v$.

Sketch the problem

The following figure describes the problem. A point at the equator moves in a circle of radius r_E, where r_E is the Earth's radius.

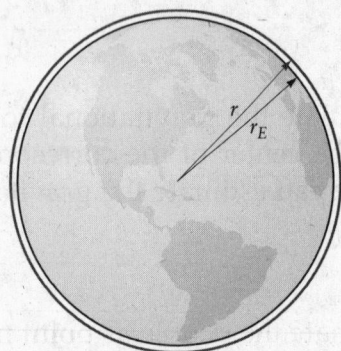

Identify the relationships and Solve

From the figure we see that a point at the Earth's equator moves a distance of $2\pi r_E$ in 1 day = 86,400 s, so the speed is

$$v_R = \frac{2\pi r_E}{T} = \frac{2\pi(6.37 \times 10^6 \text{ m})}{86,000 \text{ s}} = 470 \text{ m/s}$$

The orbital speed in low Earth orbit is given by a similar expression. As discussed in your textbook, the time is the orbital time (90 min = 5400 s), whereas the radius of the orbit is $r = 6.66 \times 10^6$ m, leading to an orbital speed of

$$v = \frac{2\pi r_{\text{orbit}}}{T_{\text{orbit}}} = \frac{2\pi(6.66 \times 10^6 \text{ m})}{5400 \text{ s}} = \boxed{7800 \text{ m/s}}$$

Comparing these two speeds shows that a satellite in orbit has a speed that is only about 17 times greater than v_R.

What does it mean?

The speed of an object "at rest" on the surface of the Earth is 470 m/s ≈ 1000 mi/h!

CP 5.1 Circular Motion Involving Many Forces

Consider the figure shown in which a mass, $m_1 = 5$ kg, is spinning around in a circular path on the end of a massless string. The string goes through a frictionless tube and is attached to another mass, $m_2 = 10$ kg, which is hanging freely. (a) Determine the speed of m_1, its angle θ with the vertical, and the tension in the string such that m_2 remains stationary.

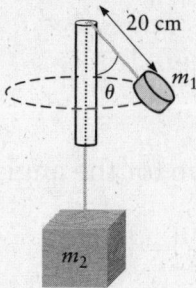

Solution

Recognize the principle

Several principles must be applied to solve this problem. First we recognize that m_1 is traveling around in a circular path so there is a centripetal acceleration. We'll need to draw a free-body diagram and apply Newton's second law to each of the masses.

Sketch the problem

The following are free-body diagrams for the two masses.

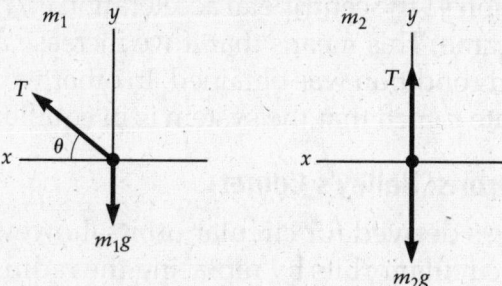

Identify the relationships

We begin by applying Newton's second law for each of the masses in both the x and y directions.

$$\sum F_{1,x} = m_1 a_{1,x} \rightarrow T \cos \theta = m_1 a_{1,x}$$

$$\sum F_{1,y} = m_1 a_{1,y} \rightarrow T \sin \theta - m_1 g = m_1 a_{1,y}$$

$$\sum F_{2,y} = m_2 a_{2,y} \rightarrow T - m_2 g = m_2 a_{2,y}$$

Since m_1 is not accelerating in the y direction, $a_{1,y} = 0$. However, m_1 is accelerating in the x direction and this acceleration is centripetal acceleration, $a_{1,x} = v^2/r$.

Since m_2 is not accelerating, $a_{2,y} = 0$.

Substituting in and rearranging the expressions we get,

$$T \cos \theta = m_1 \frac{v^2}{r} \tag{#1}$$

$$T \sin \theta = m_1 g \tag{#2}$$

$$T = m_2 g \tag{#3}$$

Finally, we recognize that since r is the radius of the circular path,

$$r = (0.20 \text{ m}) \sin \theta \tag{#4}$$

Solve

From expression #3 we determine the tension in the rope to be

$$T = m_2 g = (10 \text{ kg})(9.8 \text{ m/s}^2) = \boxed{98 \text{ N}}$$

Plugging this value for the tension into expression #2 we solve for the angle,

$$T \sin \theta = m_1 g \rightarrow \theta = \sin^{-1}\left(\frac{m_1 g}{T}\right) = \sin^{-1}\left(\frac{5 \text{ kg} \cdot 9.8 \text{ m/s}^2}{98 \text{ N}}\right) = \boxed{30°}$$

Solving for v from expression #1 and substituting in the numerical values we find,

$$v = \sqrt{\frac{r}{m_1} T \cos \theta} = \sqrt{\frac{0.2 \sin(30°) \text{ m}}{5 \text{ kg}} \cdot 98 \cos(30°) \text{ N}} = \boxed{1.3 \text{ m/s}}$$

What does it mean?

What would happen if you rotated m_1 at a greater rotational rate? Would the angle change? Would the tension change? Would r change? From expression #3 we see that the tension in the string only depends on m_2, if m_2 is to remain stationary. From expression #2 we see that the angle would also remain the same as long as the tension was the same. Finally, from expression #1 the centripetal acceleration v^2/r would remain the same since T and θ are the same. This means that if we increase v then r would change so that a new equilibrium condition was obtained. In other words, for any value of r we can find an appropriate v such that the system is in equilibrium.

CP 5.2 Kepler's Laws for Non-circular Orbits: Halley's Comet

In your textbook, Kepler's third law was derived for circular orbits, however, this law can be generalized to include non-circular orbits by replacing the radius of the circular orbit, r, by the semi-major axis, a, of the elliptical orbit. As shown in the figure, the semi-major axis is the distance from the center of an ellipse to one end of the ellipse. The closest point of an orbit to the Sun is called the perihelion and the furthest point is called the aphelion. Halley's comet has a highly eccentric orbit with a period of approximately 76 years and comes to within 8.6×10^{10} m of the Sun at its perihelion. (a) Using the modified version of Kepler's third law, determine the semi-major axis of the orbit of Halley's comet. (b) Determine the aphelion distance of Halley's comet. Compare this with Pluto's orbital radius. (c) Given the perihelion speed of Halley's comet to be approximately 54 km/s, use Kepler's third law to determine its speed at aphelion.

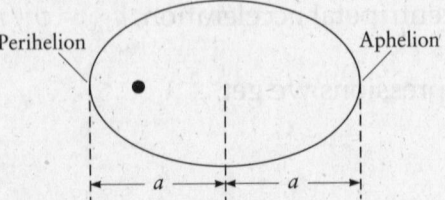

Solution

Recognize the principle

Since we know the period of Halley's comet we can use the modified form of Kepler's third law to determine the semi-major axis of the orbit. From this information and the perihelion distance it's easy to solve for the aphelion distance. Finally, we can apply Kepler's second law at these two points in the orbit to determine the speed of Halley's comet at aphelion.

Sketch the problem

Sketch is provided in problem.

Identify the relationships and Solve

(a) We begin by modifying Kepler's third law so that it applies to elliptical orbits around the Sun:

$$T^2 = \left(\frac{4\pi^2}{GM_{Sun}} \right) a^3$$

Using $T = 76$ years $= 2.4 \times 10^9$ s, and solving for a we find,

$$a = \sqrt[3]{\frac{T^2 G M_{Sun}}{4\pi^2}} = \sqrt[3]{\frac{(2.4 \times 10^9 \text{ s})^2 (6.67 \times 10^{-11} \text{ Nm}^2/\text{kg}^2)(1.99 \times 10^{30} \text{ kg})}{4\pi^2}}$$

$$= \boxed{2.7 \times 10^{12} \text{ m}}$$

(b) From the figure we see that the aphelion distance is simply twice the semi-major axis minus the perihelion distance. Therefore,

$$r_{aphelion} = 2a - r_{perihelion} = 2(2.7 \times 10^{12}) - (8.6 \times 10^{10}) = \boxed{5.3 \times 10^{12} \text{ m}}$$

This value is close to the mean orbital radius of Pluto.

(c) Knowing the orbital speed at perihelion, $r_{perihelion}$, and $r_{aphelion}$, we can apply Kepler's second law to determine the orbital speed at aphelion by comparing the area swept out by the orbit at these two points. Consider the following figure comparing the area formed by the line connecting the Sun to Halley's comet at points on either side of the perihelion and the aphelion.

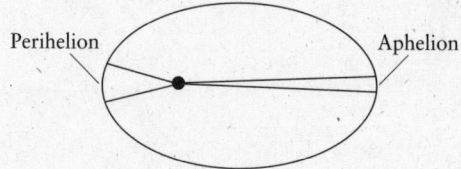

If we consider a short period of time, then the areas can be approximated by triangles with the base equal to the distance traveled through the orbit during that small time period and the height equal to the radius of the orbit during that time,

$$A \approx A_{triangle} = \frac{\text{(Distance traveled)(Orbital radius)}}{2} = \frac{d \times r}{2}$$

Now, the distance traveled through the orbit is approximately the orbital speed times the time interval,

$$d = v\Delta t$$

So, we can write the following expressions for the area of the two triangles,

$$A_{\text{perihelion}} = \frac{d_{\text{perihelion}}\, r_{\text{perihelion}}}{2} = \frac{(v_{\text{perihelion}}\, \Delta t)\, r_{\text{perihelion}}}{2}$$

$$A_{\text{aphelion}} = \frac{d_{\text{aphelion}}\, r_{\text{aphelion}}}{2} = \frac{(v_{\text{aphelion}}\, \Delta t)\, r_{\text{aphelion}}}{2}$$

Rearranging these two expressions we get,

$$\frac{A_{\text{perihelion}}}{\Delta t} = \frac{v_{\text{perihelion}}\, r_{\text{perihelion}}}{2}$$

$$\frac{A_{\text{aphelion}}}{\Delta t} = \frac{v_{\text{aphelion}}\, r_{\text{aphelion}}}{2}$$

Kepler's second law tells us that these two quantities must be the same. Therefore, we can set them equal and solve for the speed of Halley's comet at aphelion,

$$v_{\text{perihelion}}\, r_{\text{perihelion}} = v_{\text{aphelion}}\, r_{\text{aphelion}}$$

$$v_{\text{aphelion}} = \frac{v_{\text{perihelion}}\, r_{\text{perihelion}}}{r_{\text{aphelion}}} = \frac{(54 \text{ km/s})(8.6 \times 10^{10} \text{ m})}{(5.3 \times 10^{12} \text{ m})} = \boxed{0.9 \text{ km/s}}$$

What does it mean?

If fact, Halley's comet spends some of its time outside the orbit of Pluto due to the eccentricity of both their orbits. Also, since Halley's comet is so eccentric its orbital speed at aphelion is 900 m/s (2000 mi/h). This is around the maximum speed of a fighter jet!

Part E. MCAT Review Problems and Solutions

PROBLEMS

1. A 0.2-kg ball is tied to the end of a string and rotated in a horizontal circular path of radius 25 cm. If the speed is 10π cm/s, what is the centripetal acceleration of the ball?
   - (a) $2.5\pi^2$ cm/s^2
   - (b) $4.0\pi^2$ cm/s^2
   - (c) $0.40\pi^2$ cm/s^2
   - (d) $25\pi^2$ cm/s^2

2. A 70-kg woman standing at the equator rotates with the Earth around its axis at a speed of about 500 m/s. If the radius of the Earth is approximately 6×10^6 m, which is the best estimate of the centripetal acceleration experienced by the woman?
   - (a) 4×10^{-2} m/s^2
   - (b) 4 m/s^2
   - (c) 10^{-4} m/s^2
   - (d) 24 m/s^2

3. What is the centripetal acceleration of a 20-kg object traveling at a uniform speed of 40 m/s around a circular path of radius 10 m?
   (a) 320 m/s²
   (b) 160 m/s²
   (c) 80 m/s²
   (d) 20 m/s²

4. Car A has a mass twice that of car B. If both are traveling at the same uniform speed around a circular racetrack, the centripetal acceleration of car A is
   (a) twice that of car B.
   (b) half that of car B.
   (c) the same as that of car B.
   (d) four times that of car B.

5. A ball rolls with uniform speed around a frictionless flat horizontal circular track. If the speed of the ball is doubled, the centripetal acceleration is
   (a) quadrupled.
   (b) doubled.
   (c) halved.
   (d) unchanged.

6. The force of gravity between two objects is
   (a) inversely proportional to the distance between them.
   (b) directly proportional to the distance between them.
   (c) inversely proportional to the square of the distance between them.
   (d) directly proportional to the square of the distance between them.

7. Two objects of equal mass are separated by a distance of 2 m. If the mass of one object is doubled, the force of gravity between the two objects will
   (a) be half as great.
   (b) be twice as great.
   (c) be one-fourth as great.
   (d) be 4 times as great.

8. The distance between a spaceship and the center of the Earth increases from one Earth radius to three Earth radii. What happens to the force of gravity acting on the spaceship?
   (a) It becomes 1/9 as great.
   (b) It becomes 9 times as great.
   (c) It becomes 1/3 as great.
   (d) It becomes 3 times as great.

9. A 100-kg astronaut lands on a planet with a radius 3 times that of Earth and a mass 9 times that of Earth. The force of gravity experienced by the astronaut will be

   (a) 9 times the value on Earth.

   (b) 3 times the value on Earth.

   (c) the same value as on Earth.

   (d) one-third the value on Earth.

10. According to Kepler's third law of planetary motion if you double the distance to an orbiting planet as measured from the center of the Sun, the period of its orbit will

    (a) double as well.

    (b) increase by a factor of $2^{3/2}$.

    (c) increase by a factor of $2^{2/3}$.

    (d) remain the same.

SOLUTIONS

1. MCAT strategies

No answers can be easily eliminated so we proceed with our problem-solving strategy.

Recognize the principle

We must use the expression for the centripetal acceleration of an object traveling in a circular path.

Sketch the problem

None required.

Identify the relationships

The expression of centripetal acceleration is

$$a_c = \frac{v^2}{r}$$

Solve

Substituting in the numerical values and solving we find,

$$a_c = \frac{v^2}{r} = \frac{(10\pi \text{ cm/s})^2}{25 \text{ cm}} = \frac{100\,\pi^2}{25} \text{ cm/s}^2 = 4\pi^2 \text{ cm/s}^2$$

Therefore the correct answer is (b).

What does it mean?

Note that the mass of the ball was not needed.

2. MCAT strategies

No answers can be easily eliminated so we proceed with our problem-solving strategy.

Recognize the principle

We must use the expression for the centripetal acceleration of an object traveling in a circular path.

Sketch the problem

None required.

Identify the relationships

The expression of centripetal acceleration is

$$a_c = \frac{v^2}{r}$$

Solve

Substituting in the numerical values and solving we find,

$$a_c = \frac{v^2}{r} = \frac{(500 \text{ m/s})^2}{6 \times 10^6 \text{ m}} = \frac{25 \times 10^4}{6 \times 10^6} \text{ m/s}^2 \approx 4 \times 10^{-2} \text{ m/s}^2$$

Therefore the correct answer is (a).

What does it mean?

Again, that the mass of the woman was not needed.

3. MCAT strategies

No answers can be easily eliminated so we proceed with our problem-solving strategy.

Recognize the principle

We must use the expression for the centripetal acceleration of an object traveling in a circular path.

Sketch the problem

None required.

Identify the relationships

The expression of centripetal acceleration is

$$a_c = \frac{v^2}{r}$$

Solve

Substituting in the numerical values and solving we find,

$$a_c = \frac{v^2}{r} = \frac{(40 \text{ m/s})^2}{10 \text{ m}} = \frac{1600}{10} \text{ m/s}^2 = 160 \text{ m/s}^2$$

Therefore the correct answer is (b).

What does it mean?

Once again, the mass was provided in the question but not needed in the solution. It's common among MCAT tests to provide you with information which is not needed in the solution to the problem. You need to identify and not be confused by this superfluous information.

4. MCAT strategies

Since the centripetal acceleration of the cars is dependent on their speeds and the radius of the track, not their masses, the centripetal acceleration of the cars will be the same. Therefore the correct answer is (c).

5. MCAT strategies

Since the centripetal acceleration depends on the square of the speed of the object, then holding all else constant doubling the speed of the ball will produce 4 times the centripetal acceleration. Therefore the correct answer is (a).

6. MCAT strategies

This problems simply requires us to know the functional dependence of Newton's law of gravity on the distance of separation of the two objects. Since $F_{grav} = Gm_1m_2/r^2$, then F_{grav} is inversely proportional to the square of the distance between the two objects. Therefore the correct answer is (c).

7. MCAT strategies

Since the force of gravity between two objects is directly proportional to the mass of each object, then by doubling one of the masses the force will also double. Therefore the correct answer is (b). Note that the distance of separation was unnecessary information.

8. MCAT strategies

Since the force of gravity is inversely proportional to the square of the distance between the two objects, tripling the distance of separation will produce a force which is 1/9th the magnitude of the original force. Therefore the correct answer is (a).

9. MCAT strategies

Solving this problem requires a little more algebra than the previous few problems so we'll proceed with our usual problem-solving strategy.

Recognize the principle

This problem involves the manipulation of Newton's law of gravity.

Sketch the problem

None required.

Identify the relationships

Newton's law of gravity is given by $F_{grav} = Gm_1m_2/r^2$. Now if we let m_2 be $9m_2$ and r be $3r$, we can figure out how the force of gravity compares to the force of gravity on Earth.

Solve

Substituting in the values we find,

$$F_{grav,\,planet} = \frac{Gm_1\,(9m_2)}{(3r)^2} = \frac{9}{9}\frac{Gm_1m_2}{r^2} = \frac{Gm_1m_2}{r^2} = F_{grav,\,Earth}$$

Therefore the force of gravity on the astronaut remains the same. The correct answer is (c).

What does it mean?

Note that the mass of the astronaut was not necessary to solve this problem, and we didn't actually need to calculate the force.

10. MCAT strategies

This problem simply requires us to know the functional form of Kepler's third law. Kepler's third law tells us that the square of the period of an orbit is proportional to the cube of the orbital radius.

$$T^2 \propto a^3$$

Solving for T we find,

$$T \propto \sqrt{a^3} = a^{3/2}$$

So if we double the orbital radius then the period will change by a factor of $2^{3/2}$. Therefore the correct answer is (b).

6 Work and Energy

Part A. Summary of Key Concepts and Problem-Solving Strategies

KEY CONCEPTS

Work

When a force acts on an object, the force does *work* on the object:

$$W = F(\Delta r)\cos\theta$$

where F is the force acting on the object, Δr is the displacement of the object, and θ is the angle between the force and the displacement. Only if there is a component of the force in the direction of the displacement, does the force do work on the object.

The work done by an applied force can be interpreted in a graphical manner as the area under the force–displacement curve.

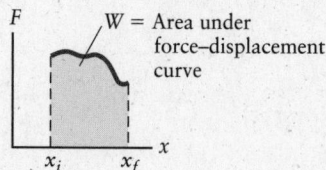

Kinetic energy

Kinetic energy is energy associated with the motion of an object and is given by the expression,

$$KE = \frac{1}{2} mv^2$$

where m is the mass and v is the speed of the object.

Work and kinetic energy are related through the *work–energy theorem*

$$W = \Delta KE$$

where W is the work done on the object. This expression implies that if net work is done on an object then it will result in a change in the object's kinetic energy, therefore, its speed will change.

Potential energy

Potential energy is stored energy and is associated with a particular force or forces. The potential energy associated with a force is related to the work done by this force:

$$\Delta PE = PE_f - PE_i = -W$$

The work done by a *conservative force* is independent of the path.

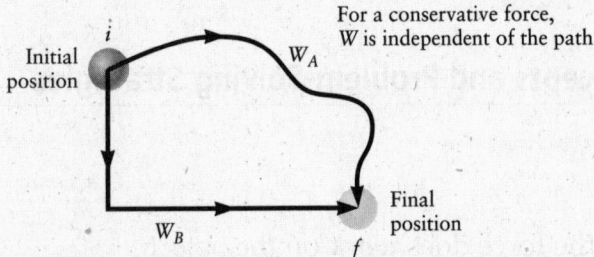

Conservation of energy

The principle of conservation of energy can be written as

$$KE_i + PE_i + W_{noncon} = KE_f + PE_f$$

where W_{noncon} is the work done on the system by nonconservative forces. Since work done on an object due to a nonconservative force removes energy from the object, this work is always negative. The principle of conservation of energy allows us to analyze complicated processes which we would not be able to solve using Newton's laws of motion.

APPLICATIONS

Hooke's law

The force exerted by a spring or similar elastic object is

$$F = -kx$$

where k is the "spring constant" and x is the amount that the spring is stretched or compressed. The negative sign provides information about the direction of the force which is always toward the equilibrium point of the system.

Potential energy functions

Gravitation potential energy of an object near Earth's surface is given by the expression,

$$PE_{grav} = mgy$$

where m is the mass of the object, and y is the vertical position of the object relative to where we've defined the origin of our coordinate system.

A more general expression for the gravitational potential energy of two masses is given by the expression,

$$PE_{grav} = -\frac{Gm_1m_2}{r}$$

where m_1 and m_2 are the masses, $G = 6.67 \times 10^{-11}\ \text{Nm}^2/\text{kg}^2$, and r is the distance of separation of the two masses measured from the center of one object to the center of the other object.

Elastic potential energy such as what is associated with a spring is given by the expression,

$$PE_{spring} = \frac{1}{2}kx^2$$

where k is the "spring constant" and x is the amount that the "spring" is stretched or compressed.

PROBLEM-SOLVING STRATEGIES

In this chapter we were introduced to our first conservation principle. The principle of conservation of energy provides us with a powerful tool for analyzing complex mechanical problems. The following schematic outlines the thought process we should follow when applying this concept.

Problem Solving: Applying the Principle of Conservation of Energy

Recognize the principle

Start by finding the object (or system of objects) whose mechanical energy (kinetic plus potential energy) is conserved. Note that there may be several forms of mechanical energy due to several forces and energy losses due to the work done by nonconservative force.

Sketch the problem

In your sketch, show the initial and final states of the object. This sketch should contain a coordinate system, including an origin, with which to measure the potential energy. Also, you should label the forces involved and the speeds of the object at the points of interest.

Identify the relationships

Find expressions for the initial and final kinetic and potential energies. One or more of these energies may involve unknown quantities. If there are nonconservative forces involved such as friction, determine the work done by these forces. Use the appropriate expression for energy conservation:

Without nonconservative forces → $KE_i + PE_i = KE_f + PE_f$

With nonconservative forces → $KE_i + PE_i + W_{noncon} = KE_f + PE_f$

Solve

Substitute in the known quantities and solve for the unknown quantities.

What does it mean?

Always *consider what your answer means* and check that it makes sense.

Part B. Frequently Asked Questions

1. *Since I'm able to define a potential energy associated with the force of gravity and the force of a spring, why can't I define a potential energy for the force of friction?*

There are several ways to think about this question. We can only associate a potential energy with a force which is conservative. The work done by a conservative force is independent of the path followed. Another way to think about this is to look at what we mean by potential energy. Potential energy is that form of energy which depends on position. It is stored energy. When you let go of a compressed spring or an object above the ground, the object gains kinetic energy as potential energy is lost. The same cannot be said for the force of friction. As you slide an object around on the floor, energy is lost due to the frictional force. This energy is not stored for us to use later. If you let go of the object, the energy lost due to friction does not convert back into kinetic energy of the object.

2. *Is it important where I define the origin of my coordinate axis when determining the gravitational potential energy of an object?*

If you are using the expression for the gravitational potential energy of an object near the surface of the Earth, $PE_{grav} = mgy$, then it doesn't matter where you define the origin of your coordinate system. As it turns out, the useful quantity is the change in the potential energy which depends on the change in the position, Δy. This is independent of where you define $y = 0$. If you use the general expression for the gravitational potential energy, then the origin of your coordinate system is defined for you. See the next FAQ for more on this question.

3. *When using the expression $PE_{grav} = -\frac{Gm_1m_2}{r}$ to calculate the gravitational potential energy for an object-Earth system, can I define $PE_{grav} = 0$ at the surface of the Earth?*

No, certainly not. By using this more general expression, $PE_{grav} = 0$ occurs when the objects are infinitely far apart, $r = \infty$. Remember, gravitational potential energy is associated with the configuration of the system. The system includes both objects m_1 and m_2, and the force of gravity between them only approaches zero when they are very far apart.

4. *When calculating the gravitational potential energy associated with an object-Earth system, when do I need to use the expression $PE_{grav} = -\frac{Gm_1m_2}{r}$ instead of $PE_{grav} = mgy$? Does it really matter?*

That's a good question, the answer of which really depends on the degree of precision we want in our final answer. Certainly if we're dealing with objects thousands of kilometers above the surface of the Earth then we should use the more general expression. What about 10 km, or 100 km? Let's calculate the change in gravitational potential energy moving an object from the surface to 100 km above the surface of the Earth using the two expressions to see how much they differ.

$$\Delta PE_{grav} = mg\Delta y = (1 \text{ kg})(9.8 \text{ m/s}^2)(100{,}000 - 0 \text{ m}) = \boxed{9.8 \times 10^5 \text{ J}}$$

$$\Delta PE_{grav} = -\frac{GM}{R_E + h} - \left(-\frac{GM}{R_E}\right) = GM\left(\frac{1}{R_E} - \frac{1}{(R_E + h)}\right)$$

$$= \left(6.67 \times 10^{-11}\,\frac{N \cdot m^2}{kg}\right)(5.98 \times 10^{24}\,kg)$$

$$\times \left(\frac{1}{6.37 \times 10^6\,m} - \frac{1}{(6.37 \times 10^6 + 1 \times 10^5)\,m}\right)$$

$$= \boxed{9.7 \times 10^5\,J}$$

Comparing these two answers we see that the simpler expression overestimates the potential energy by about 10%. So, if this degree of precision is sufficient then you can use the expression $PE_{grav} = mgy$ up to 100 km above the surface of the Earth. If greater precision is required, then you should stick with the more general expression.

5. *The minus sign in Hooke's law really confuses me. Does it mean that the force is always negative?*

Although the minus sign in Hooke's law tells us the direction of the force, it does not mean that the force is always negative. The force of a spring on an object is always directed toward the equilibrium point (where the spring is neither stretched nor compressed). If we stretch the spring in the positive direction then $x > 0$ and the minus sign tells us the force of the spring is in the negative direction. If we compressed the spring then $x < 0$ and the resulting force is in the positive direction.

6. *If the force on an object varies with position, how do we determine the work done by the force?*

As was discussed in section 6.1 in the textbook, the work done by a variable force can be determined graphically as the area under the force–displacement graph. However, if we know the expression which describes the force as a function of position, we can calculate the work done by this force in much the same way as the graphical technique, without having to draw the graph. The following example will help illustrate the process:

Problem: Suppose the force acting on an object varies with position according to the expression $F = (5x^2 - 2x + 3)N$. Determine the work done by this force on the object while moving the object from $x = 0$ to $x = 10$ m.

Solution: Rather than drawing the force–displacement graph and using the techniques described in the textbook, let's split the calculation up into little pieces. Each piece could be 1 m in length and we can assume the force is approximately constant over that distance. We can calculate the work done for each of these little pieces and add them together to get the total work. This is equivalent to calculating the area under the force–displacement graph by dividing it up into little pieces. Of course, the smaller we divide the little pieces, the more accurate our final answer will be. The following table shows the calculations:

Displacement $\Delta x = x_f - x_i$	Average Force $F_{average} = (F_f + F_i)/2$	Work $W = F_{average} \times \Delta x$
$\Delta x = 1 - 0 = 1$ m	$F_{average} = [5(1)^2 - 2(1) + 3]/2 = 3.0$ N	3.5 J
$\Delta x = 2 - 1 = 1$ m	$F_{average} = [5(2)^2 - 2(2) + 3 + 5(1)^2 - 2(1) + 3]/2 = 12.5$ N	12.5 J
$\Delta x = 3 - 2 = 1$ m	$F_{average} = [5(3)^2 - 2(3) + 3 + 5(2)^2 - 2(2) + 3]/2 = 30.5$ N	30.5 J
$\Delta x = 4 - 3 = 1$ m	$F_{average} = [5(4)^2 - 2(4) + 3 + 5(3)^2 - 2(3) + 3]/2 = 58.5$ N	58.5 J
$\Delta x = 5 - 4 = 1$ m	$F_{average} = [5(5)^2 - 2(5) + 3 + 5(4)^2 - 2(4) + 3]/2 = 96.5$ N	96.5 J
$\Delta x = 6 - 5 = 1$ m	$F_{average} = [5(6)^2 - 2(6) + 3 + 5(5)^2 - 2(5) + 3]/2 = 144.5$ N	144.5 J
$\Delta x = 7 - 6 = 1$ m	$F_{average} = [5(7)^2 - 2(7) + 3 + 5(6)^2 - 2(6) + 3]/2 = 202.5$ N	202.5 J
$\Delta x = 8 - 7 = 1$ m	$F_{average} = [5(8)^2 - 2(8) + 3 + 5(7)^2 - 2(7) + 3]/2 = 270.5$ N	270.5 J
$\Delta x = 9 - 8 = 1$ m	$F_{average} = [5(9)^2 - 2(9) + 3 + 5(8)^2 - 2(8) + 3]/2 = 348.5$ N	348.5 J
$\Delta x = 10 - 9 = 1$ m	$F_{average} = [5(10)^2 - 2(10) + 3 + 5(9)^2 - 2(9) + 3]/2 = 436.5$ N	436.5 J
	Total Work =	**1604 J**

So the total work done by this force in moving the object from $x = 0$ to $x = 10$ m is approximately 1604 J. It turns out that using calculus we would obtain the exact answer of 1597 J, so this approximate method works well!

7. *How do I know what angle to use in the expression $W = F(\Delta r)\cos\theta$?*

The angle θ is the angle between the direction of $\vec{F}$ and the direction of $\Delta\vec{r}$ and is measured clockwise from $\vec{F}$ to $\Delta\vec{r}$. The following figure shows where the angle is measured for different forces and a given displacement.

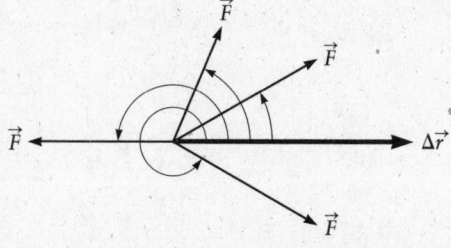

8. *When applying the principle of conservation of mechanical energy I need to define a system. How do I know what to include in the system?*

This is a good question and really gets at what we mean with the term potential energy. Often we get a little sloppy with our descriptions and say that an object has potential energy when what we really mean is that the system has potential energy. For example, by lifting an object up off the ground we are increasing the gravitational potential energy of the object-Earth system. Since the potential energy is associated with a force, and the force is between the Earth and the object, so is the potential energy. For a mass on a spring we would say the potential energy of the mass-spring system. By thinking of things in this way it's easy to know what to include in our system. If we have several objects interacting such as many cars on a roller coaster, then we can define the system to include all of the cars as well as the Earth.

9.   *Suppose a ball is tied to the end of a string and twirled around in a circle. Since the tension in the string is causing the ball to follow this path, isn't it doing work on the ball?*

No. Remember how we defined work in terms of the force and displacement, $W = F(\Delta r)\cos\theta$. Although tension is supplying a force on the ball, and the ball is moving, the angle between the force and the displacement is always 90°. Since $\cos(90°) = 0$, the work done by the tension is zero.

Part C. Selection of End-of-Chapter Answers and Solutions

QUESTIONS

Q6.6 For the bungee jumper in Example 6.9, plot the following quantities as a function of time: (a) the gravitational potential energy of the jumper, (b) the kinetic energy of the jumper, (c) the potential energy associated with the bungee cord, and (d) the total mechanical energy.

Answer

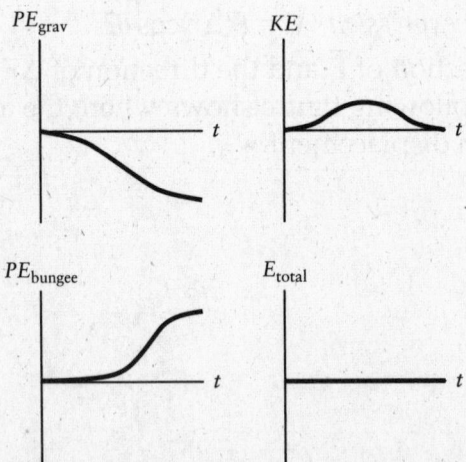

Q6.17 When a rubber ball is dropped onto a concrete floor, it bounces to a height that is slightly lower than its initial height. Compare the ball's (a) initial mechanical energy, to (b) its energy just before hitting the ground, to (c) just after bouncing off the ground, and to (d) the total energy when it reaches its final height. Is the ball's mechanical energy conserved? Explain and discuss using the principle of conservation of total energy.

Answer

(a) The ball's initial mechanical energy is slightly larger than its energy just before hitting the ground, since there are losses due to air drag.

(b) There will again be some losses in the elastic "spring" action of the ball so the mechanical energy just before hitting the ground will be slightly larger than its mechanical energy just after bouncing off the ground.

(c) On its way back up, the ball again loses energy due to the effect of air drag. This further reduces its mechanical energy making its final height less than its starting height.

(d) The ball's mechanical energy is not conserved; however, since there are nonconservative forces acting on the ball, there will be work done by these forces so that the total energy of the system is conserved.

PROBLEMS

P6.4   A hockey puck of mass 0.25 kg is sliding along a slippery frozen lake, with an initial speed of 60 m/s. The coefficient of friction between the ice and the puck is $\mu_K = 0.030$. Friction eventually causes the puck to slide to a stop. Find the work done by friction.

Solution

Recognize the principle

Work is being done on the puck by friction. From Newton's second law we can determine the acceleration of the puck. Using this acceleration and the kinematic equations we can determine the distance the puck travels before coming to rest. Finally, we can calculate the work done by the friction.

Sketch the problem

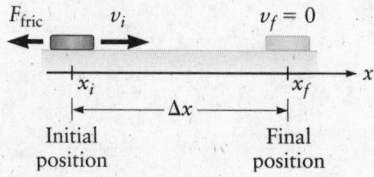

Identify the relationships

The magnitude of the force will be equal to the frictional force as given by,

$$F_{\text{friction}} = \mu_K N = \mu_K mg$$

Applying Newton's second law we obtain the acceleration of the puck,

$$F = F_{\text{fric}} = -\mu_K g = ma \rightarrow a = -\frac{\mu_K g}{m}$$

The distance traveled by the puck can be found using one of the kinematic equations discussed in Chapter 3,

$$v^2 = v_0^2 + 2a(x - x_0) \rightarrow \Delta r = x - x_0 = \frac{v^2 - v_0^2}{2a}$$

Inserting $v = 0$ and $a = -\mu_K g$,

$$\Delta r = x - x_0 = \frac{v_0^2}{2\mu_K g}$$

Inserting these expressions into our expression for work we find,

$$W = F(\Delta r)\cos\theta = \mu_K mg\left(\frac{v_0^2}{2\mu_K g}\right)\cos\theta$$

Solve

Since the force is in the opposite direction as the displacement, $\theta = 180°$ and $\cos(180°) = -1$. The work is then

$$W = \mu_K mg\left(\frac{v_0^2}{2\mu_K g}\right)(-1) = -\frac{1}{2}mv_0^2$$

Inserting the numerical values we find the work done by the friction,

$$W = -\frac{1}{2}(0.25\text{ kg})(60\text{ m/s})^2$$

$$\boxed{W = -450\text{ J}}$$

What does it mean?

Note that the work done by friction is negative. Negative work on an object results in the object slowing down, as is observed with the puck. In a later section of this chapter we'll see that it is easier to solve this problem with the work-energy theorem.

P6.28 A car of mass $m = 1500$ kg is pushed off a cliff of height $h = 24$ m (Fig. P6.28). If the car lands a distance of 10 m from the base of the cliff, what was the kinetic energy of the car the instant after it left the cliff?

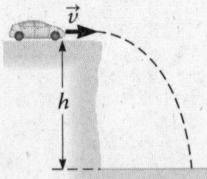

Figure P6.28

Solution

Recognize the principle

We can apply the expressions of kinematic motion discussed in Chapter 3 to determine the speed of the car the instant it left the cliff. From this speed we can calculate the kinetic energy of the car.

Sketch the problem

We can define the coordinate system as shown:

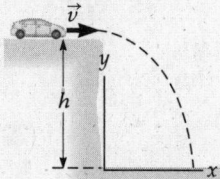

Identify the relationships

The kinetic energy of the car the instant after it left the cliff is found by knowing the speed of the car at this point in time. At this point the velocity of the car was only in the x direction. To calculate the velocity we first need to know the time the car was in the air. Applying the kinematic equations in the y direction we find,

$$y = y_0 + v_{0,y}t - \frac{1}{2}gt^2$$

The initial height, y_0, is h, with the final height of 0. The initial velocity in the y direction is also zero. Therefore,

$$0 = h + 0 - \frac{1}{2}gt^2$$

Solving for time we find,

$$t = \sqrt{\frac{2h}{g}}$$

The distance traveled in the x direction, from the base of the cliff, x, is then the initial velocity multiplied by t. Combining these equations and solving for the initial velocity in the x direction,

$$x = v_{0,x}t$$
$$v_{0,x} = \frac{x}{t} = \frac{x}{\sqrt{\frac{2h}{g}}}$$

Inserting this speed into the kinetic energy equation,

$$KE = \frac{1}{2}m\left(\frac{x}{\sqrt{\frac{2h}{g}}}\right)^2$$

$$KE = \frac{1}{2}mx^2\left(\frac{g}{2h}\right)$$

Solve

Inserting the numerical values we find,

$$KE = \frac{1}{2}(1500 \text{ kg})(10 \text{ m})^2 \left(\frac{(9.8 \text{ m/s}^2)}{2(24 \text{ m})} \right)$$

$$\boxed{KE = 15,000 \text{ J}}$$

What does it mean?

As the car falls from the cliff, it loses gravitational potential energy and gains kinetic energy. By the time it hits the ground below, its kinetic energy will be much greater than 15,000 J so it will have a speed much greater than what it had when it left the cliff. The horizontal speed, however, will be the same.

P6.37 A roller coaster (Fig. P6.37) starts at the top of its track (point A) with a speed of 12 m/s. If it reaches point B traveling at 16 m/s, what is the vertical distance (h) between A and B? Assume friction is negligible and ignore the kinetic energy of the wheels.

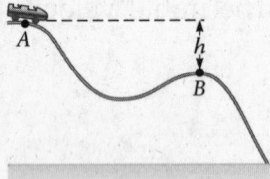

Figure P6.37

Solution

Recognize the principle

Apply concepts of gravitational potential energy and conservation of mechanical energy.

Sketch the problem

See problem figure.

Identify the relationships

Gravity is a conservative force, so the total energy will be conserved. The initial total energy at point A will equal the final total energy at point B.

$$KE_i + PE_i = KE_f + PE_f$$

$$\frac{1}{2} mv_i^2 + mgh_i = \frac{1}{2} mv_f^2 + mgh_f$$

Solve

Let $h = y_i - y_f$ and solve for h,

$$\frac{1}{2} mv_i^2 + mgy_i = \frac{1}{2} mv_f^2 + mgy_f$$

$$g(y_i - y_f) = \frac{1}{2}\left(v_f^2 - v_i^2\right)$$

$$h = \frac{1}{2g}\left(v_f^2 - v_i^2\right)$$

Inserting values,

$$h = \frac{1}{2(9.8 \text{ m/s}^2)}\left[(16 \text{ m/s})^2 - (12 \text{ m/s})^2\right]$$

$$\boxed{h = 5.7 \text{ m}}$$

What does it mean?

If no other forces are involved, a change in velocity due to a change in gravitation potential energy only depends on the change in vertical position.

P6.42 A rock of mass 12 kg is tied to a string of length 2.4 m, with the other end of the string fastened to the ceiling of a tall room (Fig. P6.42). While hanging vertically, the rock is given an initial horizontal velocity of 2.5 m/s. (a) Add a coordinate system to the sketch in Figure P6.42. Where is a convenient place to choose the origin of the vertical (y) axis? (b) What are the initial kinetic and potential energies of the rock? (c) If the rock swings to a height h above its initial point, what is its potential energy? What is its total mechanical energy at that point? (d) How high will the rock swing? Express your answer in terms of the angle θ that the string makes with the vertical when the rock is at its highest point. (e) Make qualitative sketches of the kinetic energy and potential energies as functions of height h.

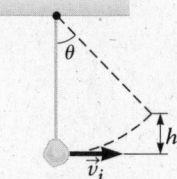

Figure P6.42

Solution

Recognize the principle

Since gravity is a conservative force, the energy of this system will be conserved.

Sketch the problem

Defining our coordinate system with the origin at the rock's lowest point we have,
(a)

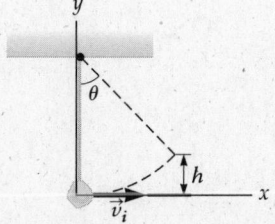

The most convenient place to choose the origin of the y axis is at the bottom of the trajectory of the rock since the rock swings up from this point.

Identify the relationships

Since the total mechanical energy is conserved, we can use the expression,

$$KE_i + PE_i = KE_f + PE_f$$

Substituting in the expressions for kinetic energy and gravitational potential energy we get,

$$\frac{1}{2}mv_i^2 + mgh_i = \frac{1}{2}mv_f^2 + mgh_f$$

Solve

(b) From the diagram we see that the lowest point in the trajectory is defined as 0, so $h_i = 0$. The initial potential energy is then

$$PE_i = mgh_i$$

$$\boxed{PE_i = 0}$$

The initial kinetic energy is given by,

$$KE_i = \frac{1}{2}mv_i^2$$

Inserting in the numerical values,

$$KE_i = \frac{1}{2}(12\,\text{kg})(2.5\,\text{m/s})^2$$

$$\boxed{KE_i = 38\,\text{J}}$$

(c) At a given height h, the potential energy will be given by,

$$\boxed{PE_{\text{at }h} = mgh}$$

The total mechanical energy will remain constant, which will be equal to the initial total mechanical energy of $\boxed{38\,\text{J}}$.

(d) To find the maximum height the rock will swing we can apply the principle of conservation of mechanical energy. We know the total initial mechanical energy is 38 J. At its maximum height, the speed of the rock will be zero; therefore, the final kinetic energy will also be zero.

$$KE_i + PE_i = KE_f + PE_f$$

$$\frac{1}{2}mv_i^2 + mgh_i = \frac{1}{2}mv_f^2 + mgh_f$$

$$\frac{1}{2}mv_i^2 + 0 = 0 + mgh_f$$

Solving for the final height h,

$$\frac{1}{2}mv_i^2 = mgh$$

$$h = \frac{v_i^2}{2g}$$

Inserting values,

$$h = \frac{(2.5 \text{ m/s})^2}{2(9.8 \text{ m/s}^2)}$$

$$h = 0.32 \text{ m}$$

The angle can then be found by looking at the diagram,

$$\cos\theta = \frac{L - h}{L}$$

$$\theta = \cos^{-1}\left(\frac{L - h}{L}\right)$$

Inserting in the numerical values,

$$\theta = \cos^{-1}\left(\frac{2.4 \text{ m} - 0.32 \text{ m}}{2.4 \text{ m}}\right)$$

$$\boxed{\theta = 30°}$$

(e)

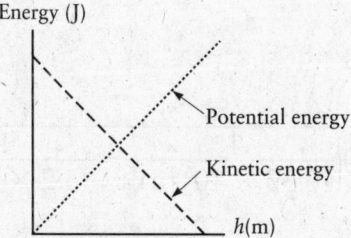

What does it mean?

Energy is conserved in an ideal pendulum, switching from all kinetic energy at the lowest point to all potential energy at the highest point.

P6.54 Calculate the velocity needed for an object starting at Earth's surface to just barely reach a satellite in a geosynchronous orbit. Ignore air drag and assume the object has a speed of zero when it reaches the satellite.

Solution

Recognize the principle

Since we can ignore air drag, the total mechanical energy is conserved. Therefore, we can equate the total mechanical energy of an object at Earth's surface to that at the radius of a geosynchronous satellite.

Sketch the problem

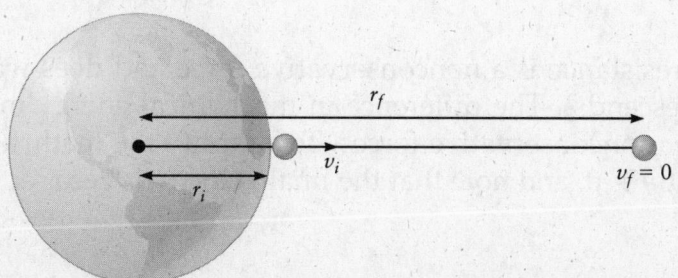

Identify the relationships

Noting that the initial radius is the radius of the Earth, $r_i = r_E$, and starting with the conservation of mechanical energy, we have,

$$\frac{1}{2}m_o v_i^2 - \frac{GM_{Earth}m_o}{r_E} = \frac{1}{2}m_o v_f^2 - \frac{GM_{Earth}m_o}{r_f}$$

The final velocity will be zero, and solving for the initial velocity,

$$\frac{1}{2}m_o v_i^2 - \frac{GM_{Earth}m_o}{r_E} = 0 - \frac{GM_{Earth}m_o}{r_f}$$

$$\frac{1}{2}m_o v_i^2 = \frac{GM_{Earth}m_o}{r_E} - \frac{GM_{Earth}m_o}{r_f}$$

$$v_i = \sqrt{\frac{2GM_{Earth}}{r_E} - \frac{2GM_{Earth}}{r_f}}$$

The final distance, r_f, will be the distance from the center of Earth to a geosynchronous orbit, and can be found in Example 5.10 in your textbook.

Solve

Inserting the numerical values,

$$v_i = \sqrt{2GM_{Earth}\left(\frac{1}{r_E} - \frac{1}{r_f}\right)}$$

$$v_i = \sqrt{2(6.67 \times 10^{-11}\,\text{N} \cdot \text{m}^2/\text{kg}^2)(6.0 \times 10^{24}\,\text{kg})\left(\frac{1}{(6.37 \times 10^6\,\text{m})} - \frac{1}{(4.2 \times 10^7\,\text{m})}\right)}$$

$$\boxed{v_i = 1.0 \times 10^4\,\text{m/s}}$$

What does it mean?

This is only about 10% lower than the escape velocity from Earth!

P6.67 A tennis ball ($m = 57$ g) is projected vertically with an initial speed of 8.8 m/s. (a) If the ball rises to a maximum height of 3.7 m, how much kinetic energy was dissipated by the drag force of air resistance? (b) How much higher would the ball have gone in a vacuum?

Solution

Recognize the principle

Apply conservation of mechanical energy and the concept of work done by nonconservative forces.

Sketch the problem

No sketch needed for this problem.

Identify the relationships

(a) The drag force of air resistance is a nonconservative force and does work on the ball as it rises (and descends). The difference in mechanical energy must be equal to the work done by nonconservative forces. To simplify the math, assume the displacement begins at $y = 0$, and note that the final velocity is zero.

$$(KE_i + PE_i) - (KE_f + PE_f) = W_{noncon}$$
$$(KE_i + 0) - (0 + PE_f) = W_{noncon}$$
$$W_{noncon} = \frac{1}{2}mv_i^2 - mgh_f$$

(b) In a vacuum there would be conservation of mechanical energy:

$$(KE_i + PE_i) = (KE_f + PE_f)$$
$$(KE_i + 0) = (0 + PE_f)$$
$$\frac{1}{2}mv_i^2 = mgh_f$$

Solve

(a) So we can calculate the energy lost to the work done by air drag:

$$W_{noncon} = \frac{1}{2}(0.057 \text{ kg})(8.8 \text{ m/s})^2 - (0.057 \text{ kg})(9.8 \text{ m/s}^2)(3.7 \text{ m})$$
$$W_{noncon} = KE_{dis} = \boxed{0.14 \text{ J}}$$

(b) The height attained without air drag is then,

$$h_f = \frac{v_i^2}{2g} = \frac{(8.8 \text{ m/s})^2}{2(9.8 \text{ m/s}^2)} = 4.0 \text{ m}$$

So the ball would have attained 4.0 m − 3.7 m = $\boxed{0.3 \text{ m}}$ of additional vertical displacement.

What does it mean?

We can cross-check our answer to part (b) by considering how much additional height the ball would get from 0.14 J of kinetic energy:

$$KE_{dis} = mgh_{add} = 0.14 \text{ J}$$

$$h_{add} = \frac{0.14 \text{ J}}{(0.057 \text{ kg})(9.8 \text{ m/s}^2)} = 0.3 \text{ m}$$

We see that the attained height in air added to this additional displacement gives a total of 4.0 m which agrees with our prior calculation.

P6.83 In Section 6.8, we calculated the force produced by a myosin molecular motor and found that it is approximately 10 piconewtons. Compare this force to the weight of a typical amino acid.

Solution

Recognize the principle

Apply the concept of weight.

Sketch the problem

No sketch needed for this problem.

Identify the relationships

The weight of an object near the surface of the Earth is given by $F_{grav} = mg$. Looking up the amino acid phenylalanine in the *CRC Handbook of Chemistry*, we find that it has a molar mass of 165.19 g/mol = 0.16519 kg/mol. Using Avogadro's number, 6.02×10^{23}, we find that a single molecule has a mass of 2.7×10^{-25} kg.

Solve

The weight is then mg or about $\boxed{3 \times 10^{-24}\,\text{N}}$.

What does it mean?

This is about 12 orders of magnitude smaller than the force a myosin molecular motor can apply!

P6.84 Conceptualizing units. A small frozen burrito, like those bought at your local fast-food restaurant, has a mass of approximately 100 g and contains about 300 C. (The capital "C" indicates that this unit is a "food Calorie.") From what height would you need to drop the burrito to give it a kinetic energy equal to its dietary "energy"? Note that 1 food Calorie is equal to 1000 calories = 4186 J. Ignore air drag.

Solution

Recognize the principle

Apply concepts of gravitational potential energy and conservation of mechanical energy.

Sketch the problem

No sketch is needed for this problem.

Identify the relationships

Using conservation of mechanical energy, and assuming air drag is negligible:

$$KE_i + PE_i = KE_f + PE_f$$
$$0 + PE_i = KE_f + 0$$
$$KE_f = mgh$$

Solve

(a) The unit of energy, the joule, is equivalent to $(1\,\text{N})(1\,\text{m}) = 1\,\text{N} \cdot \text{m}$

Here, for the burrito, $mg = 1\,\text{N}$, so we would drop it from a height of 1 m to obtain 1 J of kinetic energy. (How much would it hurt to have the burrito land on your toe and deliver 1 J of energy?)

$$h = \boxed{1\,\text{m}}$$

(b) The kinetic energy of the arrow in Example 6.8 is:

$$KE = \frac{1}{2}mv^2 = \frac{1}{2}(0.05 \text{ kg})(100 \text{ m/s})^2 = 250 \text{ J}$$

$$= \boxed{250 \text{ burritos dropped from 1 m}}$$

(c) The kinetic energy of the snowboarder in Figure 6.12 gives:

$$KE = \frac{1}{2}mv_i^2 + mgh_i = \frac{1}{2}(65 \text{ kg})(15 \text{ m/s})^2 + (65 \text{ kg})(9.8 \text{ m/s}^2)(15 \text{ m})$$

$$KE = 17 \text{ kJ} \approx \boxed{17,000 \text{ burritos dropped from 1 m}}$$

What does it mean?

We find that 250 burritos dropped from 1 m would have the equivalent kinetic energy of the arrow, and more than 17,000 burritos would need to be dropped from 1 m to have an equivalent kinetic energy of the snowboarder!

P6.87 A doughnut contains roughly 350 C (1.5×10^6 J) of potential energy locked in chemical bonds. Find the ratio of the potential energy in the doughnut to that in an equivalent volume of TNT (trinitrotoluene), which has a density of 1.65 g/cm³ and releases 2.7×10^6 J per kg of explosive. Aren't you glad the doughnut does not release all its potential energy at once?

Solution

Recognize the principle

Apply concepts of potential energy, density, and conservation of energy.

Sketch the problem

No sketch is needed for this problem.

Identify the relationships

First find the approximate mass of a doughnut-sized volume of TNT. The volume of a disk is given by the expression,

$$V = \pi r^2 h$$

The energy within the volume is proportional to the mass of TNT that could occupy that volume given by $m = \rho V$.

Solve

For a common doughnut, $r = 5$ cm and $h = 4$ cm, giving an approximate volume of $V = 314$ cm³. Thus a doughnut-shaped mass of TNT is:

$$m = \rho V = (1.65 \text{ g/cm}^3)(314 \text{ cm}^3) = 518 \text{ g} \approx 0.5 \text{ kg}$$

The equivalent energy released by detonating this mass is:

$$E = (0.5 \text{ kg})(2.7 \times 10^6 \text{ J/kg}) = 1.4 \times 10^6 \text{ J}$$

Compare this with the energy in the doughnut's chemical bonds:

$$\frac{E_{\text{doughnut}}}{E_{\text{TNT}}} = \frac{1.5 \times 10^6 \, \text{J}}{1.4 \times 10^6 \, \text{J}} = 1.07 \approx \boxed{1}$$

What does it mean?

The doughnut has a chemical potential energy equal to an equivalent volume of TNT! Good thing the doughnut does not release the energy very quickly!

P6.101 For a car moving with speed v, the force of air drag is proportional to v^2. If the power output of the car's engine is doubled, by what factor does the speed of the car increase?

Solution

Recognize the principle

There are two horizontal forces acting on the car, the force from air drag and the force that results from the engine rotating the wheels which are in contact with the road. Assuming that all the power generated by the engine transfers to the force of moving the car forward, we can find the force of the engine using our expression for power,

$$P_{\text{engine}} = F_{\text{engine}} \, v$$

$$F_{\text{engine}} = \frac{P_{\text{engine}}}{v}$$

Sketch the problem

No sketch needed.

Identify the relationships

Assuming the car is traveling at a constant speed, its acceleration is zero so the engine force must be equal and opposite to the force of air drag. In terms of Newton's second law, we have

$$\Sigma F = F_{\text{engine}} + F_{\text{drag}} = ma = 0$$

This drag force is given approximately by $F_{\text{drag}} = \frac{1}{2}\rho v^2$ (see Chapter 3). The negative sign in the drag force expression indicates that the drag force is directed opposite to the velocity. Inserting F_{engine} and F_{drag} gives,

$$F_{\text{engine}} + F_{\text{drag}} = \frac{P_{\text{engine}}}{v} - \frac{1}{2}\rho A v^2 = 0$$

Rearranging this expression to solve for v, we have

$$v = \left(\frac{2P_{\text{engine}}}{\rho A}\right)^{1/3}$$

Solve

Therefore, if the power output of the engine doubles the speed of the car increases by a factor of

$$(2)^{1/3} = 1.26 \approx \boxed{1.3}$$

What does it mean?

This means you only obtain a 30% speed increase by doubling the power output of a car's engine.

Part D. Additional Worked Examples and Capstone Problems

The following worked examples and capstone problems provide you with additional practice with work, energy, and energy conservation. Also, working these problems will give you greater insight into these concepts. Although these five problems do not incorporate all the material discussed in this chapter, they do highlight several of the key concepts. If you can successfully solve these problems then you should have confidence in your understanding of these key concepts, so use these problems as a test of your understanding of the chapter material.

WE 6.1 Potential Energy: How Big Is It?

A car of mass $m = 1200$ kg is lifted by a large crane to a height h in preparation for dropping it in a junkyard. Find the change in potential energy if the car is dropped from a height $h = 20$ m. If all this potential energy were somehow converted into the kinetic energy of the car, what would the car's speed be?

Solution

Recognize the principle

Since the initial and final heights are given, we can use $PE_{grav} = mgy$ to find the change in potential energy of the car.

Sketch the problem

The figure shows the initial state of the car (just before it is released by the crane) and the final state (just before it hits the ground).

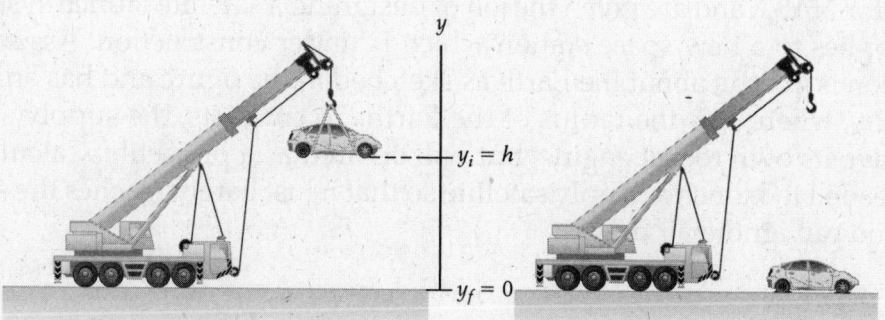

Identify the relationships

If we take ground level as the origin, the car begins at $y_i = h$ and ends at $y_f = 0$. Therefore, we can calculate the change in the potential energy,

$$\Delta PE = mg(y_f - y_i) = mg(0 - h) = -mgh$$

Solve

Inserting our values for m and h, we have

$$\Delta PE = -mgh = -(1200 \text{ kg})(9.8 \text{ m/s}^2)(20 \text{ m})$$

$$\Delta PE = -2.4 \times 10^5 \text{ kg} \cdot \text{m}^2/\text{s}^2 = -2.4 \times 10^5 \text{ J}$$

Notice that the answer is in units of joules. The value is negative because we have removed some of the stored gravitational potential energy. If this energy were all added to the car's kinetic energy, the kinetic energy would increase by

$$\Delta KE = +2.4 \times 10^5 \text{ J}$$

If the car is initially at rest, this added kinetic energy equals the total (final) kinetic energy. Using our expression for KE and solving for v, we find

$$\Delta KE = \frac{1}{2}mv^2$$

$$v^2 = \frac{2(\Delta KE)}{m}$$

Inserting the given value of m and taking ΔKE from above leads to

$$v = \sqrt{\frac{2(2.4 \times 10^5 \text{ J})}{1200 \text{ kg}}} = 20 \text{ m/s}$$

This speed is approximately 40 mi/h, so the final speed of the car is quite substantial.

What does it mean?

Since the gravitational potential energy and the kinetic energy are both proportional to the mass of the car, the final speed is independent of m.

WE 6.2 Launching into Low Earth Orbit

You work for NASA and are given the job of designing a satellite launch system for taking supplies to a new space station which is under construction. Assume this space station is in orbit about the Earth as sketched in the figure and has an orbital radius of $2r_E$, where r_E is the radius of the Earth. To cut costs, the supply satellite will not have its own rocket engine, but will be fired as a projectile. Calculate the velocity needed to launch a supply satellite so that it just barely reaches the station. Assume you can ignore air drag.

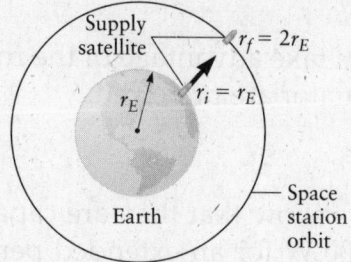

Solution

Recognize the principle

This problem is similar to our calculation of the escape velocity, except that now $r_f = 2r_E$. The final potential energy is thus not zero because the supply satellite does not completely escape from the Earth's gravitational attraction. The supply satellite is assumed to just barely reach the station; hence, its velocity on arrival will still be $v_f = 0$.

Sketch the problem

Figure provided.

Identify the relationships

We begin with the conservation of energy condition for a satellite of mass m_s:

$$\frac{1}{2}m_s v_i^2 - \frac{GM_{Earth}m_s}{r_E} = \frac{1}{2}m_s v_f^2 - \frac{GM_{Earth}m_s}{r_f}$$

The satellite is initially at the Earth's surface ($r_i = r_E$) with velocity v_i. Inserting our values for the final velocity and r_f, and solving for v_i, we get

$$\frac{1}{2}m_s v_i^2 - \frac{GM_{Earth}m_s}{r_E} = -\frac{GM_{Earth}m_s}{2r_E}$$

$$v_i^2 = \frac{2GM_{Earth}}{r_E} - \frac{GM_{Earth}}{r_E} = \frac{GM_{Earth}}{r_E}$$

Solve

Inserting the known values for the mass and radius of the Earth and for the constant G leads to

$$v_i = \sqrt{\frac{GM_{Earth}}{r_E}} = \sqrt{\frac{(6.67 \times 10^{-11}\,\text{N} \cdot \text{m}^2/\text{kg}^2)(6.0 \times 10^{24}\,\text{kg})}{6.4 \times 10^6\,\text{m}}} = \boxed{8000\,\text{m/s}}$$

What does it mean?

This result is approximately 71% of the speed required for a projectile to completely escape from the Earth. Because the initial kinetic energy is proportional to v_i^2, the kinetic energy required to reach this space station is half the kinetic energy required to escape completely from the Earth. In practice, not all of this speed needs to be provided by the launch system. We can take advantage of the rotation of the Earth which is why launch sites near the equator are preferred.

WE 6.3 Top Speed of a Bicycle Racer

Physiological tests with professional bicycle racers show that they are capable of expending a power output of typically $P_{cyclist} = 400$ W for an extended period of time. What is the maximum velocity that such a bicyclist can achieve on a level road? *Hint 1*: You must include the effect of air drag; recall from Chapter 3 that the force of air drag in this case is given approximately by

$$F_{drag} = -\frac{1}{2}\rho A v^2$$

where the density of air is $\rho = 1.3$ kg/m^3 and A is the frontal area of the bicyclist. This relation is not exact—it does not account for the aerodynamic profile of the bicyclist—but it is an adequate approximation for this problem. *Hint 2*: Assume all friction in the wheels is negligible (which is actually a good approximation).

Solution

Recognize the principle

There are two horizontal forces acting on the bicyclist, the force from air drag and the force that results from his pedaling. Calculating the force from pedaling is an interesting problem that we discuss more when we consider rolling motion in Chapter 8. For now, we can simply say that the bicyclist's pedaling leads to a horizontal force. Because there is no friction in the bearings of the wheels, there is no energy lost to friction, and all the power expended by the bicyclist goes into producing this pedaling force that moves him along the road. We can find the pedaling force using our expression for power,

$$P_{cyclist} = F_{pedal} v$$

$$F_{pedal} = \frac{P_{cyclist}}{v}$$

Sketch the problem

The following sketch helps to illustrate the problem.

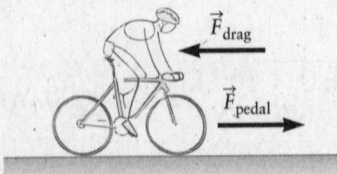

Identify the relationships

When the bicyclist has reached top speed, his acceleration is zero, so the pedaling force must be equal and opposite to the force of air drag. In terms of Newton's second law, we have

$$\Sigma F = F_{pedal} + F_{drag} = ma = 0$$

This drag force is given approximately by expression above. The negative sign in the drag force expression indicates that the drag force is directed opposite to the velocity. Inserting F_{pedal} and F_{drag} gives,

$$F_{pedal} + F_{drag} = \frac{P_{cyclist}}{v} - \frac{1}{2}\rho Av^2 = 0$$

Solve

Solving for v, we find

$$\frac{P_{cyclist}}{v} = \frac{1}{2}\rho Av^2$$

$$v^3 = \frac{2P_{cyclist}}{\rho A}$$

Here, $P_{cyclist}$ is given and we know the density of air, but we need to also know the value of A. For a bicyclist in racing position, we appeal to our intuition (and experience) and estimate $A \approx 0.3 \text{ m}^2$. So,

$$v = \left(\frac{2P_{cyclist}}{\rho A}\right)^{1/3} = \left[\frac{2(400 \text{ W})}{(1.3 \text{ kg/m}^3)(0.3 \text{ m}^2)}\right]^{1/3} = 13 \text{ m/s}$$

What does it mean?

This speed is approximately 30 mi/h. This result is actually quite close to the speeds found at the Tour de France and other elite bicycle races. Physics does work!

CP 6.1 Springs, Gravity, Friction, and More: Combining Multiple Forms of Energy and Work

Consider the hypothetical amusement park ride shown in the figure. At the start of the ride the car and passengers ($m = 700$ kg) are launched using a very strong spring ($k = 1000$ N/m). After traveling down the 80-m high track you travel upside down around a circular loop-the-loop, 30 m in diameter. The track is frictionless except for a 40-m long segment near the end of the ride used to help slow you down before you bump into an identical spring at point C. You then bounce off of this spring and travel back through the loop-the-loop and up the hill to the starting point. (a) What minimum compression of the starting spring is required to ensure you make it back up the 80-m high hill? (b) Assuming this minimum compression is used, how much will the spring at point C be compressed? (c) Determine the speed of the car at point B for both ways of the trip. (d) Are these speeds sufficient to keep the car on the track without the need for additional mechanisms/restraints?

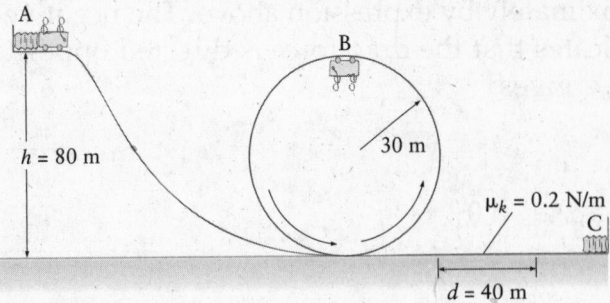

Solution

Recognize the principle

The main principle we can apply is conservation of energy. We'll also need to calculate the work done by nonconservative forces (friction in this case) and take it into account when adding up the total energy.

Sketch the problem

Refer to the figure in the problem.

Identify the relationships

We can begin by defining a few subscripts on our variables. Let the subscript "A1" refer to the starting point, "A2" for when the car makes it back up to point A, "C" is for point C, and "B1" and "B2" for when the car is at point B on the way down and way back, respectively. Also, we will use the symbol E_{lost} to indicate the energy lost as the cart goes across the frictional part of the track. Using this notation and conserving energy we can write the following expression:

$$E_{A1} = E_{B1} = E_C + E_{lost} = E_{B2} + 2E_{lost} = E_{A2} + 2E_{lost}$$

In words, we can describe it as follows. We start off with a certain amount of mechanical energy at point A and still have all this energy at point B1. By the time we get to point C we have lost some of this energy due to the work done by friction. On the way back we again lose energy due to friction but whatever energy remains is the same at points B2 and A2.

Now, if we define the origin of our vertical axis at the bottom of the hill, then we can define the gravitational potential energy in terms of the height above this point. Thinking about all the forms of energy at each point and the energy lost due to friction we can write the following expressions:

$$E_{lost} = -W_{fric} = (\mu_K mg)d = (0.2)(700 \text{ kg})(9.8 \text{ m/s}^2)(40 \text{ m}) = 54,880 \text{ J}$$

$$E_{A1} = PE_{spring} + PE_{grav} = \frac{1}{2}kx_{A1}^2 + mgh$$

$$= \frac{1}{2}(1000 \text{ N/m})x_{A1}^2 + (700 \text{ kg})(9.8 \text{ m/s}^2)(80 \text{ m})$$

$$= (500x_{A1}^2 + 548,800) \text{ J}$$

$$E_{B1} = KE_{B1} + PE_{grav, B1} = \frac{1}{2}mv_{B1}^2 + mgy_B$$

$$= \frac{1}{2}(700 \text{ kg})v_{B1}^2 + (700 \text{ kg})(9.8 \text{ m/s}^2)(60 \text{ m})$$

$$= (350 v_{B1}^2 + 411,600) \text{ J}$$

$$E_C = PE_{spring} = \frac{1}{2}kx_C^2 = \frac{1}{2}(1000 \text{ N/m})x_C^2 = 500x_C^2 \text{ J}$$

$$E_{B2} = KE_{B2} + PE_{grav, B2} = \frac{1}{2}mv_{B2}^2 + mgy_B$$

$$= \frac{1}{2}(700 \text{ kg})v_{B2}^2 + (700 \text{ kg})(9.8 \text{ m/s}^2)(60 \text{ m})$$

$$= (350 v_{B2}^2 + 411{,}600) \text{ J}$$

$$E_{A2} = PE_{grav} = (700 \text{ kg})(9.8 \text{ m/s}^2)(80 \text{ m})$$

$$= 548{,}800 \text{ J}$$

We have only included gravitational potential energy in the expression for E_{A2} since we want the cart to just make it back up the hill and stop without compressing the spring.

Solve

Now we can solve for all of our unknowns by substituting these expressions into our general expression for energy conservation, and suppressing units

$$E_{A1} = E_{B1} = E_C + E_{lost} = E_{B2} + 2E_{lost} = E_{A2} + 2E_{lost}$$

(a)

$$E_{A1} = E_{A2} + 2E_{lost}$$
$$500x_{A1}^2 + 548{,}800 = 548{,}800 + 2(54{,}880)$$
$$x_{A1} = \boxed{15 \text{ m}}$$

(b)

$$E_{A1} = E_C + E_{lost}$$
$$500x_{A1}^2 + 548{,}800 = 500x_C^2 + 54{,}880$$
$$500(15)^2 + 548{,}800 = 500x_C^2 + 54{,}880$$
$$x_C = \boxed{35 \text{ m}}$$

(c)

$$E_{A1} = E_{B1} = E_{B2} + 2E_{lost}$$
$$661{,}300 = 350v_{B1}^2 + 411{,}600 = 350v_{B2}^2 + 411{,}600 + 2(54{,}880)$$
$$v_{B1} = \boxed{27 \text{ m/s}} \quad \text{and} \quad v_{B2} = \boxed{20 \text{ m/s}}$$

(d) We can determine the minimum speed required to keep the car on the track by applying Newton's second law. Let's start by drawing the free-body diagram of the car at point B.

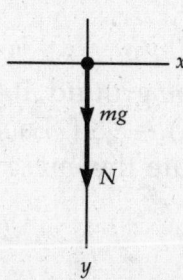

Applying Newton's second law in the y direction and recognizing that the acceleration of the car at this point is centripetal acceleration we have,

$$\Sigma F = ma = ma_c$$

$$N + mg = m\left(\frac{v^2}{r}\right)$$

$$v = \left(\frac{r(N + mg)}{m}\right)^{1/2}$$

The minimum speed required for the car to stay on the track is that for which $N = 0$ at the top. Substituting in the numerical values we find,

$$v_{\min} = \left(\frac{r(N + mg)}{m}\right)^{1/2} = \left(\frac{(30 \text{ m})[0 + (700 \text{ kg})(9.8 \text{ m/s}^2)]}{700 \text{ kg}}\right)^{1/2} = 17 \text{ m/s}$$

Therefore, the car is moving fast enough in both directions to stay on the track.

What does it mean?

Although this seemed like a difficult problem, the key concept was energy conservation. By keeping track of the different forms of energy and the energy lost due to friction we were able to solve for various unknowns.

CP 6.2 The Energy of a Bouncing Ball: Representing Energy on a Graph

Consider a ball ($m = 1$ kg) which is dropped from a height of 10 m above the ground. This ball is quite elastic so that each time it collides with the ground it only loses 10% of its total mechanical energy at that time. (a) Assuming that the ball will not bounce off the floor once its total mechanical energy drops below 1.0 J, determine the number of times the ball bounces off the ground. To simplify the calculations, treat the ball as a point mass and neglect air drag. (b) On the same graph, create a plot of the gravitational potential energy, the kinetic energy, and the total energy versus time. (c) Is the total mechanical energy of the ball conserved? If not, where does this energy go?

Solution

Recognize the principle

This is an application of conservation of energy. We'll need to keep track of the energy lost each time the ball makes contact with the ground, but the energy it has during a particular time in the air will remain constant.

Sketch the problem

No sketch required.

Identify the relationships and Solve

(a) Since the ball starts from rest at a height of 10 m above the ground, it begins with a total energy of $PE_{\text{grav}} = mgh = (1 \text{ kg})(9.8 \text{ m/s}^2)(10 \text{ m}) = 98$ J. Each time it bounces it loses 10% of the energy it has so we can tabulate the energy after each bounce:

Starting energy	=	9.8 J
Energy after 1 bounce	=	9.8 J − 0.98 J = 8.8 J
Energy after 2 bounces	=	8.8 J − 0.88 J = 7.9 J
Energy after 3 bounces	=	7.9 J − 0.79 J = 7.1 J
Energy after 4 bounces	=	7.1 J − 0.71 J = 6.4 J
Energy after 5 bounces	=	6.4 J − 0.64 J = 5.8 J
Energy after 6 bounces	=	5.8 J − 0.58 J = 5.2 J
Energy after 7 bounces	=	5.2 J − 0.52 J = 4.7 J
Energy after 8 bounces	=	4.7 J − 0.47 J = 4.2 J
Energy after 9 bounces	=	4.2 J − 0.42 J = 3.8 J
Energy after 10 bounces	=	3.8 J − 0.38 J = 3.4 J
Energy after 11 bounces	=	3.4 J − 0.34 J = 3.1 J
Energy after 12 bounces	=	3.1 J − 0.31 J = 2.8 J
Energy after 13 bounces	=	2.8 J − 0.28 J = 2.5 J
Energy after 14 bounces	=	2.5 J − 0.25 J = 2.3 J
Energy after 15 bounces	=	2.3 J − 0.23 J = 2.1 J
Energy after 16 bounces	=	2.1 J − 0.21 J = 1.9 J
Energy after 17 bounces	=	1.9 J − 0.19 J = 1.7 J
Energy after 18 bounces	=	1.7 J − 0.17 J = 1.5 J
Energy after 19 bounces	=	1.5 J − 0.15 J = 1.4 J
Energy after 20 bounces	=	1.4 J − 0.14 J = 1.3 J
Energy after 21 bounces	=	1.3 J − 0.13 J = 1.2 J
Energy after 22 bounces	=	1.2 J − 0.12 J = 1.1 J
Energy after 23 bounces	=	1.1 J − 0.11 J = 1.0 J

So, the ball bounces 22 times. On the 23rd bounce its energy drops to 1.0 J so it doesn't leave the ground.

(b) While in the air the energy of the ball will remain constant and be divided between kinetic energy and gravitational potential energy. As the ball comes off the ground, its energy is in the form of kinetic energy. By the time it reaches its maximum height the energy has been converted into gravitational potential energy. This process repeats itself each time the ball bounces while steadily decreasing the total energy as given in the table. Each time the ball is in contact with the ground it loses energy so the kinetic energy immediately after contact will be 10% less than immediately before contact. Now, the difficult part of this problem is determining the shape of the curves for the potential and kinetic energies. For this we need to think of how the height of the ball and the speed of the ball vary with time. Using the kinematic equations from Chapter 3 we have,

$$y = y_0 + v_0 t + \frac{1}{2} a t^2$$
$$v = v_0 + at$$

Choosing up as positive, $y_0 = h$ and $v_0 = 0$ we can determine how y and v vary with time while the ball is falling,

$$y = h - \frac{g}{2} t^2$$
$$v = -gt$$

Therefore, the potential energy and kinetic energy as the ball falls vary with time according to the following expressions,

$$PE_{grav} = mgy = mg(h - gt^2) = mgh - mgt^2$$
$$KE = \frac{1}{2} mv^2 = \frac{1}{2} m(-gt)^2 = \frac{1}{2} mg^2 t^2$$

So the bottom line is the potential energy decreases like t^2 and the kinetic energy increases like t^2 as the ball falls. So, the graphs for the gravitational potential energy, kinetic energy, and total energy of the ball versus time would look as follows:

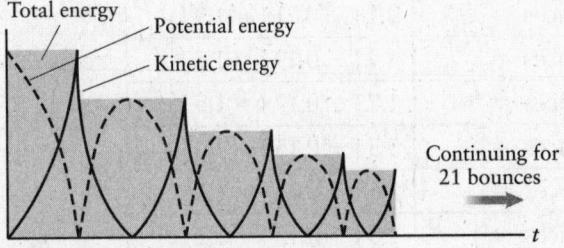

(c) The total mechanical energy of the ball is not conserved since it steadily decreases each time the ball bounces off the ground. The energy is lost to heat due to the internal friction of the ball.

What does it mean?

Note that while the ball is in the air the total mechanical energy remains constant. If we had included air drag this would not be the case.

Part E. MCAT Review Problems and Solutions

PROBLEMS

1. If the speed at which a car is traveling is tripled, by what factor does its kinetic energy increase?
   (a) $3^{1/2}$
   (b) 3
   (c) 6
   (d) 9

2. What is the total amount of work done by a 100-N force when it pushes a box up an incline 5.0 m in length and then another 5.0 m along a horizontal surface?

   (a) 5.0×10^2 J

   (b) 7.5×10^2 J

   (c) 1.0×10^3 J

   (d) 1.5×10^3 J

3. What is the speed of a car if its engine is rated at 100 kW and provides a constant force of 5.0×10^3 N?

   (a) 0.05 m/s

   (b) 0.02 m/s

   (c) 20 m/s

   (d) 50 m/s

4. How much work is done when a 0.50-kg mass is pushed by a 20-N force over a distance of 10.0 m?

   (a) 5 J

   (b) 10 J

   (c) 49 J

   (d) 200 J

5. Object A has a mass of 1 kg and a constant speed of 4 m/s. Object B has a mass of 1.5 kg and a constant speed of 2 m/s. Which of the following statements is true?

   (a) Object A has a greater kinetic energy.

   (b) Object B has a greater kinetic energy.

   (c) Object A has a greater acceleration.

   (d) Object B has a greater acceleration.

6. What is the kinetic energy of a 10.0-kg mass with a speed of 2.0 m/s?

   (a) 20 J

   (b) 10 J

   (c) 5 J

   (d) 2.5 J

7. The work done in raising an object must

   (a) increase the kinetic energy of the object.

   (b) decrease the total mechanical energy of the object.

   (c) decrease the internal energy of the object.

   (d) increase the gravitational potential energy of the object.

8. A frictionless incline has a length of 5.0 m and a height of 4 m. How much work must be done to move a 50-N box from the bottom to the top of the incline?

   (a) 100 J

   (b) 150 J

   (c) 200 J

   (d) 250 J

9. What is the average power output of a 50-kg boy who climbs a 2.0-m step ladder in 10 s?

   (a) 10 W

   (b) 49 W

   (c) 98 W

   (d) 250 W

10. Consider the following figure in which a 10-kg box slides 2 m down an incline where the coefficient of kinetic friction between the box and the incline is $\mu_K = 0.1$. How much work is done on the box by friction during this process?

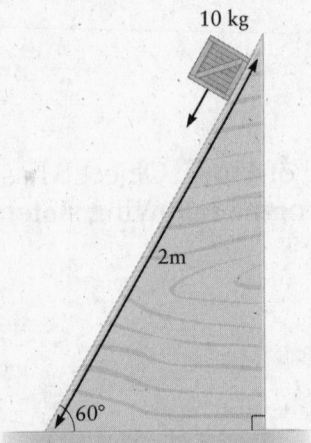

   (a) −98 J

   (b) 196 J

   (c) 9.8 J

   (d) −9.8 J

SOLUTIONS

1. *MCAT strategies*

Since the kinetic energy of an object is proportional to the square of the speed of the object, tripling the speed of the object will result in 9 times the kinetic energy. Therefore, the correct answer is (d).

2. MCAT strategies

No answers can be easily eliminated so we proceed with our problem-solving strategy.

Recognize the principle

The key principle involved in solving this problem is work and the relationship between work, force, and displacement. During the first part of the process the force is directed along the incline in the direction of the displacement, so the angle between the force and displacement is zero. Along the horizontal surface the force is again along the direction of the displacement so the angle between the force and displacement is once again zero.

Sketch the problem

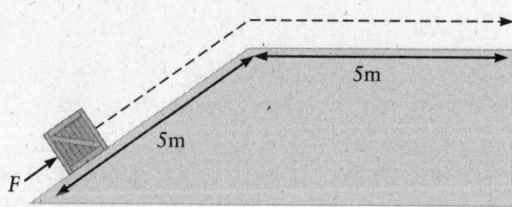

Identify the relationships

We begin with the relationship for work, $W = F(\Delta r)\cos\theta$, and split the problem into the two segments.

$$W = W_1 + W_2 = F(\Delta r_1)\cos\theta_1 + F(\Delta r_2)\cos\theta_2$$

Solve

Substituting the numerical values and solving we find,

$$W = F(\Delta r_1)\cos\theta_1 + F(\Delta r_2)\cos\theta_2 = 100(5)\cos(0) + 100(5)\cos(0)$$
$$= 1000\ \text{J} = 1 \times 10^3\ \text{J}$$

Therefore, the correct answer is (c).

What does it mean?

Although the displacement of the object for the two segments was along different directions, the work done by the force was the same, since in each case the force was in the same direction as the displacement.

3. MCAT strategies

No answers can be easily eliminated so we proceed with our problem-solving strategy.

Recognize the principle

The main principle involved in this problem is power and its dependence on force and speed.

Sketch the problem

None required.

Identify the relationships

We begin with the relationship for power in terms of force and speed, $P = Fv$. Rearranging this expression for the speed we find, $v = P/F$.

Solve

Substitution in the numerical values and solving the expression we find,

$$v = \frac{P}{F} = \frac{100 \times 10^3 \, W}{5 \times 10^3 \, N} = 20 \, m/s$$

Therefore the correct answer is (c).

What does it mean?

Just to be sure we have the correct expression we should check the units to be sure if they are indeed in m/s.

$$\left[\frac{W}{N}\right] = \left[\frac{J/s}{N}\right] = \left[\frac{N \cdot m}{s/N}\right] = [m/s]$$

4. MCAT strategies

No answers can be easily eliminated so we proceed with our problem-solving strategy.

Recognize the principle

The key principle involved in solving this problem is work and the relationship between work, force, and displacement. The force is directed along the direction of the displacement so the angle between the force and displacement is zero.

Sketch the problem

None required.

Identify the relationships

We begin with the relationship for work, $W = F(\Delta r)\cos\theta$.

Solve

Substituting the numerical values and solving we find,

$$W = F(\Delta r)\cos\theta = (20 \, N)(10 \, m)\cos(0) = 200 \, J$$

Therefore, the correct answer is (d).

What does it mean?

Note that the mass of the object was given in the problem but not needed in the solution.

5. MCAT strategies

The solution to this problem requires us to understand the relationship between kinetic energy, mass, and speed. Kinetic energy is given by

$$KE = \frac{1}{2}mv^2$$

Therefore substituting in the values for the two objects we find,

$$KE_A = \frac{1}{2}(1 \text{ kg})(4 \text{ m/s})^2 = 8 \text{ J}$$

$$KE_B = \frac{1}{2}(1.5 \text{ kg})(2 \text{ m/s})^2 = 3 \text{ J}$$

$$KE_A > KE_B$$

Therefore the correct answer is (a).

6. MCAT strategies

Solving this problem requires a quick calculation using the expression for kinetic energy.

$$KE = \frac{1}{2}mv^2 = \frac{1}{2}(10 \text{ kg})(2 \text{ m/s})^2 = 20 \text{ J}$$

Therefore the correct answer is (a).

7. MCAT strategies

The key word in the phrasing of this question is "must". We are not looking for parameters which <u>can</u> change, only the parameter which <u>must</u> change as we raise an object. For the kinetic energy of the object to change, its speed must change. Although lifting the object may result in changing its speed, that does not have to be the case, so we can eliminate answer (a). A change in the internal energy may also occur if there is friction involved in the process, but again that does not need to happen. Therefore answer (c) can be eliminated. As we lift the object we are increasing its gravitational potential energy which may or may not affect the total mechanical energy of the object. Clearly, the only parameter which must change is the gravitational potential energy. Therefore the correct answer is (d).

8. MCAT strategies

No answers can be easily eliminated so we proceed with our problem-solving strategy.

Recognize the principle

There are two ways we could tackle this problem. The first involves determining the angle of the incline from the information provided and calculating the work done by this force as the object moves up the incline. The second, and easier, way is to recognize that the work done on the object goes into changing the gravitation potential energy of the system. By calculating the change in the potential energy, we effectively determine the work done on the object.

Sketch the problem

None required.

Identify the relationships

The change in height of the object is 4 m, given in the problem. The force of gravity on the object is its weight, $F_{grav} = mg = 50 \text{ N}$. The expression for the gravitational potential energy is $PE_{grav} = mgy$.

Solve

Putting this all together we can calculate the change in the gravitational potential energy of the system,

$$\Delta PE_{\text{grav}} = mg \, \Delta y = (50 \text{ N})(4 \text{ m}) = 200 \text{ J}$$

Therefore, the correct answer is (c).

What does it mean?

Since there was no friction involved in this process, the work needed was independent of the path (the 5-m long incline) and only depended on the change in height.

9. *MCAT strategies*

No answers can be easily eliminated so we proceed with our problem-solving strategy.

Recognize the principle

There are two principles we'll need to apply. First, we'll need to calculate the work done while climbing the step ladder and then we can determine the power associated with this work.

Sketch the problem

None required.

Identify the relationships

Since the boy is climbing a step ladder, his gravitational potential energy is changing and is given by the expression,

$$\Delta PE_{\text{grav}} = mg \, \Delta y$$

To change the potential energy by this amount the boy must do an equivalent amount of work, therefore, $W = \Delta PE_{\text{grav}}$. The power the boy must generate is given by the expression $P = W/t$.

Solve

Combining these expressions and substituting in the numerical values we find,

$$P = \frac{W}{t} = \frac{\Delta PE_{\text{grav}}}{t} = \frac{mg\Delta y}{t} = \frac{(50 \text{ kg})(9.8 \text{ m/s}^2)(2 \text{ m})}{10 \text{ s}} = 98 \text{ W}$$

Therefore the correct answer is (c).

What does it mean?

As in the previous problem, the answer only depended on the change in height of the boy since no forces other than gravity were important.

10. *MCAT strategies*

Since the work done by kinetic friction on an object is always negative, we can eliminate answers (b) and (c). No answers can be easily eliminated so we proceed with our problem-solving strategy.

Recognize the principle

To solve this problem we'll need to determine the force of friction on the box by first determining the normal force of the incline on the box. Next we can calculate the work done by the force of friction.

Sketch the problem

For this problem it's helpful to start with a free-body diagram for the box.

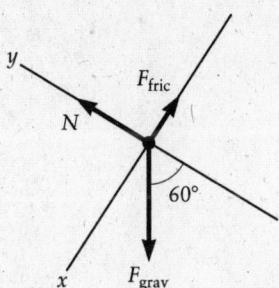

Identify the relationships

Applying Newton's second law along the y direction we can solve for the normal force on the box:

$$\Sigma F_y = N - mg\cos(60°) \rightarrow N = mg\cos(60°)$$

The frictional force is given by $F_{fric} = \mu_K N$. Finally, the work done by a force is given by the expression $W = F(\Delta r)\cos\theta$.

Solve

Combining these expressions and substituting in the numerical values we find,

$$W = F_{fric}(\Delta r)\cos(\theta) = \mu_K N(\Delta r)\cos(\theta) = \mu_K(mg\cos(60°))(\Delta r)\cos(\theta)$$

$$= (0.1)\left[(10 \text{ kg})(9.8 \text{ m/s}^2)\left(\frac{1}{2}\right)\right](2 \text{ m})\cos(180°)$$

$$= -9.8 \text{ N}$$

Therefore the correct answer is (d).

What does it mean?

Since friction was involved in this process, we could not use the expression for the change in the gravitational potential to analyze the problem. Fortunately, our experience with Newton's second law proved useful.

7 Momentum, Impulse, and Collisions

CONTENTS

Part A. Summary of Key Concepts and Problem-Solving Strategies

KEY CONCEPTS

Momentum

The ***momentum*** of a particle is given by the expression

$$\vec{p} = m\vec{v}$$

where m is the mass of the particle and $\vec{v}$ is the velocity.

The total momentum of a system of particles is equal to the sum of the momenta of the individual particles. Since momentum is a vector quantity we need to be sure to add momenta as vectors.

Impulse theorem

When a constant force acts on an object, it imparts an ***impulse*** to the object that is equal to $\vec{F}\Delta t$. Here, $\vec{F}$ is the force and Δt is the time interval over which the force is applied. When the force varies with time, calculating the impulse is more difficult. If we can graph the force as a function of time, then the impulse is equal to the area under the force–time curve. The impulse imparted to an object is equal to the change in the momentum of the object:

$$\text{impulse} = \vec{F}\Delta t = \Delta\vec{p}$$

Conservation of momentum

If there is no net external force on a system of particles, the total momentum of the system is constant. In such a case, momentum is *conserved*. This means that during the interaction of these particles the total momentum of the system remains constant. Momentum conservation is a powerful tool that allows us to analyze the many complicated interactions without the need to fully understand the forces involved in these interactions.

APPLICATIONS

Collisions

The objects involved in a collision can be thought of as a "system." During a collision, the forces external to the system are usually very small, and the total momentum of the system after the collision is equal to the total momentum of the system before the collision. Therefore the momentum is conserved. In an *elastic collision*, both momentum and kinetic energy are conserved. In an *inelastic collision*, the momentum is conserved, but kinetic energy is not conserved. In *a completely inelastic collision*, the two objects stick together after the collision, however, momentum is still conserved.

Center of mass

Most real objects have a size and shape, and they can be thought of as being composed of many separate pieces. Such extended objects possess a *center of mass*, and the motion of this point is very simple. For any object, no matter how complicated, the center of mass moves in response to the total external force acting on the object. This motion follows Newton's laws as if all the mass were located at the center of mass. For a system of two particles, the coordinates of the center of mass are given by

$$x_{CM} = \frac{m_1 x_1 + m_2 x_2}{m_1 + m_2}, \; y_{CM} = \frac{m_1 y_1 + m_2 y_2}{m_1 + m_2}, \; z_{CM} = \frac{m_1 z_1 + m_2 z_2}{m_1 + m_2}$$

For a system of many particles, we can generalize these expressions to be

$$x_{CM} = \frac{\sum_i m_i x_i}{\sum_i m_i} = \frac{\sum_i m_i x_i}{M_{tot}}, \; y_{CM} = \frac{\sum_i m_i y_i}{M_{tot}}, \; z_{CM} = \frac{\sum_i m_i z_i}{M_{tot}}$$

where M_{tot} is the total mass of the system.

PROBLEM-SOLVING STRATEGIES

In this chapter we were introduced to our second conservation principle. The principle of conservation of linear momentum provides us with a powerful tool for analyzing the dynamics of collisions. The following schematic outlines the thought process we should follow when applying this concept.

Problem Solving: Analyzing a Collision

Recognize the principle

The momentum of a system is conserved only when the total external force is zero. If the net external force in just one direction is zero, then momentum in only that direction is conserved. The principle of momentum conservation can be applied when the force between the particles within the system is much larger than the external forces.

Sketch the problem

Identify the system and make a sketch of the system, showing the coordinate axes and (where possible) the initial and final velocities of the particles in the system.

Identify the relationships

What type of collision is it? Remember, explosions and objects pushing off each other can be treated as completely inelastic collisions in reverse!

Elastic Collisions

Momentum and kinetic energy are conserved.

$$\vec{p}_{\text{total},\,i} = \vec{p}_{\text{total},\,f} \text{ and } KE_{\text{total},\,i} = KE_{\text{total},\,f}$$

Inelastic Collisions

Momentum is conserved but kinetic energy is not conserved.

$$\vec{p}_{\text{total},\,i} = \vec{p}_{\text{total},\,f}$$

Completely Inelastic Collisions

Momentum is conserved but kinetic energy is not conserved. Particles within the system stick together after the collision or begin together before pushing apart.

$$\vec{p}_{\text{total},\,i} = \vec{p}_{\text{total},\,f}$$

If the collision involves more than one dimension, be sure to conserve momentum in *all* directions.

Solve

Use the equations for momentum conservation (and kinetic energy conservation if the collision is elastic) to solve for the unknowns. This may involve solving multiple equations for multiple unknowns.

What does it mean?

Always *consider what your answer means* and check that it makes sense.

Part B. Frequently Asked Questions

1. *I'm confused about impulse and force. Aren't they really the same thing?*

No, force and impulse are not the same. The impulse is the product of the force and the time interval over which the force is applied, $\vec{F}\Delta t$. The same impulse can result from two different forces. For example, a large force times a small time interval can result in the same impulse as a small force times a longer time interval.

2. *Is the impulse of an object the same as its momentum, or the change in its momentum? What's the difference?*

The impulse theorem is described by the expression, impulse $= \vec{F}\Delta t = \Delta\vec{p}$. It states that the impulse, which is the product of the force and the time interval over which the force is applied, is equal to the change in the momentum. This expression is derived from Newton's second law and can be interpreted in much the same way. Newton's second law states that a force will result in an acceleration. The impulse theorem states that an impulse (force times time) will result in a change in momentum. Remember momentum and the change in momentum are not the same. Momentum is defined as $\vec{p} = m\vec{v}$ whereas the change in momentum is given by $\Delta\vec{p} = \vec{p}_f - \vec{p}_i$.

3. *When analyzing a collision, I'm never sure when to choose the before and after times to calculate the momentum. Does it really make a difference?*

Typically you want to analyze the momentum (and perhaps kinetic energy) of the system immediately before and immediately after the interaction (collision). This is particularly important when there are external forces acting on the objects being analyzed. For example, consider the collision of two cars. After the collision the cars usually skid to a stop, so friction is an external force to the system. We can still conserve momentum in this collision as long as we analyze the collision immediately before and immediately after the two cars make contact. If we wait until the cars skid to a stop, then their momentum has been lost and we can no longer apply the conservation of momentum.

4. *What's the difference between elastic, inelastic, and perfectly inelastic interactions?*

The difference between these interactions is really the degree to which the energy, more specifically the kinetic energy, is lost. As long as there is no or negligible external force on the system, then the momentum will be conserved for all these interactions. For elastic interactions the kinetic energy is also conserved. During an inelastic interaction some of the kinetic energy of the objects is lost to deforming the objects, heat, sound, etc. In a perfectly inelastic collision the objects remain stuck together after the collision. Consider the following figure which relates kinetic energy loss to the types of interactions:

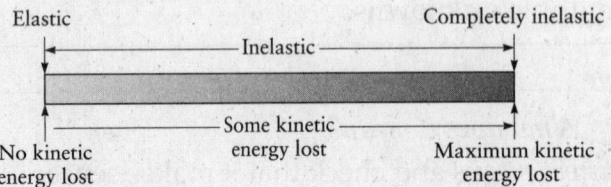

5. *Under what conditions is momentum conserved? What about kinetic energy?*

As stated in the previous FAQ, the total momentum of a system of particles is conserved as long as the system is isolated. By isolated, we mean that there is no net external force acting on the system. In practice we can still conserve momentum as long as the external force is small compared to the forces of interaction between the particles. Kinetic energy is only conserved in the case of an elastic collision. During elastic collisions the energy is momentarily stored in the elastic properties of the interacting objects and converted back into kinetic energy. For all other types of interactions kinetic energy is not conserved.

6. *Suppose I want to knock over an object with a ball. Should I use a sticky, clay ball or a rubber ball with the same mass?*

Although your intuition probably tells you that you should use a sticky, clay ball, you would be wrong. The interaction between the clay ball and the object is inelastic whereas the interaction between the rubber ball and the object is elastic. Consider the following example where a ball collides with a block and we're interested in comparing the final velocity of the block after the collision for two types of collisions, an elastic and a completely inelastic collsion.

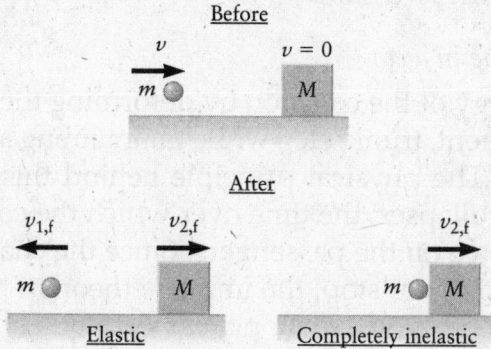

Applying the conservation of momentum to these two situations we get the following expressions for the final velocity of the block:

Elastic

$$mv = -mv_{1,f} + Mv_{2,f} \rightarrow v_{2,f} = \frac{m(v + v_{1,f})}{M}$$

Completely Inelastic

$$mv = mv_{2,f} + Mv_{2,f} \rightarrow v_{2,f} = \frac{mv}{(m + M)}$$

Comparing these two expressions it is easy to see that the final velocity of the block for the elastic collision will always be larger than for the completely inelastic collision. In other words, more momentum is transferred to the block when the collision is elastic. So, you use a rubber ball if you are trying to knock over an object.

7. *Since it is almost impossible to completely isolate a system from external forces, is momentum ever really conserved?*

That's a good question. Certainly out in space far from the gravitational effects of stars and planets, the net external force on our system is zero. In practice, however, this is seldom the case. Fortunately the forces involved in the interactions within our system (collision forces) are usually very large compared to these external forces and only change the total momentum of the system by very small amount. For this reason, momentum conservation can still be applied and is considered to be a very useful and powerful tool in physics.

8. *When applying the conservation of momentum, how do I choose my system?*

When defining a system, we need to include the objects of interest and the forces of interaction which are important. Also, the total external force on the system should be minimized so that we can apply the conservation of momentum. For example, consider two balls colliding in the air. The objects of interest are the two balls and the interaction forces between the two balls must be included in the system. Although the gravitational force from the Earth on the two balls is not zero, it is small compared to the interaction forces. So, we would define our system as the two balls and include their interaction forces. As long as we limit our analysis to immediately before and after they collide, the effect of the Earth's gravity is negligible and will have little effect on the total momentum of our system. Although it is tempting to include the Earth in the system (and you can!), this will have little effect on our analysis of the collision of the two balls.

9. *What is the purpose of the crumple zone in a car?*

Crumple zones absorb some of the energy of the collision by deforming the external portions of the car (engine compartment, trunk, etc.) while maintaining a structurally intact passenger compartment. The physical principle behind this is the same as air bags. As the crumple zone collapses, the time over which the collision occurs is extended thus reducing the force on the passengers. Since the change in momentum is the same regardless of how you stop, the impulse theorem tells us that the impulse must also be the same. So, you can stop a car very quickly with a very large force or over a longer period of time using a smaller force. Smaller forces are less damaging to the passengers. Of course, the trade off for this safety feature is that there is more damage to your car! See Problem 6.64 in your textbook for a calculation comparing cars with and without a crumple zone.

10. *When playing pool I've noticed that the cue ball and the ball it strikes always go off at 90° from each other. Is this always the case in an elastic collision?*

No. In fact, this is only the case for an elastic collision between objects of equal mass. The following example will help illustrate this situation:

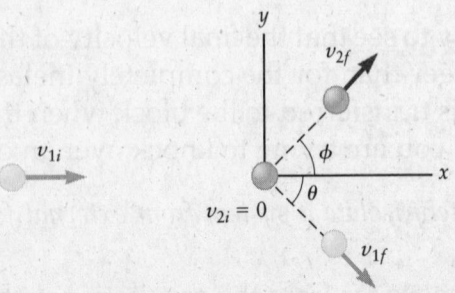

Applying momentum conservation in the x direction we find:

$$mv_{1i} = mv_{1f} \cos \theta + mv_{2f} \cos \phi \rightarrow v_{1i} = v_{1f} \cos \theta + v_{2f} \cos \phi \qquad (\#1)$$

Similarly, applying momentum conservation in the y direction we find:

$$0 = -mv_{1f} \sin \theta + mv_{2f} \sin \phi \rightarrow v_{1f} \sin \theta = v_{2f} \sin \phi \qquad (\#2)$$

Since this is an elastic collision, kinetic energy is also conserved:

$$\frac{1}{2} mv_{1i}^2 = \frac{1}{2} mv_{1f}^2 + \frac{1}{2} mv_{2f}^2 \rightarrow v_{1i}^2 = v_{1f}^2 + v_{2f}^2 \qquad (\#3)$$

Squaring (#1) and substituting this into equation (#3) we get,

$$0 = 2v_{1f} v_{2f} \cos \theta \cos \phi - v_{1f}^2 (1 - \cos^2 \theta) - v_{2f}^2 (1 - \cos^2 \phi)$$

Using the trig identity $1 - \cos^2 \theta = \sin^2 \theta$ and $1 - \cos^2 \phi = \sin^2 \phi$ and substituting in using equation (#2) this expression simplifies to,

$$0 = 2v_{1f}^2 \frac{\sin \theta \cos \theta \cos \phi}{\sin \phi} - v_{1f}^2 \sin^2 \theta - v_{1f}^2 \frac{\sin^2 \theta}{\sin^2 \phi} \sin^2 \phi$$

Upon further simplification we get,

$$\frac{\sin \theta}{\cos \theta} = \frac{\cos \phi}{\sin \phi} \rightarrow \tan \theta = \cot \phi \rightarrow \tan \theta = \tan (90° - \phi)$$

Therefore,

$$\theta = 90 - \phi \rightarrow \boxed{\theta + \phi = 90°}$$

So the balls go off 90° from each other. This is only true for an elastic collision between two equal masses where one of the masses is initially at rest.

Part C. Selection of End-of-Chapter Answers and Solutions

QUESTIONS

Q7.1 A bomb that is initially at rest breaks into several pieces of approximately equal mass, two of which are shown in Figure Q7.1. Use conservation of momentum to determine if there might be other pieces of the bomb not shown in the figure. Assume there is only one missing piece and estimate the direction it is traveling after the explosion.

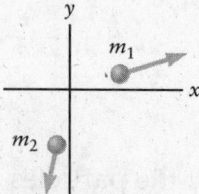

Figure Q7.1

Answer

The sum of the momenta of the particles involved must be equal to zero. Since each mass is about the same, we can use the size of the velocity vectors to represent the magnitudes of the momenta of the particles. Adding the velocity vectors as in the figure, it is apparent that they do not add to zero. So, there must be more mass moving that is not shown. If we assume that there is only one missing piece, then it must be traveling in a direction opposite the sum of the two existing vectors as shown.

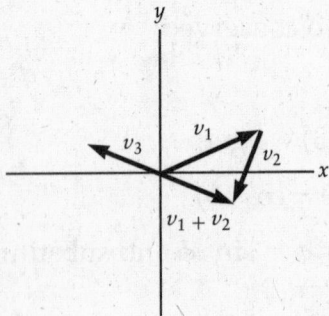

Q7.14 A tennis ball and a ball of soft clay are dropped from the same height onto the floor below. The force from the floor produces an impulse on each ball. If the balls have the same mass, which impulse is larger? Explain.

Answer

Since the tennis ball bounces off the floor, its change in momentum is greater than the change in momentum for the ball of clay. From the impulse theorem we see that the impulse is equal to the change in momentum. Therefore, the tennis ball will experience the greater impulse.

PROBLEMS

P7.4 Two particles of mass $m_1 = 1.2$ kg and $m_2 = 2.9$ kg are traveling as shown in Figure P7.4. What is the total momentum of this system? Be sure to give the magnitude and direction of the momentum.

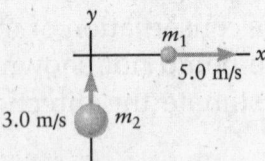

Figure P7.4

Solution

Recognize the principle

Since momentum is a vector, we need to add the momenta of the particles as vectors, taking into account their x and y components.

Sketch the problem

Refer to Figure P7.4.

Identify the relationships

From the diagram, m_2 is moving in the positive y direction; and m_1 is moving in the positive x direction. Therefore, we can easily write the components of the momenta along each axis:

$$p_x = m_1 v_x$$
$$p_y = m_2 v_y$$

The magnitude and direction of the resultant momentum can be found from these components,

$$|\vec{p}| = \sqrt{p_x^2 + p_y^2}$$
$$\theta = \tan^{-1}\left(\frac{p_y}{p_x}\right)$$

Solve

First, we find the values for components of the total momentum:

$$p_x = (1.2 \text{ kg})(5.0 \text{ m/s}) = 6.0 \text{ kg}\cdot\text{m/s}$$
$$p_y = (2.9 \text{ kg})(3.0 \text{ m/s}) = 8.7 \text{ kg}\cdot\text{m/s}$$

Then, inserting these component values into our expressions for the magnitude and the angle we find,

$$|\vec{p}| = \sqrt{(6.0 \text{ kg}\cdot\text{m/s})^2 + (8.7 \text{ kg}\cdot\text{m/s})^2}$$

$$\boxed{|\vec{p}| = 11 \text{ kg}\cdot\text{m/s}}$$

$$\theta = \tan^{-1}\left(\frac{8.7 \text{ kg}\cdot\text{m/s}}{6.0 \text{ kg}\cdot\text{m/s}}\right)$$

$$\boxed{\theta = 55° \text{ above the } x \text{ axis}}$$

What does it mean?

The net momentum of the system is the vector sum of each particle's momentum.

P7.11 A baseball player hits a baseball ($m = 0.14$ kg) as shown in Figure P7.11. The ball is initially traveling horizontally with speed of 40 m/s. The batter hits a fly ball as shown, with a speed $v_f = 55$ m/s. (a) What is the magnitude and direction of the impulse imparted to the ball? (b) If the ball and bat are in contact for a time of 8.0 ms, what is the magnitude of the average force of the bat on the ball? Compare this answer to the weight of the ball. (c) What is the impulse imparted to the bat?

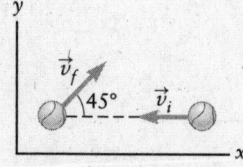

Figure P7.11

Solution

Recognize the principle

We can use our expression relating impulse and momentum to determine the impulse imparted to the ball as well as the average force of the bat on the ball. This is a two-dimensional problem so we will need to analyze the problem in terms of the x and y components.

Sketch the problem

Refer to Figure P7.11.

Identify the relationships

(a) From the diagram we can determine the components of the velocity of the ball before and after it is struck by the bat.

$$v_{ix} = -40 \text{ m/s}$$
$$v_{iy} = 0$$
$$v_{fx} = (55 \text{ m/s})\cos 45° = 38.89 \text{ m/s}$$
$$v_{fy} = (55 \text{ m/s})\sin 45° = 38.89 \text{ m/s}$$

From these we can calculate the change in momentum of the ball which is the impulse imparted to the ball.

(b) The magnitude of the average force can be found using the following expression,

$$F = \frac{\Delta p}{\Delta t}$$

(c) Newton's third law tells us that the force on the bat is equal in magnitude and in the opposite direction to the force on the ball. Also, the time of contact of the bat and ball remains the same so the impulse on the bat is equal in magnitude but in the opposite direction to the impulse imparted to the ball.

Solve

(a) The impulse imparted to the ball will be

$$\text{impulse} = \Delta \vec{p}$$

For each direction,

$$\Delta p_x = p_{fx} - p_{ix} = mv_{fx} - mv_{ix}$$
$$\Delta p_y = p_{fy} - p_{iy} = mv_{fy} - mv_{iy}$$

Inserting values,

$$\Delta p_x = (0.14 \text{ kg})(38.89 \text{ m/s}) - (0.14 \text{ kg})(-40 \text{ m/s})$$
$$\Delta p_x = 11 \text{ kg} \cdot \text{m/s}$$
$$\Delta p_y = (0.14 \text{ kg})(38.89 \text{ m/s}) - (0.14 \text{ kg})(0)$$
$$\Delta p_y = 5.4 \text{ kg} \cdot \text{m/s}$$

The magnitude of the impulse is then,

$$\Delta p = \sqrt{\Delta p_x^2 + p_y^2}$$

$$\Delta p = 12 \text{ kg} \cdot \text{m/s}$$

The angle above the horizontal is then

$$\theta = \tan^{-1}\left(\frac{\Delta p_y}{\Delta p_x}\right)$$

$$\theta = 26°$$

(b) Inserting values,

$$F = \frac{12 \text{ kg} \cdot \text{m/s}}{0.0080 \text{ s}}$$

$$F = 1500 \text{ N}$$

The contact force is more than 1000 times the balls weight:

$$W = mg = 0.14 \text{ kg} \, (9.8 \text{ m/s}^2) = 1.4 \text{ N}$$

(c) The magnitude of the impulse is then,

$$\Delta p = 12 \text{ kg} \cdot \text{m/s}$$

The angle from the x axis going counterclockwise is then $180° + 26°$

$$\theta = 206°$$

What does it mean?

The bat receives an impulse equal in magnitude to the impulse on the ball. It doesn't change its velocity as much as the ball because it has a much larger mass. Also, the batter grips the bat.

P7.17   A golf ball is hit from the tee and travels a distance of 300 yards. Estimate the magnitude of the impulse imparted to the golf ball. Ignore air drag in your analysis.

Solution

Recognize the principle

The distance a golf ball travels depends on its initial velocity (speed and angle). The magnitude of the impulse determines this initial velocity.

Sketch the problem

No sketch needed.

Identify the relationships

In order to calculate the impulse imparted to the golf ball, we need to determine the velocity of the ball when it is hit from the tee. From Chapter 4 we know,

$$x = \text{range} = \frac{v_0^2 \sin(2\theta)}{g}$$

Assuming the golfer is trying to obtain the maximum range, we can set θ equal to 45°. Solving for the velocity,

$$x = \text{range} = \frac{v_0^2 \sin(2\theta)}{g} = \frac{v_0^2}{g}$$

$$v_0 = \sqrt{xg}$$

This will be the speed of the ball after being struck. Since the ball is at rest initially, the impulse imparted to the golf ball will be the mass times this speed. That is:

$$\text{impulse} = \Delta p = mv_f - mv_i = mv_f$$

Solve

Inserting the value for the range converted to meters,

$$v_0 \approx \sqrt{\left(300 \text{ yards} \times \frac{0.9144 \text{ m}}{1 \text{ yard}}\right)(9.8 \text{ m/s}^2)}$$

$$v_0 \approx 50 \text{ m/s}$$

The mass of a golf ball is about 50 g. Inserting the velocity, into our expression for impulse:

$$\text{impulse} \approx (0.05 \text{ kg})(50 \text{ m/s})$$

$$\boxed{\text{impulse} \approx 3 \text{ N} \cdot \text{s}}$$

What does it mean?

A relatively modest impulse sends a golf ball 300 yards when aimed for maximum range.

P7.25   Consider again the collision between two hockey pucks in Figure P7.24, but now they do not stick together. Their speeds before the collision are $v_{1i} = 20$ m/s and $v_{2i} = 15$ m/s. It is found that after the collision one of the pucks is moving along x with a speed of 10 m/s. What is the final velocity of the other puck?

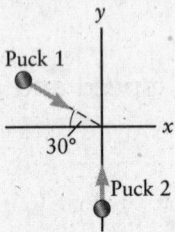

Figure P7.24

Solution

Recognize the principle

The pucks do not stick together, however we cannot assume it is an elastic collision. We proceed by applying the conservation of momentum in both the x and y directions.

Sketch the problem

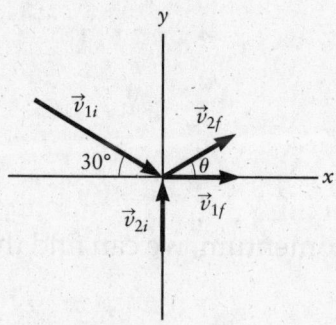

Identify the relationships

We express the initial and final velocities in terms of components. For puck 1 the initial velocity is,

$$v_{1ix} = v_{1i}\cos 30° = (20\text{ m/s})\cos 30° = 17.3\text{ m/s}$$
$$v_{1iy} = -v_{1i}\sin 30° = -(20\text{ m/s})\sin 30° = -10\text{ m/s}$$

For puck 2 the initial velocity is,

$$v_{2ix} = 0$$
$$v_{2iy} = v_{2i} = 15\text{ m/s}$$

For puck 1 the final velocity is,

$$v_{1fx} = v_{1f} = 10\text{ m/s}$$
$$v_{1fy} = 0$$

For puck 2 the final velocity is,

$$v_{2fx} = v_{2f}\cos\theta$$
$$v_{2fy} = v_{2f}\sin\theta$$

Assuming the pucks have identical masses and applying the conservation of momentum for each axis we find:

Along the x axis,

$$mv_{1ix} + mv_{2ix} = mv_{1fx} + mv_{2fx}$$
$$mv_{1ix} + 0 = mv_{1fx} + mv_{2fx}$$
$$v_{1ix} = v_{1fx} + v_{2fx}$$

Along the y axis,

$$mv_{1iy} + mv_{2iy} = mv_{1fy} + mv_{2fy}$$
$$mv_{1iy} + mv_{2iy} = 0 + mv_{2fy}$$
$$v_{1iy} + v_{2iy} = v_{2fy}$$

Solve

From the first equation for conservation of momentum, we can find the final velocity of puck 2 in the x direction,

$$v_{1ix} = v_{1fx} + v_{2fx}$$
$$v_{2fx} = v_{1ix} - v_{1fx}$$
$$v_{2fx} = v_{1i} \cos 30° - v_{1f}$$
$$v_{2fx} = 7.3 \text{ m/s}$$

From the second equation for the conservation of momentum, we can find the final velocity of puck 2 in the y direction,

$$v_{2fy} = v_{1iy} + v_{2iy}$$
$$v_{2fy} = -v_{1i} \sin 30° + v_{2i}$$
$$v_{2fy} = 5.0 \text{ m/s}$$

The magnitude and angle of the final velocity is given by,

$$v_{2f} = \sqrt{v_{2fx}^2 + v_{2fy}^2}$$
$$\boxed{v_{2f} = 8.9 \text{ m/s}}$$

$$\theta = \tan^{-1}\left(\frac{v_{2fy}}{v_{2fx}}\right)$$

$$\boxed{\theta = 34° \text{ above the } x \text{ axis}}$$

What does it mean?

Note that kinetic energy is not conserved in this collision. If we had assumed that it was conserved, we would have obtained the wrong answer. In a closed system, momentum is always conserved.

P7.30   Two cars of equal mass are traveling as shown in Figure P7.30 just before undergoing a collision. Before the collision one of the cars has a speed of 18 m/s along $+x$ while the other has a speed of 25 m/s along $+y$. The cars lock bumpers and then slide away together after the collision. What is the magnitude and direction of their final velocity?

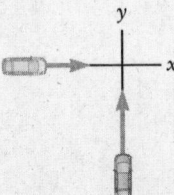

Figure P7.30

Solution

Recognize the principle

This is a collision in two dimensions. Since the two cars stick together this is an inelastic collision and only the conservation of momentum will apply.

Sketch the problem

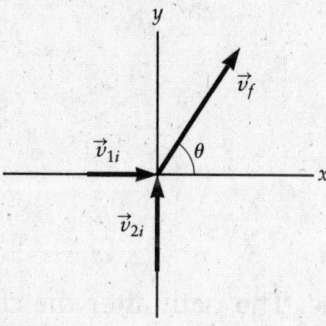

Identify the relationships

Expressing the velocities in terms of components we find:

For car one,

$$v_{1ix} = 18 \text{ m/s}$$
$$v_{1iy} = 0$$

For car two,

$$v_{2ix} = 0$$
$$v_{2iy} = 25 \text{ m/s}$$

Since the cars stick together and have identical mass, the conservation of momentum for each axis is given by:
Along the x axis,

$$mv_{1ix} + mv_{2ix} = (m_1 + m_2) v_{fx}$$
$$mv_{1ix} + 0 = 2 mv_{fx}$$
$$v_{1ix} = 2v_{fx}$$

Along the y axis,

$$mv_{1iy} + mv_{2iy} = (m_1 + m_2) v_{fy}$$
$$0 + mv_{2iy} = 2 mv_{fy}$$
$$v_{2iy} = 2 v_{fy}$$

Solve

To find the magnitude of the velocity we use,

$$v_f = \sqrt{v_{fx}^2 + v_{fy}^2}$$

Inserting values,

$$v_f = \sqrt{\left(\frac{1}{2}\,(18\text{ m/s})\right)^2 + \left(\frac{1}{2}\,(25\text{ m/s})\right)^2}$$

$$\boxed{v_f = 15\text{ m/s}}$$

The angle θ is given by,

$$\theta = \tan^{-1}\left(\frac{v_{fy}}{v_{fx}}\right)$$

$$\theta = \tan^{-1}\left(\frac{25\text{ m/s}}{18\text{ m/s}}\right)$$

$\boxed{\theta = 54°}$ above the x axis

What does it mean?

This type of analysis is often done at a crash scene. The path after the collision gives information about the speeds of the cars before the collision.

P7.40   A railroad car containing explosive material is initially traveling south at a speed of 5.0 m/s on level ground. The total mass of the car plus explosives is 3.0×10^4 kg. An accidental spark ignites the explosive, and the car breaks into two pieces, which then roll away along the same track. If one piece has a mass of 2.0×10^4 kg and a final speed of 2.5 m/s toward the south, what is the final speed of the other piece?

Solution

Recognize the principle

This event can be treated like a collision in one dimension along the horizontal north-south axis. Momentum will be conserved.

Sketch the problem

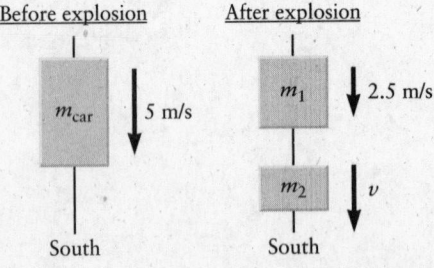

Identify the relationships

Applying the conservation of momentum we find,

$$m_{\text{car}}\,v_i = m_1 v_{1f} + m_2 v_{2f}$$

Solve

Solving for the final speed of mass 2,

$$m_{car}\, v_i = m_1 v_{1f} + m_2 v_{2f}$$
$$m_2 v_{2f} = m_{car}\, v_i - m_1 v_{1f}$$
$$v_{2f} = \frac{m_{car}\, v_i - m_1 v_{1f}}{m_2}$$

Assuming the explosives were very small compared to the mass of the car, $m_2 = m_{car} - m_1$. Inserting values,

$$v_{2f} = \frac{m_{car}\, v_i - m_1 v_{1f}}{m_{car} - m_1}$$

$$v_{2f} = \frac{(3.0 \times 10^4\ \text{kg})(5.0\ \text{m/s}) - (2.0 \times 10^4\ \text{kg})(2.5\ \text{m/s})}{3.0 \times 10^4\ \text{kg} - 2.0 \times 10^4\ \text{kg}}$$

$$\boxed{v_{2f} = 10\ \text{m/s}}$$

What does it mean?

Note that this positive velocity indicates that both pieces are still moving southward. The explosion slowed the heavier piece down to half its original velocity, and doubled the lighter piece's velocity.

P7.44 Consider the motion of the two ice skaters in Figure 7.28 and assume they have masses of 60 kg and 100 kg. If the larger skater has moved a distance of 12 m from his initial position, where is the smaller skater?

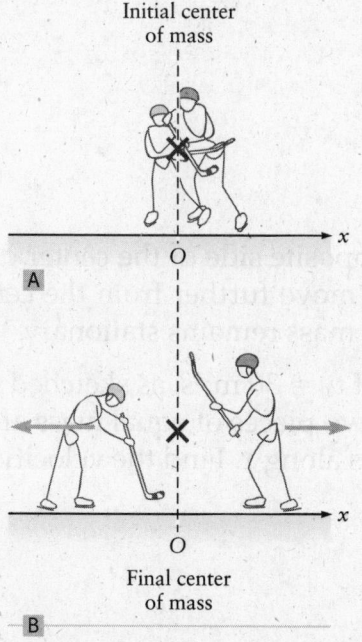

Figure 7.28

Solution

Recognize the principle

To find how far the smaller skater moved, we need to find the center of mass. Regardless of how far they move, the center of mass will remain the same.

Sketch the problem

The sketch is provided in the problem (Figure 7.28).

Identify the relationship

For the two skaters the center of mass is given by the expression,

$$x_{CM} = \frac{m_1 x_1 + m_2 x_2}{m_1 + m_2}$$

We will assume the skater with the smaller mass is on the left and the skater with the larger mass is on the right. If we label the center of mass of the system as the $x = 0$, then we can solve for the distance the smaller mass is away from the center this point,

$$x_{CM} = 0 = \frac{m_1 x_1 + m_2 x_2}{m_1 + m_2}$$

$$0 = m_1 x_1 + m_2 x_2$$

$$x_2 = \frac{-m_1 x_1}{m_2}$$

Solve

Inserting the values,

$$x_2 = \frac{-(100 \text{ kg})(12 \text{ m})}{60 \text{ kg}}$$

$$\boxed{x_2 = -20 \text{ m}}$$

or 20 m to the left of the center of mass.

What does it mean?

The lighter (60 kg) skater must be 20 m on the opposite side of the center of mass from the heavier skater. The lighter skater must move further from the center of mass than the heavier skater so that the center of mass remains stationary.

P7.51   A hand grenade is thrown with a speed of $v_0 = 30$ m/s, as sketched in Figure P7.51, just prior to exploding. It breaks into two pieces of equal mass after the explosion. One piece has a final velocity of 40 m/s along x. Find the velocity of the other piece after the explosion.

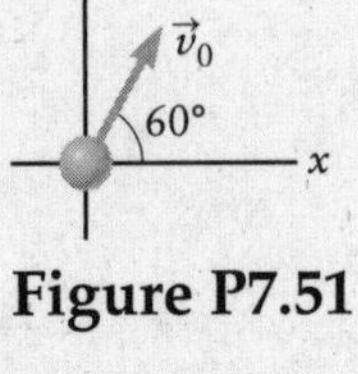

Figure P7.51

Solution

Recognize the principle

Momentum will be conserved in both the x and y directions. Since the bomb is whole (both pieces together) before it explodes, the two pieces have the same initial velocity.

Sketch the problem

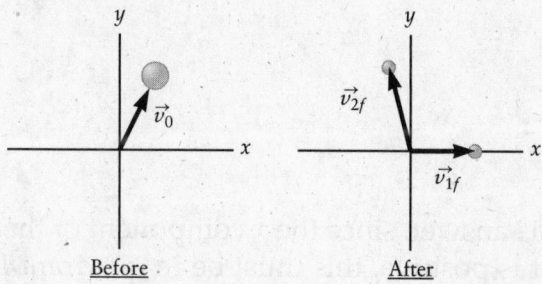

Identify the relationships

Since the two pieces are of equal mass, let m be the mass. Applying the conservation of momentum along both the x and y directions,

$$(m + m)v_{ix} = mv_{1fx} + mv_{2fx} \qquad (m + m)v_{iy} = mv_{1fy} + mv_{2fy}$$
$$2v_{ix} = v_{1fx} + v_{2fx} \qquad \text{and} \qquad 2v_{iy} = v_{1fy} + v_{2fy}$$

Solving for the velocity of mass two in the x direction,

$$2v_{ix} = v_{1fx} + v_{2fx}$$
$$v_{2fx} = 2v_{ix} - v_{1fx}$$

And solving for the velocity of mass two in the y direction,

$$2v_{iy} = v_{1fy} + v_{2fy}$$
$$v_{2fy} = 2v_{iy} - v_{1fy}$$

Solve

From the diagram, the initial velocity in the x direction is $v_0 \cos 60°$,

$$2v_{ix} = v_{1fx} + v_{2fx}$$
$$v_{2fx} = 2v_0 \cos 60° - v_{1fx}$$

Inserting values,

$$v_{2fx} = 2(30 \text{ m/s}) \cos 60° - 40 \text{ m/s}$$
$$v_{2fx} = -10 \text{ m/s}$$

From the diagram, the initial velocity in the y direction is $v_0 \sin 60°$ and mass #1 doesn't have a final velocity in the y direction,

$$2v_{iy} = v_{1fy} + v_{2fy}$$
$$v_{2fy} = 2v_0 \sin 60° - 0$$

Inserting values,

$$v_{2fy} = 2(30 \text{ m/s})\sin 60°$$
$$v_{2fy} = 52 \text{ m/s}$$

The magnitude and direction of mass #2 is then,

$$v_{2f} = \sqrt{v_{2fx}^2 + v_{2fy}^2}$$
$$\boxed{v_{2f} = 53 \text{ m/s}}$$

$$\theta = \tan^{-1}\left(\frac{v_{2fy}}{v_{2fx}}\right)$$
$$\theta = -79°$$

What does it mean?

Even though our calculator gives this answer, since the x component of the velocity is negative and the y component is positive, this must be in quadrant II. This means it is $\boxed{101° \text{ clockwise}}$ from the x axis.

P7.71 Rocket engines work by expelling gas at a high speed as illustrated in Figure P7.71. We wish to design an engine that expels an amount of gas $m_g = 50$ kg each second, and we want the engine to exert a force $F = 20{,}000$ N on a rocket. At what speed v_g should the gas be expelled? Assume the mass of the rocket is very large.

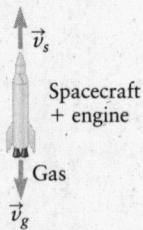

Figure P7.71

Solution

Recognize the principle

The impulse provided in pushing the mass of gas is (by Newton's third law) equal and opposite to the impulse on the rocket.

Sketch the problem

Refer to Figure P7.71.

Identify the relationships

The impulse theorem states,

$$F\Delta t = \Delta p$$

If we assume the gas starts at rest and choose down as positive, the change in momentum of the gas will be:

$$\Delta p = p_f - p_i$$
$$\Delta p = m_g v_f - m_g v_i = m_g v_f - 0 = m_g v_f$$

Inserting this into the above equation,

$$F\Delta t = m_g v_f$$

Solve

Solving for the speed of the gas,

$$v_f = \frac{F\Delta t}{m_g}$$

Inserting the values for a length of time of 1.0 s, to have a force of 20,000 N by expelling 50 kg of gas,

$$v_f = \frac{(20,000 \text{ N})(1.0 \text{ s})}{50 \text{ kg}}$$

$$\boxed{v_f = 400 \text{ m/s}}$$

What does it mean?

The gas has to be expelled at a high speed (greater than the speed of sound) in order to provide the force required.

Part D. Additional Worked Examples and Capstone Problems

The following worked examples and capstone problems provide you with practice determining the center of mass of an object, dealing with momentum conservation, and understanding the dynamics of collisions. Although these four problems do not incorporate all the material discussed in this chapter, they do highlight several of the key concepts. If you can successfully solve these problems then you should have confidence in your understanding of these key concepts, so use these problems as a test of your understanding of the chapter material.

WE 7.1 Center of Mass of a Bracket

Suppose you work for a carpenter and are building a bracket composed of two straight sections of wood as sketched in the figure. You want to show your boss that physics can be useful, so you decide to calculate the position of the center of mass of the bracket. (Knowing this position might help in supporting the bracket during a construction project.) If the two sections of wood have lengths $L_1 = 1.2$ m and $L_2 = 2.0$ m and masses $m_1 = 0.50$ kg and $m_2 = 1.0$ kg, what is the location of the center of mass of the entire bracket?

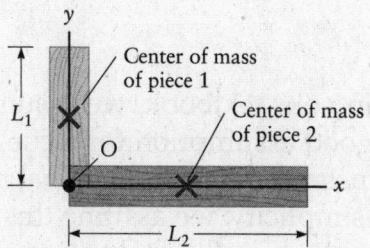

Solution

Recognize the principle

We could find the center of mass of this system of particles by applying the definition and summing over all the pieces that make up the bracket. A simpler approach, though, uses the notion of the center of mass as a balance point. With that in mind, we first find the center of mass of each straight section of the bracket. We can then treat these two pieces as particles of mass m_1 and m_2 located at their respective centers of mass and then compute the location of the overall center of mass of this two-particle system.

Sketch the problem

The figure shows the problem.

Identify the relationships

For each piece of wood, we use the balance-point notion of center of mass to tell us that the center of mass of each piece will be at each respective center as indicated in the figure. Thus, the center of mass of piece 1 is at

$$x_{1,\,CM} = 0 \text{ and } y_{1,\,CM} = \frac{L_1}{2}$$

and for piece 2, we have

$$x_{2,\,CM} = \frac{L_2}{2} \text{ and } y_{2,\,CM} = 0$$

Solve

We can now treat this problem as a system of two particles of mass m_1 and m_2 at these two locations and use the relations for the center of mass coordinates to find the center of mass of the entire bracket. So,

$$x_{CM} = \frac{m_1\,x_{1,\,CM} + m_2\,x_{2,\,CM}}{m_1 + m_2} = \frac{(0.50 \text{ kg})(0) + (1.0 \text{ kg})(L_2/2)}{(0.50 \text{ kg}) + (1.0 \text{ kg})} = \boxed{0.67 \text{ m}}$$

$$y_{CM} = \frac{m_1\,y_{1,\,CM} + m_2\,y_{2,\,CM}}{m_1 + m_2} = \frac{(0.50 \text{ kg})(L_1/2) + (1.0 \text{ kg})(0)}{(0.50 \text{ kg}) + (1.0 \text{ kg})} = \boxed{0.20 \text{ m}}$$

The center of mass is outside the bracket.

What does it mean?

When an object has a simple shape (e.g., a wheel, ball, or a straight piece of wood), the center of mass is at the center of symmetry of the object.

WE 7.2 Bouncing from a Wall

In our example with the cue ball in Figure 7.30 in your textbook, we assumed an elastic collision with the rail. Although that is a good assumption for a cue ball, it fails for many other types of balls. Consider, for instance, a rubber ball that reflects from a wall as sketched in following figure. For simplicity, we assume this ball is rolling on a horizontal table and the wall forms one edge of the table (i.e., there is no effect from gravity on the ball's trajectory). Even with a ball that is extremely elastic, the kinetic energy after colliding with the wall is always smaller than the initial kinetic energy. This situation can be described mathematically by a quantity called the coefficient of restitution, denoted by α. For the collision in the figure, the initial and final components of the velocity along y are related by

$$v_{fy} = -\alpha v_{iy} \tag{1}$$

Suppose a ball with a coefficient of restitution $\alpha = 0.80$ hits the wall at an initial angle of $\theta_i = 45°$. Find the angle θ_f that the outgoing velocity makes with the y direction. Ignore any effects due to the spin of the ball.

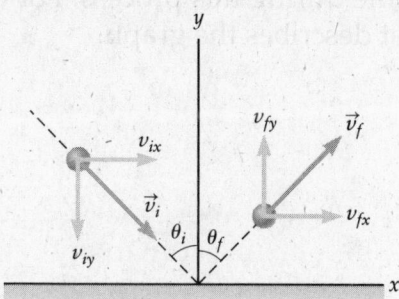

Figure 7.30

Solution

Recognize the principle

We can follow an approach similar to the one taken to analyze the cue ball in Figure 7.30 of your textbook. The ball's momentum along x is conserved because there are no forces on the ball in this direction, and this condition will give us the final component of the velocity along x.

Sketch the problem

The figure describes the problem and shows the initial and final velocities and angles.

Identify the relationships

Momentum along the x direction is conserved, so we get

$$mv_{ix} = mv_{fx}$$
$$v_{ix} = v_{fx} \tag{2}$$

The outgoing angle is (see Fig. 7.30)

$$\theta_f = \tan^{-1}\left|\frac{v_{fx}}{v_{fy}}\right|$$

Using the results from Equations (1) and (2) gives

$$\theta_f = \tan^{-1}\left|\frac{v_{fx}}{v_{fy}}\right| = \tan^{-1}\left|\frac{v_{ix}}{\alpha v_{iy}}\right|$$

Solve

The incoming angle $\theta_i = 45°$, so $v_{ix} = v_{iy}$. Using this result along with $\alpha = 0.80$ leads to

$$\theta_f = \tan^{-1}\left|\frac{v_{ix}}{\alpha v_{iy}}\right| = \tan^{-1}\left(\frac{1}{\alpha}\right) = \tan^{-1}\left(\frac{1}{0.80}\right) = \boxed{51°}$$

What does it mean?

This angle is larger than the incident angle, so the ball will travel closer to the wall after the collision than before.

CP 7.1 Understanding Collision Dynamics using Graphs

Consider two carts moving toward each other on a frictionless track. Initially cart A is moving in the x direction while cart B is moving in the $-x$ direction. They bounce off each other in an elastic collision. Each of the graphs (1–6) represents a particular physical quantity as a function of time during this process. For each of these graphs match up an item (a–f) which best describes the graph.

(a) The force on cart A.

(b) The force on cart B.

(c) The momentum of cart A.

(d) The momentum of cart B.

(e) The position of cart A.

(f) The position of cart B.

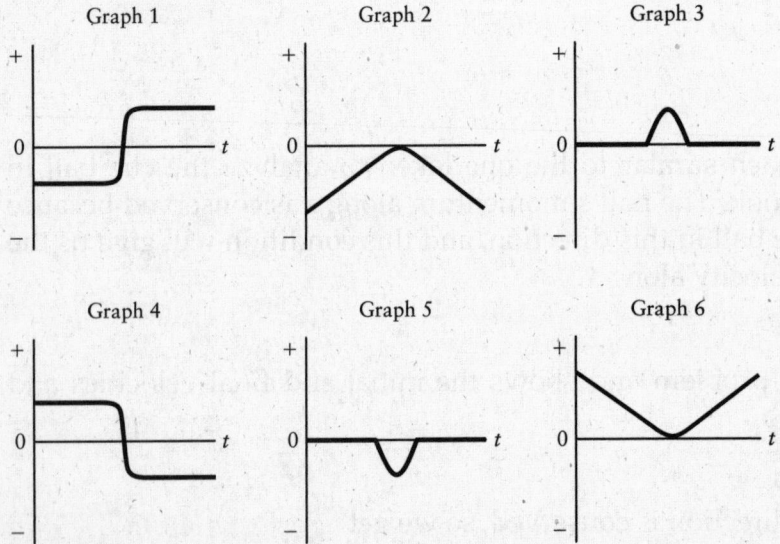

Solution

Recognize the principle

We will need to understand how the position, momentum, and force vary as a function of time as the carts move before, during, and after the collision.

Sketch the problem

The sketch of the problem is provided in the question.

Identify the relationships

Several key relationships are important. Before and after the collision the carts move at a constant velocity since there is no force acting on the carts. For these portions of the motion the following is true:

- The force is zero.
- The velocity is constant, therefore, the position-time graph should be a straight line with constant slope.

- Since the velocity is constant for the carts, so is their momentum. So, the momentum-time graphs should be straight, horizontal lines.

During the actual collision of the carts the following is true.

- The carts change direction so the slope of the position-time graphs should reflect a change in sign.
- The velocity will change direction so the momentum-time graphs should reflect a change in sign.
- The force will increase from zero to some maximum value and then decrease to zero during the collision.

Solve

Since cart A is initially moving in the x direction its position-time graph will initially have a positive slope and its momentum-time graph will be positive. During the collision, cart A changes direction, therefore is accelerated in the $-x$ direction. So, the force on cart A during the collision must be negative. Following this same logic for cart B, we match up an item (a–f) which best describes the graphs (1–6).

(a) $\rightarrow$ (5)
(b) $\rightarrow$ (3)
(c) $\rightarrow$ (4)
(d) $\rightarrow$ (1)
(e) $\rightarrow$ (2)
(f) $\rightarrow$ (6)

What does it mean?

Note that the area under the force-time graphs for the two carts is the same. This is a consequence of Newton's third law! Also, according to the impulse theorem, the areas under the force-time graphs equal the change in momentum observed in the momentum-time graphs.

CP 7.2 Applying the Conservation of Momentum to an Exploding Bomb

Consider a bomb launched at 200 m/s at an angle of 30° above horizontal ground. At 15 s into the trajectory the bomb detonates and breaks into two fragments of equal mass which fly directly away from each other. One fragment lands 4000 m from the original launch point. Where does the other fragment land?

Solution

Recognize the principle

There are three parts to this problem. During the first part of the trajectory we can apply our expressions for two-dimensional motion under constant acceleration (projectile motion) to determine the position and velocity of the bomb just before it explodes. During the explosion momentum is conserved, thus relating the velocities of the fragments after the explosion to the velocity of the bomb just before the explosion. Finally, we can use our expressions for projectile motion to analyze the trajectories of the fragments after the explosion.

Sketch the problem

It is useful to draw a sketch of the entire process as well as a separate sketch of the explosion.

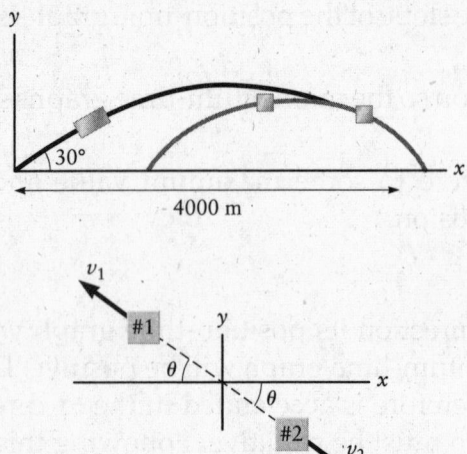

Identify the relationships and Solve

Let's begin by determining the position and velocity of the bomb just before it explodes.

x direction

$$v_{0x} = v_0 \cos \theta = 200 \text{ m/s} \cos 30° = 173 \text{ m/s}$$
$$x = v_{0x}\, t = 173 \text{ m/s} (15) = 2595 \text{ m}$$

y direction

$$v_{0y} = v_0 \sin \theta = 200 \text{ m/s} \sin 30° = 100 \text{ m/s}$$
$$y = y_0 + v_{0y}t + \frac{1}{2} a_y t^2 = 100 \text{ m/s} (15 \text{ s}) - 4.9 \text{ m/s}^2 (15 \text{ s})^2 = 398 \text{ m}$$
$$v_y = v_{0y} + a_y t = 100 \text{ m/s} - 9.8 \text{ m/s}^2 (15 \text{ s}) = -47 \text{ m/s}$$

So, just before the bomb explodes it is at a position $x = 2595$ m, $y = 398$ m and has a velocity of $v_x = 173$ m/s, $v_y = -47$ m/s.

Now we can use the conservation of momentum to determine the velocities of the two bomb fragments after the explosion. Again, from the diagram and the information we just obtained about the position and velocity of the bomb before it explodes, we can split the problem up into both directions and conserve momentum it both directions.

x direction

$$p_{ix} = p_{fx}$$
$$p_{ix} = p_{1x} + p_{2x}$$
$$m v_x = -\left(\frac{m}{2}\right) v_1 \cos \theta + \left(\frac{m}{2}\right) v_2 \cos \theta$$
$$346 \text{ m/s} = -v_1 \cos \theta + v_2 \cos \theta \rightarrow v_2 - v_1 = \frac{346 \text{ m/s}}{\cos \theta}$$

y direction

$$p_{iy} = p_{fy}$$

$$p_{iy} = p_{1y} + p_{2y}$$

$$\cancel{m}v_y = \left(\frac{\cancel{m}}{2}\right) v_1 \sin\theta - \left(\frac{\cancel{m}}{2}\right) v_2 \sin\theta$$

$$-94 \text{ m/s} = v_1 \sin\theta - v_2 \sin\theta \rightarrow v_2 - v_1 = \frac{94 \text{ m/s}}{\sin\theta}$$

Equating these two expressions we can solve for θ:

$$\frac{346 \text{ m/s}}{\cos\theta} = \frac{97 \text{ m/s}}{\sin\theta} \rightarrow \tan\theta = \left(\frac{94 \text{ m/s}}{346 \text{ m/s}}\right) \rightarrow \theta = 15°$$

Now that we know the angle at which the fragments are launched, we can use the information on where fragment #2 lands to determine what its velocity must have been immediately after the explosion.

x direction

$$v_{2x} = v_2 \cos\theta$$

$$x = x_0 + v_{2x}t + \frac{1}{2}a_x t^2$$

$$a_x = 0$$

$$x - x_0 = v_{2x}t \rightarrow t = \frac{x - x_0}{v_{2x}} = \frac{4000 \text{ m} - 2595 \text{ m}}{v_2 \cos 15°}$$

$$t = \frac{1454 \text{ m}}{v_2}$$

y direction

$$v_{2y} = -v_2 \sin\theta$$

$$y = y_0 + v_{2y}t + \frac{1}{2}a_y t^2$$

$$0 = 398 \text{ m} - v_2 \sin(15°)t - (4.9 \text{ m/s}^2)t^2$$

$$398 \text{ m} = 0.26 \, v_2 t + (4.9 \text{ m/s}^2)t^2$$

Substituting in for t from the x direction and solving for v_1 we find,

$$398 \text{ m} = 0.26 \, \cancel{v}_2 \left(\frac{1454 \text{ m}}{\cancel{v}_2}\right) + 4.9 \text{ m/s}^2 \left(\frac{1454 \text{ m}}{v_2}\right)^2$$

$$v_2 = 720 \text{ m/s}$$

Substituting back into the momentum conservation expression we find v_1,

$$v_2 - v_1 = \frac{94 \text{ m/s}}{\sin\theta} \rightarrow v_1 = v_2 - \frac{94 \text{ m/s}}{\sin 15°} = 357 \text{ m/s}$$

From this information and our expressions for projectile motion we can determine where fragment #1 lands.

x direction

$$v_{1x} = -v_1 \cos\theta = -(357 \text{ m/s}) \cos(15°) = -345 \text{ m/s}$$

$$x - x_0 = v_{1x}t \rightarrow x = x_0 - (345 \text{ m/s})t$$

$$x = 2595 \text{ m} - (345 \text{ m/s})t$$

y direction

$$v_{1y} = v_1 \sin \theta = 357 \text{ m/s} \sin 15° = 92 \text{ m/s}$$

$$y = y_0 + v_{1y}t + \frac{1}{2}a_y t^2$$

$$0 = 398 \text{ m} + (92 \text{ m/s})t - (4.9 \text{ m/s}^2)t^2$$

$$t = 22.4 \text{ s}$$

Substituting this into the expression for x we find,

$$x = 2595 \text{ m} - 345 \text{ m/s} (22.4 \text{ s}) = \boxed{-5133 \text{ m}}$$

What does it mean?

Clearly this is a very dangerous situation since one of the fragments lands behind the original launch point of the bomb. Fortunately, bombs seldom explode in this way!

Part E. MCAT Review Problems and Solutions

PROBLEMS

1. A car with a mass of 800 kg is stalled on the road. A truck with a mass of 1200 kg comes around the curve at 20 m/s and hits the car. The two vehicles remain locked together after the collision. What is their combined speed after the impact?

   (a) 3 m/s
   (b) 6 m/s
   (c) 12 m/s
   (d) 24 m/s

2. A 1000-kg car traveling at 5.0 m/s overtakes and collides with a 3000-kg truck traveling in the same direction at 1.0 m/s. During the collision, the two vehicles couple together and continue to move as one unit. What is the speed of the coupled vehicles immediately after the collision?

   (a) 2.0 m/s
   (b) 4.0 m/s
   (c) 5.0 m/s
   (d) 6.0 m/s

3. A rifle with a mass of 0.20 kg fires a 0.50 g bullet with an initial speed of 100 m/s. What is the recoil speed of the rife?

   (a) 0.25 m/s
   (b) 0.50 m/s
   (c) 1.0 m/s
   (d) 10 m/s

4. A 0.20-kg ball is bounced against a wall. It hits the wall with a speed of 20 m/s and rebounds elastically. What is the magnitude of the total change in momentum of the ball?

   (a) $0 \text{ kg} \cdot \text{m/s}$

   (b) $4.0 \text{ kg} \cdot \text{m/s}$

   (c) $8.0 \text{ kg} \cdot \text{m/s}$

   (d) $10.0 \text{ kg} \cdot \text{m/s}$

5. A tennis ball is hit with a racket and the change in momentum of the ball is $4 \text{ kg} \cdot \text{m/s}$. If the collision time of the ball and the racket is 0.01 s, what is the magnitude of the average force exerted on the ball by the racket?

   (a) $2.5 \times 10^{-3} \text{ N}$

   (b) $4.0 \times 10^{-3} \text{ N}$

   (c) 4.0 N

   (d) 400 N

6. How fast must a 2000-kg object travel in order to have the same momentum as a 200-g object traveling at 200 m/s?

   (a) 0.02 m/s

   (b) 0.05 m/s

   (c) 5.0 m/s

   (d) 20 m/s

7. A spacecraft with a total mass of 10,000 kg is at rest in space, far from any gravitational forces. The rocket engines are then fired. Its rockets burn for 15 s and produce a thrust of 3.0×10^5 N. Assuming that the net mass of the spacecraft does not change as the rocket fuel is consumed, what is the final speed of the spacecraft?

   (a) $2.0 \times 10^2 \text{ m/s}$

   (b) $4.5 \times 10^2 \text{ m/s}$

   (c) $3.0 \times 10^2 \text{ m/s}$

   (d) $5.0 \times 10^2 \text{ m/s}$

8. Gas is burned in a combustion engine. The resulting explosion produces a force that drives the pistons in the engine. The force of the explosion on the piston is due to the change in the momentum of the gas molecules as they collide with the piston. A 0.40-g sample of gas produces a force of 2400 N in an explosion that lasts 10^{-3} s. Assuming the gas is at rest after colliding with the piston, what must be the speed of the gas before it collides with the piston?

   (a) $6.0 \times 10^3 \text{ m/s}$

   (b) 6.0 m/s

   (c) $3.0 \times 10^3 \text{ m/s}$

   (d) 3.0 m/s

9. A 30-kg cart traveling due north at 5 m/s collides with a 50-kg cart that has been traveling due south. Both carts immediately come to rest after the collision. What must have been the speed of the southbound cart?

    (a) 3.0 m/s

    (b) 5.0 m/s

    (c) 6.0 m/s

    (d) 10.0 m/s

10. An eastbound car traveling at 10 m/s collides with a northbound car of equal mass. The two cars stick together and slide in a direction 60° north of east. What must have been the speed of the northbound car immediately before the collision?

    (a) $\dfrac{20}{\sqrt{3}}$ m/s

    (b) $\dfrac{10}{\sqrt{3}}$ m/s

    (c) $10\sqrt{3}$ m/s

    (d) $\dfrac{10\sqrt{3}}{2}$ m/s

SOLUTIONS

1. *MCAT strategies*

We know that the combined velocity after the collision for this situation cannot be greater than the initial velocity of the moving car (20 m/s), so answer (d) can be eliminated. No other answer can be easily eliminated so we proceed with our problem-solving strategy.

Recognize the principle

This is an example of a one-dimensional completely inelastic collision. Momentum is conserved in this situation.

Sketch the problem

Begin by sketching the situation before and after the collision.

Identify the relationships

We can apply the conservation of momentum to this situation. Since it is only in one dimension, the expression is simply,

$$p_i = p_f$$

$$p_{1i} + p_{2i} = p_{1+2f}$$

$$m_1 v_1 + m_2 v_2 = (m_1 + m_2) v$$

$$1200 \text{ kg} (20 \text{ m/s}) + 0 = (2000 \text{ kg}) v$$

Solve

Solving for the final velocity of the combined mass we find,

$$v_{1+2} = \frac{(1200 \text{ kg}) (20 \text{ m/s})}{2000 \text{ kg}} = 12 \text{ m/s}$$

So the correct answer is (c).

What does it mean?

Since the combined mass is less than twice the mass of the initially moving car, the final combined velocity should be more than half the initial velocity of the moving car so this answer makes sense.

2. MCAT strategies

We know that the combined velocity after the collision for this situation cannot be less than the truck's initial velocity nor greater than the initial velocity of the car, so answers (c) and (d) can be eliminated. No other answer can be easily eliminated so we proceed with our problem-solving strategy.

Recognize the principle

This is an example of a one-dimensional completely inelastic collision. Momentum is conserved in this situation.

Sketch the problem

Begin by sketching the situation before and after the collision.

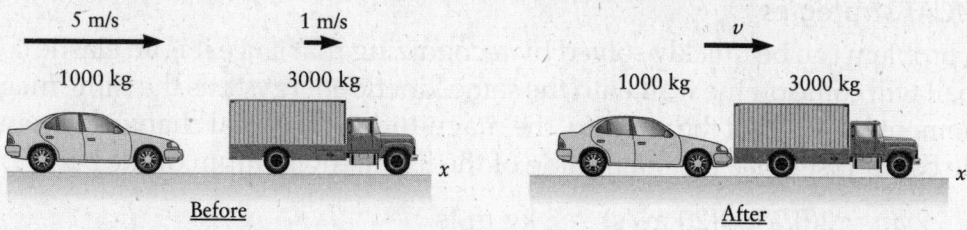

Identify the relationships

We can apply the conservation of momentum to this situation. Since it is only in one dimension, the expression is simply,

$$p_i = p_f$$
$$p_{1i} + p_{2i} = p_{1+2f}$$
$$m_1 v_1 + m_2 v_2 = (m_1 + m_2)\, v$$
$$(1000 \text{ kg})(5 \text{ m/s}) + (3000 \text{ kg})(1 \text{ m/s}) = (4000 \text{ kg})\, v$$

Solve

Solving for the final velocity of the combined mass we find,

$$v = \frac{(1000 \text{ kg})(5 \text{ m/s}) + (3000 \text{ kg})(1 \text{ m/s})}{4000 \text{ kg}} = 2 \text{ m/s}$$

So the correct answer is (a).

What does it mean?

Since the truck is much more massive than the car, the final combined velocity should be closer to the initial velocity of the truck than that of the car so this answer makes sense.

3. MCAT strategies

This problem can be quickly solve by recognizing that the momentum of the bullet must be equal in magnitude to the momentum of the rifle. This must be true since momentum is conserved in this situation and the initial momentum of the rifle plus the bullet is zero.

$$p_{\text{bullet}} = p_{\text{rifle}}$$

$$m_{\text{bullet}}\, v_{\text{bullet}} = m_{\text{rifle}}\, v_{\text{rifle}}$$

$$v_{\text{rifle}} = \frac{m_{\text{bullet}}\, v_{\text{bullet}}}{m_{\text{rifle}}} = \frac{(0.5 \times 10^{-3} \text{ kg})(100 \text{ m/s})}{0.20 \text{ kg}} = 0.25 \text{ m/s}$$

So the correct answer is (a).

4. MCAT strategies

This problem can be quickly solved by recognizing that since it is an elastic collision, the ball will come off the wall with the same kinetic energy, thus the same magnitude of momentum, as it hit the wall. So, the magnitude of the total change in momentum of the ball is just twice the magnitude of the initial momentum of the ball,

$$2mv = 2(0.2 \text{ kg})(20 \text{ m/s}) = 8 \text{ kg m/s}$$

So the correct answer is (c).

5. MCAT strategies

This is a straightforward application of the impulse theorem. Since we are given the collision time and the change in the momentum, the magnitude of the force is easily obtained,

$$\text{impulse} = \vec{F}\,\Delta t = \Delta \vec{p}$$

$$F = \frac{\Delta p}{\Delta t} = \frac{4 \text{ kg m/s}}{0.01 \text{ s}} = 400 \text{ N}$$

So the correct answer is (d).

6. MCAT strategies

This is a straightforward calculation knowing the definition of momentum and being careful to use consistent units.

$$p_1 = p_2$$

$$m_1 v_1 = m_2 v_2$$

$$v_1 = \frac{m_2 v_2}{m_1} = \frac{(0.2 \text{ kg})(200 \text{ m/s})}{(2000 \text{ kg})} = 0.02 \text{ m/s}$$

So the correct answer is (a).

7. MCAT strategies

No answers can be easily eliminated so we proceed with our problem-solving strategy.

Recognize the principle

This is an application of the impulse theorem.

Sketch the problem

None required.

Identify the relationships

We begin with the impulse theorem and substitute in for the definition of momentum.

$$F\Delta t = \Delta p = p_f - p_i$$

$$F\Delta t = mv_f - mv_i$$

$$v_f = \frac{F\Delta t}{m} + v_i$$

Solve

Substituting in the given values we find,

$$v_f = \frac{F\Delta t}{m} + v_i = \frac{(3.0 \times 10^5 \text{ N})(15 \text{ s})}{10,000 \text{ kg}} + 0 = 450 \text{ m/s}$$

So the correct answer is (b).

What does it mean?

This problem could have also been solved using the equations for motion with constant acceleration by first finding the acceleration of the rocket and then calculating its velocity after 15 s.

8. MCAT strategies

No answers can be easily eliminated so we proceed with our problem-solving strategy.

Recognize the principle

This is an application of the impulse theorem.

Sketch the problem

None required.

Identify the relationships

We begin with the impulse theorem and substitute in for the definition of momentum.

$$F\Delta t = \Delta p = p_f - p_i$$
$$F\Delta t = mv_f - mv_i$$
$$v_i = \frac{-F\Delta t}{m} + v_f$$

Solve

Substituting in the given values we find,

$$v_i = \frac{-F\Delta t}{m} + v_f = \frac{-(-2400\ \text{N})(10^{-3}\ \text{s})}{0.4 \times 10^{-3}\ \text{kg}} + 0 = 6.0 \times 10^3\ \text{m/s}$$

So the correct answer is (a).

What does it mean?

Note how the approach to solving this problem was exactly the same as the previous problem. Also, we were careful to include a negative sign with our force since the direction of the force on the gas is in the opposite direction to the initial velocity of the gas.

9. MCAT strategies

Since the carts immediately come to rest after the collision, conservation of momentum tells us that the two carts must have had the same momentum but in opposite directions prior to the collision. With this in mind we can quickly solve this problem.

$$p_1 = p_1$$
$$m_1 v_1 = m_2 v_2$$
$$v_2 = \frac{m_1\ v_1}{m_2} = \frac{(30\ \text{kg})(5\ \text{m/s})}{50\ \text{kg}} = 3\ \text{m/s}$$

So the correct answer is (a).

10. MCAT strategies

No answers can be easily eliminated so we proceed with our problem-solving strategy.

Recognize the principle

This is a two-dimensional completely inelastic collision. Momentum is conserved in both directions.

Sketch the problem

We begin by drawing a sketch of the collision.

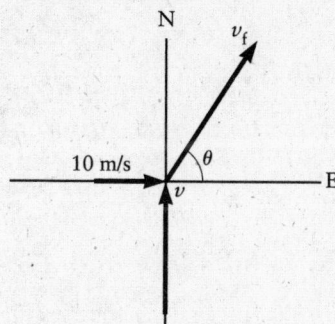

Identify the relationships

We conserve momentum in both the east and north directions. Since the mass of the two cars is the same, these expressions simplify to the following:

North direction

$$p_{iN} = p_{fN}$$

$$\cancel{m}v = 2\,\cancel{m}v_f \sin 60°$$

$$v = 2\,v_f\left(\frac{\sqrt{3}}{2}\right) = \sqrt{3}\,v_f$$

East direction

$$p_{iE} = p_{fE}$$

$$\cancel{m}(10 \text{ m/s}) = 2\,\cancel{m}v_f \cos 60°$$

$$v_f = \frac{10 \text{ m/s}}{2 \cos 60°} = \frac{10 \text{ m/s}}{2\left(\frac{1}{2}\right)} = 10 \text{ m/s}$$

Solve

Substituting v_f into the northbound expression we find,

$$v = \sqrt{3}\,v_f = 10\sqrt{3} \text{ m/s}$$

So the correct answer is (c).

What does it mean?

Since the cars are equal mass and they go off at an angle greater than 45° north of east, the speed of the northbound car must have been greater than the eastbound car. The only answer which is greater than 10 m/s is answer (c)!

8 Rotational Motion

Part A. Summary of Key Concepts and Problem-Solving Strategies

KEY CONCEPTS

Angular displacement, velocity, and acceleration

Rotational motion is described by several angular variables. *Angular displacement* θ (measured in radians) is a measure of how far an object has rotated, *angular velocity* ω (in radians per second) is a measure of how fast an object rotates, and *angular acceleration* α (in radians per second squared) is a measure of how the angular velocity is changing with respect to time. All these quantities are vectors and thus have directions. Also, they are related to the linear motion of a point on the object by the expressions

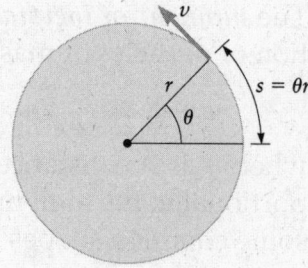

$$s = \theta r$$
$$v = \omega r$$
$$a = \alpha r$$

Torque

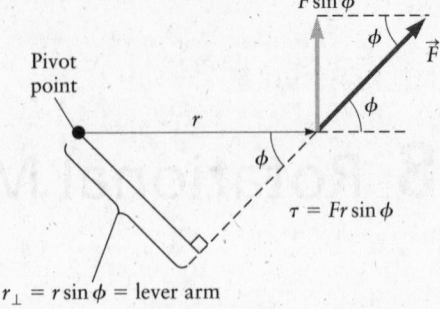

Torque plays the role of force in cases of rotational motion. In other words, to give an object angular acceleration you must apply a torque just as to get an object to accelerate you must apply a force. The torque produced by an applied force is given by

$$\tau = Fr \sin \phi$$

where F is the applied force, r is the distance measured from the pivot point, and ϕ is the angle between r and F.

This relation for the torque can be thought of in two equivalent ways. One way is

$$\tau = F_\perp r$$

where $F_\perp = F \sin \phi$ is the component of the force perpendicular to the line defined by r. The torque can also be written as

$$\tau = Fr_\perp$$

where $r_\perp = r \sin \phi$ is called the **lever arm**; it is the distance from the rotation axis to the line of action of the force, measured along a line that is perpendicular to both. You can apply either approach depending on the information provided.

Newton's second law for rotational motion

Analogous to Newton's second law for translational motion, $\Sigma \vec{F} = m\vec{a}$, for rotational motion the expression is

$$\Sigma \tau = I\alpha$$

The **moment of inertia** I of an object depends on its mass and shape. For a collection of particles of mass m_i, the moment of inertia is

$$I = \sum_i m_i r_i^2$$

where r_i is the distance from the rotation axis to m_i. The moment of inertia is proportional to the total mass and depends on the size (i.e., the radius). Values of I for some common shapes are listed in Table 8.2 in your textbook.

APPLICATIONS

Rotational equilibrium

An object is in translational equilibrium and rotational equilibrium if its linear acceleration and angular acceleration are both zero. For that to be true, both the total force and the total torque must be zero. Note that this does not mean that the object is stationary, only that its acceleration and angular acceleration are both zero.

Rotational motion

When analyzing problems which involve rotational motion, the process usually begins by applying Newton's second law for rotational motion,

$$\Sigma \tau = I\alpha$$

When the angular acceleration is constant, the solutions for the angular displacement and velocity as functions of time are analogous to the expressions for translational motion with constant acceleration:

$$\theta = \theta_0 + \omega_0 t + \frac{1}{2}\alpha t^2$$
$$\omega = \omega_0 + \alpha t$$
$$\omega^2 = \omega_0^2 + 2\alpha\,(\theta - \theta_0)$$

Rolling motion

Rolling motion is an example of combined translational and rotational motion. If an object rolls without slipping, the angular velocity and angular acceleration are related to the linear velocity and acceleration of the center of mass by

$$v = \omega R$$
$$a = \alpha R$$

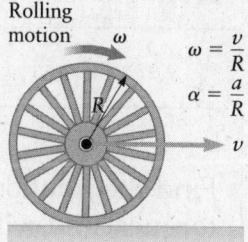

$$\omega = \frac{v}{R}$$
$$\alpha = \frac{a}{R}$$

PROBLEM-SOLVING STRATEGIES

In this chapter we've expanded our understanding of equilibrium to include rotational equilibrium. Also, we derived and applied Newton's second law for rotational motion. The following schematic outlines the thought process we should follow when applying these two main concepts.

Problem Solving: Applying the Conditions for Translational and Rotational Equilibrium

Recognize the principle

An object is in translational static equilibrium if its linear acceleration and linear velocity are both zero. An object is in rotational static equilibrium if its angular acceleration and angular velocity are both zero.

Sketch the problem

Make a drawing showing the object of interest, along with all the forces that act on it. This sketch should contain a set of coordinate axes.

Identify the relationships

Find the rotation axis and pivot point. These are necessary for calculating the torques, and will depend on the problem. The torque calculations can be broken into several steps:

- To calculate the torque from each force, first determine the lever arm, as the distance from the pivot point to point where the force acts.
- Calculate the magnitude of the torque using $\tau = Fr\sin\phi$. Here F is the magnitude of the force.
- Determine the sign of τ. If this force *acting alone* would produce a *counterclockwise* rotation, then the torque is *positive*. If the force would cause a *clockwise* rotation, the torque is *negative*.

Add the torques from each force to get the total torque. Be sure to include the *proper sign* for each of the individual torques.

Solve

Solve for the unknowns by applying the condition for rotational equilibrium $\Sigma\tau = 0$, and (if necessary) the condition for translational equilibrium $\Sigma\vec{F} = 0$.

What does it mean?

Always *consider what your answer means* and check that it makes sense.

Problem Solving: Applying Newton's Second Law for Rotational Motion

---

Recognize the principle

According to Newton's second law for rotational motion ($\Sigma\tau = I\alpha$), torque causes an angular acceleration. To find α we must therefore find the total torque ($\Sigma\tau$) and the moment of inertial I.

Sketch the problem

Make a drawing showing all of the objects of interest, along with the forces that act on them. This sketch should contain a set of coordinate axes (x and y) for translational motion.

Identify the relationships

Determine the rotation axis and pivot point for calculating the torques for objects that may rotate. These will depend on the problem. Then

- Find the total torque on the objects that are undergoing rotational motion. Calculate the torque using the method we employed in our work on rotational statics. This torque will then be used in Newton's second law for rotational motion, $\Sigma\tau = I\alpha$.

- Calculate the sum of the forces on the objects that are undergoing linear motion, for use in Newton's second law, $\Sigma\vec{F} = m\vec{a}$.

Check for a relation between the linear acceleration and the angular acceleration. There will usually be a connection between a and α through a relation such as $a = r\alpha$.

Solve

Solve for the quantities of interest (which might be a or α, or both) using Newton's second law for linear motion ($\Sigma\vec{F} = m\vec{a}$) and rotational motion ($\Sigma\tau = I\alpha$).

What does it mean?

Always *consider what your answer means* and check that it makes sense.

Part B. Frequently Asked Questions

1. *What's the difference between angular velocity and angular acceleration?*

Recall from our earlier study of translational motion, that velocity and acceleration were carefully defined. Velocity is the rate of change of an object's position with respect to time. In other words, it is a measure of how fast an object is moving and in what direction. Acceleration is the rate of change of an object's velocity with respect to time. For one dimension we have the following expressions:

$$\text{Velocity: } v = \frac{\Delta x}{\Delta t} \text{ [m/s]}$$

$$\text{Acceleration: } a = \frac{\Delta v}{\Delta t} \text{ [m/s}^2\text{]}$$

In a similar way we define the angular velocity and angular acceleration of a rotating object. Angular velocity is the rate of change of an objects angular position with respect to time. In other words, it is a measure of how fast an object is rotating and in what direction. Angular acceleration is the rate of change of an object's angular velocity with respect to time.

$$\text{Angular velocity: } \omega = \frac{\Delta \theta}{\Delta t} \text{ [rad/s]}$$

$$\text{Angular acceleration: } \alpha = \frac{\Delta \omega}{\Delta t} \text{ [rad/s}^2\text{]}$$

We can use these angular expressions to describe rotational motion in the same way as we studied translational motion.

2. *As I apply torque to an object it rotates faster and faster. Doesn't this imply that torque and angular velocity are directly related?*

Although this seems like a reasonable conclusion, it is wrong. If angular velocity and torque were directly related then as the torque on an object becomes zero, then the angular velocity should as well. We know from experience that objects will continue to rotate even after we stop pushing on them so torque and angular velocity are not proportional. Newton's second law for rotational motion describes the correct relationship between torque and rotational motion by directly relating torque and angular acceleration. If a constant torque is applied to an object, then it will angularly accelerate at a constant rate resulting in the angular velocity of the object getting faster and faster as was stated in your question.

3. *When calculating torque I have trouble knowing which angle to use. Is there an easy way to remember?*

As was discussed in the textbook, there are two ways we can think about torque. One involves using the component of the applied force which is perpendicular to the distance *r* from the pivot to the point where the force acts on the object. The other way uses the component of the distance *r* which is perpendicular to the applied force. Either way we need to use the correct angle when determining components. The following figure will help illustrate:

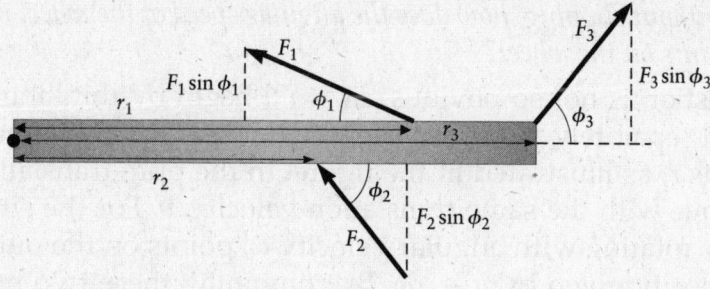

We can determine the torque due to each of these forces by using the perpendicular component of the force.

$$\tau_1 = r_1 F_{1,\perp} = r_1 F_1 \sin \phi_1$$
$$\tau_2 = r_2 F_{2,\perp} = r_2 F_2 \sin \phi_2$$
$$\tau_3 = r_3 F_{3,\perp} = r_3 F_3 \sin \phi_3$$

4. *What's the difference between velocity and angular velocity? How are they related?*

Velocity and angular velocity are two separate yet related concepts. Consider a wheel rotating about its axle as illustrated. Suppose the wheel is rotating at an angular velocity ω, as shown. Now suppose you are a bug on the wheel. No matter where you are on the wheel you rotate with an angular velocity of ω. Your velocity, however, depends on your radius from the axle. The further you are from the axle, the greater your velocity. This suggests a relationship between angular velocity, velocity, and radius. The expression is given by $v = r\omega$.

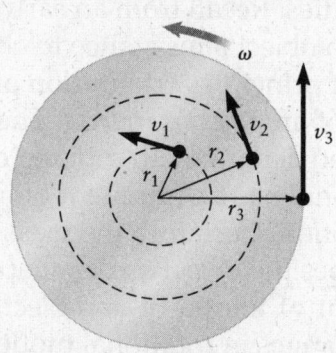

5. *When setting up a problem involving torques, where should I choose my axis of rotation?*

That's a good question. Often the choice is obvious, if a hinge or pivot point is given in the problem. For situations that do not have an obvious axis of rotation the choice is yours. In the end, it will not affect the outcome of the problem; however, a wise choice can lead to simpler mathematics. For example, by choosing the axis of rotation at a point where several forces are applied, you can eliminate the torque due to these forces since the distance from the pivot to the point they're applied is zero. This greatly simplifies the calculation. On the other hand, if you are trying to determine an unknown force then it is probably best not to choose your axis of rotation at the point where that force is applied.

6. *When a wheel rolls without slipping, how does the angular speed of the wheel relate to the linear speed of points on the wheel?*

The answer to this question is not so obvious, so we'll begin by thinking about rotation and translation separately. Consider separately the translation and rotation of a wheel of radius r, as illustrated in the figure. In the pure translation, all points on the wheel move with the same translation velocity, v. For the situation where the wheel is only rotating with angular velocity ω, points on the outer rim of the wheel have a velocity given by $v = r\omega$. By combining these two motions (a rolling wheel without slipping) we can simply add the velocities from the two separate motions to determine the combined motion. As we see from the figure, the point in contact with the ground has (momentarily) zero velocity, the point at the center of the wheel has a velocity of v, and the point at the top of the wheel is moving with a velocity of $2v$.

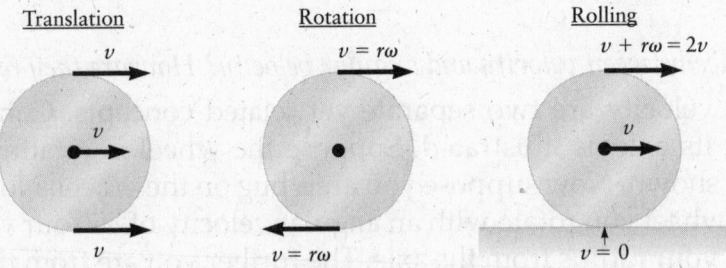

Translation	Rotation	Rolling
v	$v = r\omega$	$v + r\omega = 2v$
v		v
v	$v = r\omega$	$v = 0$

7. *What's the difference between mass and moment of inertia? Are they the same?*

Mass and moment of inertia are not the same quantities. Recall from an early chapter how inertia and mass were defined. Inertia is an object's resistance to changes in motion. Although it's not possible to give a "first principles" definition of mass we can think of mass as the quantitative measure of an object's inertia. The more mass an object has, then the more inertia it has. When dealing with rotating objects we can define an analogous term to inertia. Rotational inertia (moment of inertia) is an object's resistance to changes in rotational motion. Certainly the mass of the object plays a role in its moment of inertia but so does the object's size, shape, and the location of the axis of rotation. So, the moment of inertia of an object is the quantitative measure of the object's resistance to changes in rotational motion and depends on the mass of the object and how that mass is distributed about the axis of rotation. For a given mass, the more spread out the mass is from the axis of rotation, the greater the moment of inertia. Conversely, the more compact the mass is about the axis of rotation, then the less the moment of inertia.

8. *Since the moment of inertia of an object depends on the location of the axis of rotation, is there a minimum value for the moment of inertia?*

Yes. As it turns out, for a given object the minimum moment of inertia is when the axis of rotation goes through the center of mass of the object. For example, consider a rod pivoted at one end versus the same rod pivoted through its center of mass. The moment of inertia for these two configurations is quite different with the smaller one being through the center of mass.

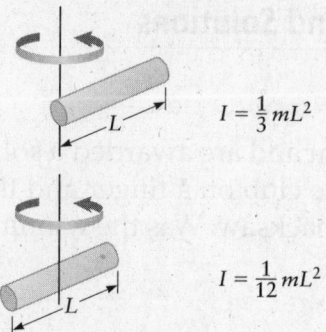

$$I = \tfrac{1}{3}mL^2$$

$$I = \tfrac{1}{12}mL^2$$

9.   *When on a merry-go-round I've noticed that the further I am from the center, the more difficult it is to hang on. Why is that?*

As we've learned, your linear velocity depends on how far you are from the axis of rotation. The further away you are from the center, the greater the linear velocity according to the expression $v = r\omega$. Recall from our previous discussion that centripetal acceleration depends on the tangential speed according to the expression $a_c = v^2/r$. Combining these two expressions we can determine how the centripetal acceleration depends on r.

$$a_c = \frac{v^2}{r} = \frac{(r\omega)^2}{r} = r\omega^2$$

So, the centripetal acceleration depends directly on the distance you are from the center of the merry-go-round. Since you need to supply the force required to cause the centripetal acceleration (by holding on!), it is more difficult to hold on the further you are from the center.

10.   *Why is it easier to balance a golf club at the handle end rather than at the club end?*

This has to do with the moment of inertia being different for these two situations and the application of Newton's second law for rotational motion. As illustrated in the figure, the moment of inertia is greater for the situation where the club head is away from your hand. This is because there is more mass distributed further from the pivot.

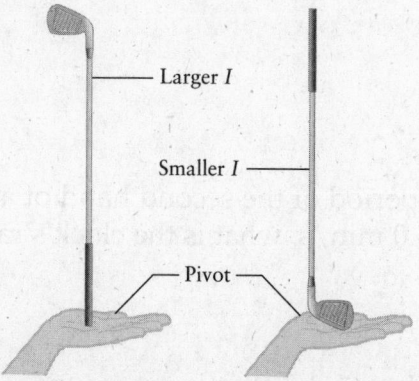

Larger *I*

Smaller *I*

Pivot

According to Newton's second law for rotational motion, for a given torque, the object with the larger moment of inertia will have the smaller angular acceleration. With the club head away from your hand the club will have a smaller angular acceleration as it starts to fall over, thus giving you more time to react and keep it balanced.

Part C. Selection of End-of-Chapter Answers and Solutions

QUESTIONS

Q8.7 Two golfers team up to win a golf tournament and are awarded a solid gold golf club. To divide their winnings, they balance the club on a finger and then cut the club into two pieces at the balance point with a hacksaw. Was the winning gold split evenly? Why or why not?

Answer

No. The person who got the club head got much more gold. The balance point is at the center of mass, which is closer to the head of the club. The handle is lighter and longer than the head, and the mass further from the balance point creates more torque than the mass close to the balance point. The smaller distributed mass of the handle balances the more massive club head which is closer to the balance point.

Q8.15 A tree has two large branches that grow out horizontally from the trunk in opposite directions. One branch has a length L and diameter d, whereas the other has a length $L/2$. If the magnitude of the torque due to gravity is the same on the two branches, what is the diameter of the shorter branch?

Answer

Since torque is proportional to the product of the force (weight of the branch in this case) and the distance from the pivot, the shorter branch will need to have twice the mass. Assuming the branches have the same density, the mass will be proportional to the volume. Treating the branches as cylinders, the volume of the longer branch is $V_1 = LA_1 = L\left(\frac{\pi d^2}{4}\right)$. The volume of the shorter branch is $V_2 = \frac{L}{2}A_2 = \frac{L}{2}\left(\frac{\pi d_2^2}{4}\right)$.

So, since the shorter branch must have twice the mass (and therefore twice the volume), we can set up the following ratio,

$$2 = \frac{V_2}{V_1} = \frac{\frac{L}{2}\left(\frac{\pi d_2^2}{4}\right)}{L\left(\frac{\pi d^2}{4}\right)} \rightarrow 4 = \frac{d_2^2}{d^2} \rightarrow \boxed{d_2 = 2d}$$

So the shorter branch has a diameter of $2d$.

PROBLEMS

P8.13 What is the angular velocity and the period of the second hand of a clock? If the linear speed of the end of the hand is 5.0 mm/s, what is the clock's radius?

Solution

Recognize the principle

Apply the equations for angular velocity and period.

Sketch the problem

No sketch needed.

Identify the relationships

The period of the second hand of a clock is $T = \boxed{60 \text{ s}}$. Using this and Equation 8.6, we can find the angular velocity of the second hand.

$$\omega = \frac{2\pi}{T}$$

Then, given the linear speed and angular velocity of the second hand, we can use Equation 8.7 to find the radius of the clock.

$$v = \omega r$$

Solve

Inserting the period,

$$\omega = \frac{2\pi}{60 \text{ s}}$$

$$\boxed{\omega = 1.0 \times 10^{-1} \text{ rad/s}}$$

Solving for r in Equation 8.7 and inserting the values for the velocity and angular velocity,

$$r = \frac{v}{\omega}$$

$$r = \frac{0.005 \text{ m/s}}{1.0 \times 10^{-1} \text{ rad/s}}$$

$$r = 0.05 \text{ m} = \boxed{5.0 \text{ cm}}$$

What does it mean?

The period is not explicitly given in this problem; however, the definition of a second hand on a clock gives the period. What other well established periods can you think of? Consider: a day, year, phases of the moon, etc.

P8.18 Consider the clock in Figure 8.17. Calculate the magnitude and sign of the torque due to gravity on the hour hand of the clock at 4 o'clock. Assume the hand has a mass of 15 kg, a length of 1.5 m, and the mass is uniformly distributed.

Solution

Recognize the principle

We will need to apply our expression for torque. That is, the torque on an object is equal to the force times the perpendicular lever arm.

Sketch the problem

Using Figure 8.17 as a guide, we have the following sketch:

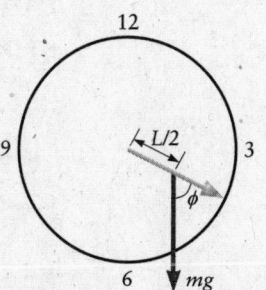

Identify the relationships

Since the force of gravity on the hand and the lever arm are not perpendicular, the magnitude of the torque for the hour hand is given by:

$$\tau = F_{applied}\, r \sin\phi$$

The force acts clockwise around the pivot point, so the torque is negative. The force will be the weight of the hour hand ($F = mg$). The length r is measured from the pivot to the center of mass of the hour hand. Therefore, r is the length of the hour hand divided by 2. If L is the length of the hour hand,

$$\tau = -mg\,\frac{L}{2}\sin\phi$$

At the 4 o'clock position, the hour hand will be 1/3 of the way between the 3 o'clock and the 6 o'clock positions. This will make the angle ϕ between the hand and the downward gravitational force equal to 60°.

Solve

Inserting values to the above equation,

$$\tau = -(15\ \text{kg})(9.8\ \text{m/s}^2)\,\frac{(1.5\ \text{m})}{2}\sin 60°$$

$$\boxed{\tau = -95\ \text{N}\cdot\text{m}}$$

What does it mean?

The magnitude of the gravitational torque on the hour hand is at a maximum at 3 o'clock and 9 o'clock, and zero at 6 o'clock and noon.

P8.23   A tree grows at an angle of 50° to the ground as shown in Figure Q8.5. If the tree is 25 m from its base to its top and has a mass of 500 kg, what is the approximate magnitude of the torque on the tree due to the force of gravity? Take the base of the tree as the pivot point. (The answer is one reason trees need roots.)

50°

Figure Q8.5

Solution

Recognize the principle

Apply the concepts of torque.

Sketch the problem

The sketch is provided in the problem.

Identify the relationships

The magnitude of the torque is given by the expression $\tau = rF \sin\phi$. The force will be the weight of the tree giving an expression for the torque,

$$\tau = mgr \sin\phi$$

We can approximate the tree as a uniform log of length $L = 25$ m. This allows us to estimate the center of mass of the tree to be at $r = \frac{1}{2}L$ from the base of the tree.

Solve

Substituting in for r in our expression for torque we find,

$$\tau = mg\frac{L}{2}\sin\phi$$

Inserting values, with the angle ϕ equal to 50°,

$$\tau \approx (500 \text{ kg})(9.8 \text{ m/s}^2)\frac{(25 \text{ m})}{2}\sin 50°$$

$$\boxed{\tau \approx 50,000 \text{ N} \cdot \text{m}}$$

What does it mean?

The trunk of the tree must be sturdy to supply the needed equal in magnitude and opposite in direction torque to keep the tree in equilibrium. It does this by using roots.

P8.40 A flagpole of length 12 m and mass 30 kg is hinged at one end, where it is connected to a wall as sketched in Figure P8.40. The pole is held up by a cable attached to the other end. Find the tension in the cable.

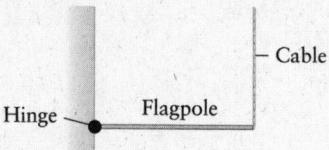

Figure P8.40

Solution

Recognize the principle

The flagpole is in translational and rotational equilibrium. Therefore, the sum of forces along any axis and the sum of torques around any axis of rotation must be zero.

Sketch the problem

We begin by drawing a sketch of the situation, being sure to define a coordinate system, labeling all known and unknown forces and where those forces act upon the flagpole, and choosing a pivot point to evaluate the torques.

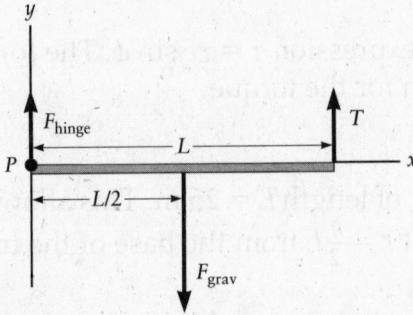

Identify the relationships

We will choose the pivot point to be at location P, with the rotation axis perpendicular to the plane of the picture. Since the flagpole is in static equilibrium, the sum of the forces in both the x and y directions must equal zero and the sum of the torques about the pivot point must equal zero. Therefore, we can write the following expressions:

$$\Sigma F_x = 0$$
$$\Sigma F_y = 0 \rightarrow F_{hinge} + T - mg = 0$$
$$\Sigma \tau = \tau_{flagpole} + \tau_{cable} = 0 \rightarrow -mgL/2 + TL = 0$$

Solve

Since neither the tension nor the vertical force of the hinge on the flagpole is known, we cannot solve the expression we obtained from our analysis of the y-direction. However, we can solve the torque expression as follows,

$$-mgL/2 + TL = 0 \rightarrow T = \frac{mg}{2}$$

Inserting the values, we find,

$$T = \frac{(30 \text{ kg})(9.8 \text{ m/s}^2)}{2} = \boxed{150 \text{ N}}$$

What does it mean?

The cable must provide about 150 N of force to support the flagpole.

P8.42   Four particles with masses $m_1 = 15$ kg, $m_2 = 25$ kg, $m_3 = 10$ kg, and $m_4 = 20$ kg sit on a very light (massless) metal sheet and are arranged as shown in Figure P8.42. Find the moment of inertia of this system with the pivot point (a) at the origin and (b) at point P. Assume the rotation axis is parallel to the z direction, perpendicular to the plane of the drawing.

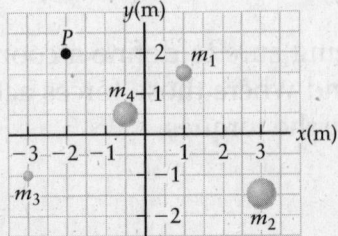

Figure P8.42

Solution

Recognize the principle

The moment of inertia depends on the mass and the square of the distance from the mass to the axis of rotation.

Sketch the problem

See Figure P8.42.

Identify the relationships

The moment of inertia for a point particle is given by,

$$I = \sum_i m_i r_i^2$$

(a) Measuring our distances to the particles from the origin we have $r = \sqrt{x^2 + y^2}$. Therefore, the moment of inertia will be

$$I = m_1\left(\sqrt{x_1^2 + y_1^2}\right)^2 + m_2\left(\sqrt{x_2^2 + y_2^2}\right)^2 + m_3\left(\sqrt{x_3^2 + y_3^2}\right)^2 + m_4\left(\sqrt{x_4^2 + y_4^2}\right)^2$$

$$I = m_1\left(x_1^2 + y_1^2\right) + m_2\left(x_2^2 + y_2^2\right) + m_3\left(x_3^2 + y_3^2\right) + m_4\left(x_4^2 + y_4^2\right)$$

(b) The coordinates of point P are $x = -2$ m, and $y = 2.0$ m. So, we can determine our distances to the particles from point P by subtracting point P's coordinates from each coordinate found in part (a).

Solve

(a) Inserting values,

$$I = (15 \text{ kg})\left[(1 \text{ m})^2 + (1.5 \text{ m})^2\right] + (25 \text{ kg})\left[(3 \text{ m})^2 + (-1.5 \text{ m})^2\right]$$
$$+ (10 \text{ kg})\left[(-3 \text{ m})^2 + (-1 \text{ m})^2\right]$$
$$+ (20 \text{ kg})\left[(-0.5 \text{ m})^2 + (0.5 \text{ m})^2\right]$$

$$\boxed{I = 440 \text{ kg} \cdot \text{m}^2}$$

(b) Subtracting the appropriate coordinate for each value and inserting them into our moment of inertia equation, we have:

$$I = (15 \text{ kg})\left[(3 \text{ m})^2 + (-0.5 \text{ m})^2\right] + (25 \text{ kg})\left[(5 \text{ m})^2 + (-3.5 \text{ m})^2\right]$$
$$+ (10 \text{ kg})\left[(-1 \text{ m})^2 + (-3.0 \text{ m})^2\right]$$
$$+ (20 \text{ kg})\left[(1.5 \text{ m})^2 + (-1.5 \text{ m})^2\right]$$

$$I = 1260 \text{ kg} \cdot \text{m}^2 \approx \boxed{1300 \text{ kg} \cdot \text{m}^2}$$

What does it mean?

The moment of inertia is very different around different points of rotation. The center of rotation must always be specified when presenting a moment of inertia.

P8.52   Figure P8.52 shows the angular displacement of an object as a function of time. (a) What is the approximate angular velocity of the object at $t = 0$? (b) What is the approximate angular velocity at $t = 0.10$ s? (c) Estimate the angular acceleration at $t = 0.050$ s.

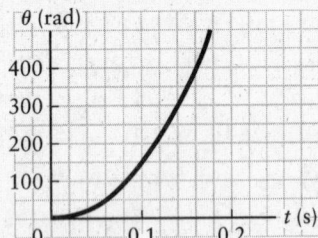

Figure P8.52

Solution

Recognize the principle

The angular velocity can be found from the slope of the angular displacement versus time graph.

Sketch the problem

No sketch needed.

Identify the relationships

(a) and (b) In both cases we can approximate the slope of the curve by drawing a tangent to the curve at the appropriate point and find the slope of this tangent line.

(c) The angular acceleration is defined as:

$$\alpha = \frac{\Delta \omega}{\Delta t}$$

Since we have estimated the angular speed at $t = 0$ and at $t = 0.10$ s, the angular acceleration at $t = 0.05$ s can be found as:

$$\alpha = \frac{\omega_{t=0.1\,s} - \omega_{t=0\,s}}{t_2 - t_1}$$

Solve

(a) From the diagram, the tangent line passing through $t = 0$ is approximately horizontal. Therefore, the approximate angular velocity at $t = 0$ is $\boxed{0 \text{ rad/s}}$.

(b) At $t = 0.10$ s, the tangential line goes through points (125 rad, 0.10 s) and (0 rad, 0.050 s). Therefore,

$$\omega = \frac{\theta_2 - \theta_1}{t_2 - t_1} \approx \frac{125 \text{ rad} - 0 \text{ rad}}{0.10 \text{ s} - 0.05 \text{ s}} = \boxed{2500 \text{ rad/s}}$$

(c) Inserting the values from parts (a) and (b):

$$\alpha \approx \frac{2500 \text{ rad/s} - 0 \text{ rad/s}}{0.10 \text{ s} - 0 \text{ s}} = \boxed{25,000 \text{ rad/s}^2}$$

What does it mean?

The angular acceleration is a very rough estimate as it depends on the accuracy of our angular velocity estimates.

P8.58 Two crates of mass $m_1 = 15$ kg and $m_2 = 25$ kg are connected by a cable that is strung over a pulley of mass $m_{pulley} = 20$ kg as shown in Figure P8.58. There is no friction between crate 1 and the table. (a) Make a sketch showing all the forces on both crates and the pulley. (b) Express Newton's second law for the crates (translational motion) and for the pulley (rotational motion). The linear acceleration a of the crates, the angular acceleration α of the pulley, and the tensions in the right and left portions of the rope are unknowns. (c) What is the relation between a and α? (d) Find the acceleration of the crates. (e) Find the tensions in the right and left portions of the rope.

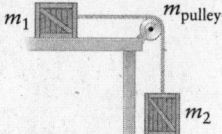

Figure P8.58

Solution

Recognize the principle

The two blocks, which are connected by a rope, must accelerate at the same rate. Knowing this, we can then use kinematics equations for both rotation and translation to find the needed quantities.

Sketch the problem

(a) The following diagram illustrates the forces on crate 1, crate 2, and the pulley. Though there are additional forces holding the pulley in place, we have omitted them from our sketch since they do not produce a torque on the pulley and thus do not play a role in the solution of this problem.

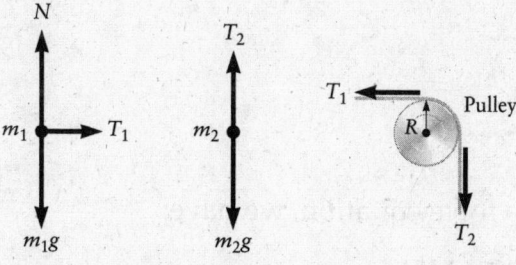

Identify the relationships

(b) For crate 1 we can apply Newton's second law in the horizontal direction and obtain the following expression,

$$\Sigma F_{m_1} = \boxed{T_1 = m_1 a}$$

Applying Newton's second law to m_2 yields,

$$\Sigma F_{m_2} = \boxed{-T_2 + m_2 g = m_2 a}$$

Applying Newton's second law for rotational motion to the pulley we get,

$$\Sigma \tau = \boxed{(T_2 - T_1)R = I_{\text{pulley}}\alpha}$$

Note that we have defined a "clockwise" torque as positive so that the sign for our angular acceleration α agrees with our sign for the acceleration a.

The moment of inertia of a pulley is given in Table 8.2 of your textbook as,

$$I_{\text{pulley}} = \frac{1}{2}m_{\text{pulley}}R^2$$

(c) The connection between the acceleration of both blocks and the angular acceleration of the pulley is given by the expression:

$$\boxed{a = \alpha R}$$

Solve

(d) Adding the two force equations, we have:

$$T_1 - T_2 + m_2 g = m_1 a + m_2 a \rightarrow T_2 - T_1 = m_2 g - (m_1 + m_2)a$$

We can then substitute this expression for the $T_2 - T_1$ in the torque equation, and using the connection between a and α to eliminate the angular acceleration, we have:

$$[m_2 g - (m_1 + m_2)a]R = \frac{1}{2}m_{\text{pulley}}R^2 \frac{a}{R}$$

Solving for the acceleration,

$$m_2 g - (m_1 + m_2)a = \frac{1}{2}m_{\text{pulley}}a$$

$$\frac{1}{2}m_{\text{pulley}}a + (m_1 + m_2)a = m_2 g$$

$$a = \frac{m_2 g}{\frac{1}{2}m_{\text{pulley}} + (m_1 + m_2)}$$

Inserting values,

$$a = \frac{(25\ \text{kg})(9.8\ \text{m/s}^2)}{\frac{1}{2}(20\ \text{kg}) + (15\ \text{kg} + 25\ \text{kg})}$$

$$\boxed{a = 4.9\ \text{m/s}^2}$$

Substituting this value back in to each force equation, we have:

$$T_1 = m_1 a = (15\ \text{kg})(4.9\ \text{m/s}^2) = \boxed{74\ \text{N}}$$

$$T_2 = m_2(g - a) = 25\ \text{kg}(9.8\ \text{m/s}^2 - 4.9\ \text{m/s}^2) = \boxed{123\ \text{N}}$$

What does it mean?

Notice that T_2 is larger than T_1 which is expected since the pulley rotates clockwise.

P8.67 Consider a tennis ball that is hit by a player at the baseline with a horizontal velocity of 45 m/s (about 100 mi/h). The ball travels as a projectile to the player's opponent on the opposite baseline, 24 m away, and makes 25 complete revolutions during this time. What is the ball's angular speed?

Solution

Recognize the principle

We'll need to use our understanding of projectile motion (motion with constant acceleration) and rotational motion with constant angular acceleration.

Sketch the problem

No sketch needed.

Identify the relationships

Since we are given the number of rotations completed (angular displacement) we can find the ball's constant angular velocity if we know how long the ball is in the air,

$$\Delta \theta = \omega t$$
$$\omega = \frac{\Delta \theta}{t}$$

To determine how long the ball is in the air, we can apply our understanding of projectile motion. Neglecting air resistance, the ball's horizontal velocity will remain constant. Knowing the distance the ball travels horizontally we can determine the time the ball is in the air using the following expression,

$$\Delta x = v_x t$$
$$t = \frac{\Delta x}{v_x}$$

Solve

Plugging in the values we find,

$$t = \frac{\Delta x}{v_x} = \frac{24 \text{ m}}{45 \text{ m/s}} = 0.53 \text{ s}$$

The angular speed is then,

$$\omega = \frac{\Delta \theta}{t} = \frac{(25 \text{ rev})(2\pi \text{ rad/rev})}{0.53 \text{ s}} = \boxed{300 \text{ rad/s}}$$

What does it mean?

The ball is spinning at a rate of more than 47 rev/s or 2800 rpm.

P8.76 Three gears of a mechanism are meshed as shown in Figure P8.76 and are rotating with constant angular velocities. The ratio of the diameters of gears 1 through 3 is 3.5/1.0/2.0. A torque of 20 N · m is applied to gear 3. If gear 2 has a radius of 10 cm, what is the torque on gears 1 and 2?

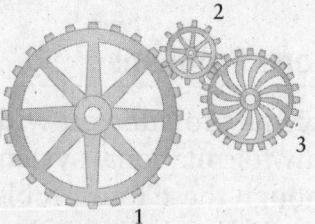

Figure P8.76

Solution

Recognize the principle

From Newton's third law we know that the forces between gears are equal and opposite. Also, we'll need to apply our expression for torque.

Sketch the problem

Refer to Figure P8.76.

Identify the relationships

The force exerted on each gear is perpendicular to the radius at the point of contact, so the angle between the force and the lever arm is 90°. This simplifies our expression for torque to:

$$\tau = Fr$$

From the given ratios we know that $r_1 = 0.35$ m, $r_2 = 0.10$ m, and $r_3 = 0.20$ m.

Solve

Since we know the torque on gear 3 we can calculate the force exerted on the teeth of gear 3, here assumed to be 0.20 m from its rotation axis.

$$\tau_3 = F_3 r_3 \rightarrow F_3 = \frac{\tau_3}{r_3} = \frac{20 \text{ N} \cdot \text{m}}{0.20 \text{ m}} = 100 \text{ N}$$

By Newton's third law the force on the teeth of gear 2 must be equal and opposite to that of gear 3. So the magnitude of the force on gear 2 is:

$$F_3 = F_2 = 100 \text{ N}$$

The torque on gear 2 is then:

$$\tau_2 = F_2 r_2 = (100 \text{ N})(0.10 \text{ m}) = \boxed{10 \text{ N} \cdot \text{m}}$$

By Newton's third law, the force on the teeth of gear 1 must be equal and opposite to that of gear 2. So the magnitude of the force on gear 1 is:

$$F_1 = F_2 = 100 \text{ N}$$

Since the same force is applied to gear 1 we can find the torque on it as well:

$$\tau_1 = F_1 r_1 = (100 \text{ N})(0.35 \text{ m}) = \boxed{35 \text{ N} \cdot \text{m}}$$

What does it mean?

The torques and forces on systems of gears can be found using Newton's third law and our application of our expression for torque.

Part D: Additional Worked Examples and Capstone Problems

The following worked examples and capstone problems provide you with practice calculating torque, applying Newton's second law for rotational systems, and determining the moment of inertia of an object. Although these four problems do not incorporate all the material discussed in this chapter, they do highlight several of the key concepts. If you can successfully solve these problems then you should have confidence in your understanding of these key concepts, so use these problems as a test of your understanding of the chapter material.

WE 8.1 Forces and Torques on a Flagpole in Equilibrium

Consider the problem illustrated in the figure in which a horizontal flagpole of mass $m = 10$ kg and length $L = 2$ m is supported by a cable at one end and by a hinge connected to a wall on the other. Determine the tension in the cable and the total force exerted by the hinge on the flagpole.

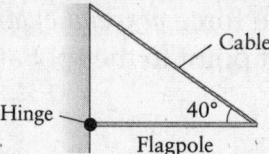

Solution

Recognize the principle

Since the flagpole is neither translating nor rotating, this is a problem where we can apply the conditions for static equilibrium.

Sketch the problem

We begin by first making a sketch showing all the forces on the flagpole along with the location at which each force acts. As shown, there are forces on the flagpole exerted by the cable, by gravity, and by the hinge. Because we don't yet know the direction of the force exerted by the hinge, we have shown this force as two separate components, one along the vertical F_V and one along the horizontal F_H. We have also written the force $\vec{T}$ exerted by the cable in terms of its components along the vertical ($T \sin 40°$) and the horizontal ($-T \cos 40°$) as shown in the figure. A convenient choice for the pivot point is at the hinge, point P. The figure also shows where the various forces act.

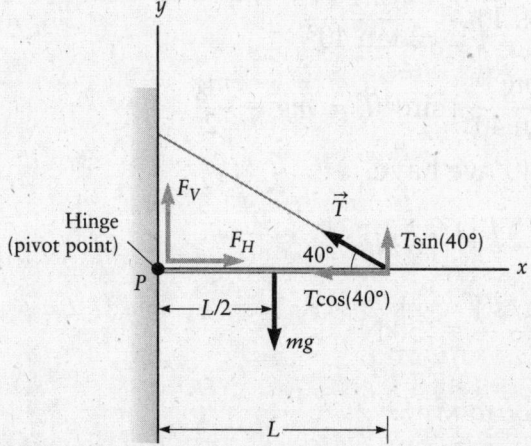

Identify the relationships

The next step is to calculate the torques from each of these forces and express the condition for rotational equilibrium $\Sigma\tau = 0$. The forces exerted by the hinge do not produce any torque because their lever arms are zero. Only the force of gravity on the pole and the force exerted by the cable on the pole produce nonzero torques. Using the forces and lever arms from the figure gives

$$\Sigma \tau = 0 = -mg\left(\frac{L}{2}\right) + T(\sin 40°)(L)$$

The first term on the right is the torque from the force of gravity on the flagpole. This torque is negative because, if acting by itself, it would make the pole rotate clockwise (the negative direction). The tension force would give a counterclockwise rotation, so its torque is positive. Also, when calculating the torque due to the tension in the cable, we must remember that this force is not directed perpendicular to the pole. This torque equals the component of the force *perpendicular* to the pole ($T \sin 40°$) multiplied by the distance from the pivot point to the spot at which this force acts on the pole.

We next apply the condition for translational equilibrium $\Sigma \vec{F} = 0$. This condition includes the forces along both the x and y axes. Along x, we have

$$\Sigma F_x = 0 = F_H - T \cos 40°$$

and along y, we have

$$\Sigma F_y = 0 = F_V + T \sin 40° - mg$$

Solve

We wish to solve for three unknowns: T, F_H, and F_V. We can get T from the torque expression; solving for T, we find

$$mg\left(\frac{L}{2}\right) = T(\sin 40°)(L)$$

$$T = \frac{mg}{2 \sin 40°}$$

We can now use this expression to find the force exerted by the hinge on the pole. We have two equations and two unknowns, F_H and F_V, so we can solve for both.

$$F_H = T \cos 40° = \left(\frac{mg}{2 \sin 40°}\right) \cos 40° = \frac{mg}{2 \tan 40°}$$

$$F_V = -T \sin 40° + mg = -\left(\frac{mg}{2 \sin 40°}\right) \sin 40° + mg = \frac{mg}{2}$$

Plugging in the values for m, g, and 40° we have,

$$T = \frac{mg}{2 \sin 40°} = \frac{(10 \text{ kg})(9.8 \text{ m/s}^2)}{2 \sin 40°} = \boxed{76 \text{ N}}$$

$$F_H = \frac{mg}{2 \tan 40°} = \frac{(10 \text{ kg})(9.8 \text{ m/s}^2)}{2 \tan 40°} = \boxed{58 \text{ N}}$$

$$F_V = \frac{mg}{2} = \frac{(10 \text{ kg})(9.8 \text{ m/s}^2)}{2} = \boxed{49 \text{ N}}$$

Therefore the total force exerted by the hinge on the flagpole is,

$$F = \sqrt{F_H^2 + F_V^2} = \sqrt{(58 \text{ N})^2 + (49 \text{ N})^2} = \boxed{76 \text{ N}}$$

$$\theta = \tan^{-1}\left(\frac{49 \text{ N}}{58 \text{ N}}\right) = \boxed{40°}$$

What does it mean?

So the force the hinge must exert on the flagpole is 76 N in a direction 40° above the horizontal. Note that the hinge and the tension exert the same force on the flagpole. This is a consequence of the fact that the center of mass of the flagpole was at the center of the pole. If we had used a non-uniform pole or hung a flag at the end of the pole, the answer would be quite different.

WE 8.2 Fancy Wheels for Your Car

It is possible to purchase "high performance" wheels for a car or other racing vehicle. Typically, the metal portion of these wheels is composed of a special (and expensive) lightweight metal alloy. These wheels are lighter than normal steel wheels, so they have a smaller moment of inertia, enabling your car to accelerate faster for a given torque applied to the axle (by the engine). Estimate the moment of inertia of such alloy wheels compared with that of a steel wheel. Assume the portion made with the lightweight metal alloy has a mass half that of a steel wheel. *Hint*: Ignore the mass of the wheel spokes and assume the mass of the tire is the same for the normal and the high-performance wheels.

Solution

Recognize the principle

A car's wheel is composed of two main pieces: an inner metal wheel and an outer tire (made of rubber). Both have a "wheel shape," and the moment of inertia of each is given in Table 8.2 of your textbook. The total moment of inertia is then the sum of I for the metal wheel plus I for the rubber tire.

Sketch the problem

The following figure shows a car's wheel and indicates the inner and outer radii of the metal wheel and the rubber tire. The metal wheel is the part that will be replaced by the special lightweight alloy.

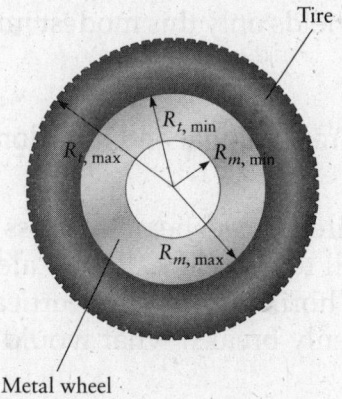

Identify the relationships

According to Table 8.2 in your textbook, the moment of inertia of a single wheel with an inner radius of R_{min} and outer radius R_{max} is

$$I_{\text{wheel}} = \frac{1}{2}m(R_{\text{max}}^2 + R_{\text{min}}^2)$$

The values of the inner and outer radii for the metal wheel and rubber tire are not given above, so we need to make estimates of their values. For the tire, we estimate $R_{t,max} = 30$ cm and $R_{t,min} = 25$ cm, and for the metal wheel, we estimate $R_{m,max} = 25$ cm and $R_{m,min} = 20$ cm. For a typical case, the mass of the rubber tire is approximately 5 kg, and for a steel wheel, we estimate the mass to also be about 5 kg. The total moment of inertia of the rubber tire plus metal wheel is thus

$$I_{total} = I_{tire} + I_{metal} = \frac{m(R_{t,max}^2 + R_{t,min}^2)}{2} + \frac{m(R_{m,max}^2 + R_{m,min}^2)}{2}$$

Solve

Inserting our estimates for the masses and radii gives

$$I_{total} = I_{tire} + I_{metal} = \frac{(5 \text{ kg})[(0.30)^2 + (0.25)^2]\text{m}^2}{2} + \frac{(5 \text{ kg})[(0.25)^2 + (0.20)^2]\text{m}^2}{2}$$

$$I_{total} = 0.38 \text{ kg} \cdot \text{m}^2 + 0.26 \text{ kg} \cdot \text{m}^2 = \boxed{0.64 \text{ kg} \cdot \text{m}^2}$$

Here we have written the values of I for the tire and steel wheel separately, showing that the rubber tire provides the larger share of the total moment of inertia. Even so, changing to a lightweight alloy wheel will still lower the overall moment of inertia. Switching to an alloy wheel decreases the mass of the metal wheel by a factor of 2, so the moment of inertia of the metal wheel will decrease by the same factor (because I is proportional to m). The result for the total moment of inertia in this case is

$$I_{total} = I_{tire} + I_{metal} = 0.38 \text{ kg} \cdot \text{m}^2 + 0.13 \text{ kg} \cdot \text{m}^2 = \boxed{0.51 \text{ kg} \cdot \text{m}^2}$$

What does it mean?

Comparing the two results shows that the total moment of inertia decreases by about 25%. Hence, a fancy (and expensive) wheel yields only this modest improvement in performance.

CP 8.1 Balancing Act: Applying the Conditions for Translational and Rotational Equilibrium

Suppose a large crate of 300 kg is hung from a uniform steel beam of mass 100 kg and 2 m in length. A cable connects from the wall to a point half way along the beam. Find (a) the tension in the cable and the (b) horizontal and (c) vertical force on the beam from the wall. (d) If the cable suddenly breaks, what would be the initial angular acceleration of the beam?

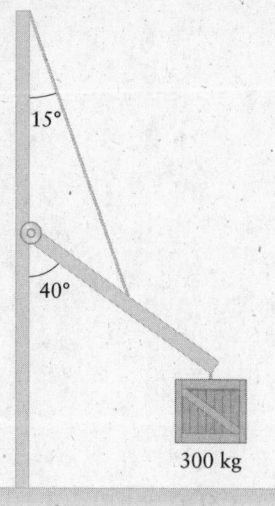

Solution

Recognize the principle

To determine the tension in the cable and the horizontal and vertical forces of the wall on the beam, we apply the conditions for translational and rotational equilibrium. Once the cable breaks the system is no longer in equilibrium, so we apply Newton's second law for rotational motion to determine the angular acceleration of the beam.

Sketch the problem

We begin by drawing the free-body diagram for the beam. Also, we can label the other angles in our drawing so that we can more easily determine the torques.

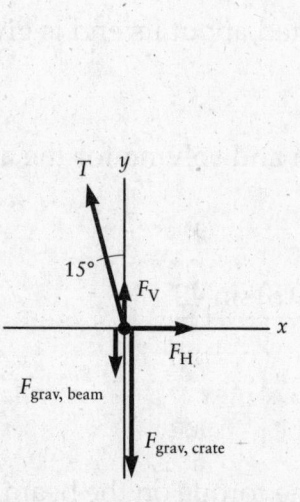

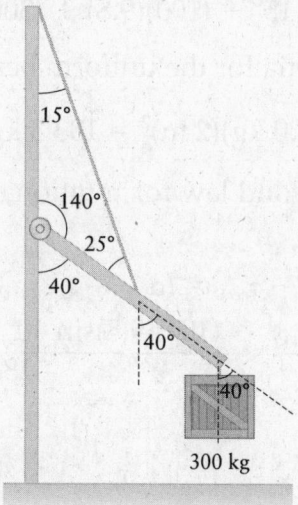

Identify the relationships

Translational Equilibrium

$$\Sigma \vec{F} = 0 \rightarrow \Sigma F_x = 0 \text{ and } \Sigma F_y = 0$$

$$\Sigma F_x = F_H - T\sin 15° = 0 \rightarrow F_H = T\sin 15°$$

$$\Sigma F_y = F_V + T\cos 15° - F_{grav, beam} - F_{grav, crate} = 0$$
$$\rightarrow F_V = -T\cos 15° + F_{grav, beam} + F_{grav, crate}$$

Rotational Equilibrium

$$\Sigma \tau = 0 \rightarrow \tau_{cable} - \tau_{beam} - \tau_{crate} = 0$$

where

$$\tau_{cable} = (1 \text{ m})T\sin 25°$$
$$\tau_{beam} = (1 \text{ m})F_{grav, beam}\sin 40°$$
$$\tau_{crate} = (2 \text{ m})F_{grav, crate}\sin 40°$$

Solve

(a) Using the values given in the problem and from the expression for rotational equilibrium we can solve for the tension in the cable.

$$\tau_{cable} - \tau_{beam} - \tau_{crate} = 0$$
$$(1 \text{ m})T\sin 25° - (1 \text{ m})F_{grav, beam}\sin 40° - (2 \text{ m})F_{grav, crate}\sin 40° = 0$$
$$T\sin 25° - (100)(9.8)\sin 40° - (600)(9.8)\sin 40° = 0$$
$$T = \boxed{10{,}400 \text{ N}}$$

(b) and (c) Plugging the value for the tension into the expressions for translational equilibrium we can solve for the horizontal and vertical forces on the beam from the wall.

$$F_H = T\sin 15° = 10{,}400 \sin 15° = \boxed{2700 \text{ N}}$$
$$F_V = -T\cos 15° + F_{grav, beam} + F_{grav, crate}$$
$$= -10{,}400\cos 15° + (100)(9.8) + (300)(9.8) = \boxed{-6100 \text{ N}}$$

(d) The moment of inertia for the uniform beam pivoted about its end is given by

$$I = \frac{1}{3} ML^2 = \frac{1}{3}(100 \text{ kg})(2 \text{ m})^2 = 133.3 \text{ kg} \cdot \text{m}^2$$

Applying Newton's second law for rotational motion and solving for the angular acceleration we find,

$$\Sigma \tau = I\alpha \rightarrow -\tau_{beam} - \tau_{crate} = I\alpha$$
$$\alpha = \frac{-\tau_{beam} - \tau_{crate}}{I} = \frac{-(100)(9.8)\sin 40° - (600)(9.8)\sin 40°}{133.3}$$
$$= \boxed{-33 \text{ rad/s}^2}$$

What does it mean?

Notice that we used the angle 25° when calculating the torque on the beam due to the cable and 40° when calculating the torques due to the beam's weight and the crate's weight. These were the correct angles to use to determine the component of these forces which were perpendicular to the beam. Often determining these angles is the most difficult part of setting up these types of problems. Also, notice that the sign on the angular acceleration is negative indicating a clockwise angular acceleration.

CP 8.2 Cables, Real Pulleys, Inclines, and Friction: Drawing Free-Body Diagrams and Applying Newton's Second Law for Translational and Rotational Motion

Consider the following figure in which one block rests on a horizontal frictionless table and the other block sits on a rough incline. The blocks are connected by a massless string which loops around three pulleys as shown. Treating the pulleys as disks, $I = \frac{1}{2}Mr^2$, determine the tensions in the string (T_1, T_2, T_3, and T_4) and the acceleration of the blocks, a, once the blocks are released. The three pulleys have the same radii and all the blocks and pulleys have the same mass, $M = 1$ kg.

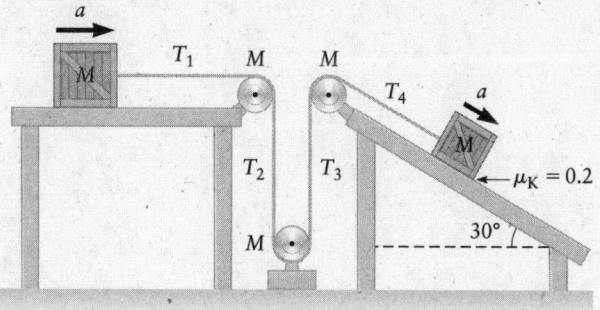

Solution

Recognize the principle

This problem involves many of the concepts discussed in previous chapters with the addition of "real" pulleys with moments of inertia. We will need to apply Newton's second law to the two blocks and Newton's second law for rotational motion to the pulleys. Also, we'll need to relate the angular acceleration of the pulleys to the translational acceleration of the blocks.

Sketch the problem

We begin by drawing a free-body diagram for each of the blocks.

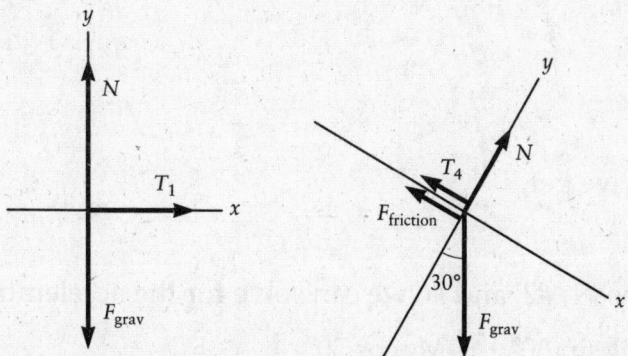

Identify the relationships

We now apply Newton's second law to the blocks and Newton's second law for rotational motion to each of the pulleys.

1st Block

$$\Sigma F_x = Ma \rightarrow T_1 = Ma \qquad (\#1)$$

2nd Block

$$\Sigma F_y = 0 \rightarrow N - Mg\cos 30° = 0 \rightarrow N = Mg\cos 30°$$

Therefore, $F_{\text{friction}} = \mu_K N = \mu_K Mg\cos 30°$

$$\Sigma F_x = Ma \rightarrow Mg\sin 30° - F_{\text{friction}} - T_4 = Ma$$
$$T_4 = -Ma + Mg\sin 30° - F_{\text{friction}}$$
$$= -Ma + Mg\sin 30° - \mu_K Mg\cos 30° \qquad (\#2)$$

1st Pulley

$$\Sigma\tau = I\alpha \rightarrow \tau_{T_2} - \tau_{T_1} = I\alpha$$
$$T_2 r - T_1 r = I\alpha \qquad (\#3)$$

2nd Pulley

$$\Sigma\tau = I\alpha \rightarrow \tau_{T_3} - \tau_{T_2} = I\alpha$$
$$T_3 r - T_2 r = I\alpha \qquad (\#4)$$

3rd Pulley

$$\Sigma\tau = I\vec{\alpha} \rightarrow \tau_{T_4} - \tau_{T_3} = I\alpha$$
$$T_4 r - T_3 r = I\alpha \qquad (\#5)$$

Note that the accelerations of the blocks will be the same. Since the pulleys have the same radius, their angular accelerations will be the same. Furthermore, the relationship between the angular acceleration of the pulleys and the linear acceleration of the blocks is given by $\alpha = \frac{a}{r}$.

Solve

Substitution of the expressions for I and α into equations #3, #4, and #5 we get,

$$T_2 - T_1 = \frac{Ma}{2}$$
$$T_3 - T_2 = \frac{Ma}{2}$$
$$T_4 - T_3 = \frac{Ma}{2}$$

Combining these expressions we get,

$$T_4 = T_1 + \frac{3Ma}{2} \qquad (\#6)$$

Finally, combining expressions #1, #2, and #6 we can solve for the acceleration a,

$$Ma + \frac{3Ma}{2} = -Ma + Mg\sin 30° - \mu_K Mg\cos 30°$$

$$\frac{7Ma}{2} = Mg\sin 30° - \mu_K Mg\cos 30°$$

$$a = \frac{2(Mg\sin 30° - \mu_K Mg\cos 30°)}{7M} = \frac{2(g\sin 30° - \mu_K g\cos 30°)}{7}$$

$$= \boxed{0.92 \text{ m/s}^2}$$

Substituting back into the previous expressions we can solve for the tensions in the string.

$$T_1 = Ma = (1 \text{ kg})(0.92 \text{ m/s}^2) = \boxed{0.92 \text{ N}}$$

$$T_2 = T_1 + \frac{Ma}{2} = \boxed{1.38 \text{ N}}$$

$$T_3 = T_2 + \frac{Ma}{2} = \boxed{1.84 \text{ N}}$$

$$T_4 = T_3 + \frac{Ma}{2} = \boxed{2.30 \text{ N}}$$

What does it mean?

Note that the tensions in the string progressively increase. This makes sense since each must supply the necessary force or torque to accelerate an additional component in the system. We can quickly check the answers by using tension T_1 and T_4 to calculate and compare the accelerations of the two blocks. These accelerations should be the same.

1st Block

$$a = \frac{T_1}{M} = \frac{0.92}{1} = 0.92 \text{ m/s}^2$$

2nd Block

$$a = \frac{-T_4 + Mg\sin 30° - \mu_K Mg\cos 30°}{M} = 0.90 \text{ m/s}^2$$

Part E. MCAT Review Problems and Solutions

PROBLEMS

1. A steel beam of uniform cross-section and composition weighs 100 N. What minimum force is required to lift one end of the beam off the ground?

   (a) 25 N

   (b) 50 N

   (c) 250 N

   (d) 500 N

2. A non-uniform wooden beam 8.0 m long is placed on a pivot 2.0 m from the lighter end of the beam. The center of gravity of the beam is located 2.0 m from the heavier end. If placing a 500 N weight on the lighter end of beam balances the beam, what must be the weight of the beam?

   (a) 125 N

   (b) 250 N

   (c) 500 N

   (d) 1000 N

3. Consider the following situation in which a massless plank has two objects placed at different positions resulting in the system being balanced. Determine the distance d of the 10-kg object.

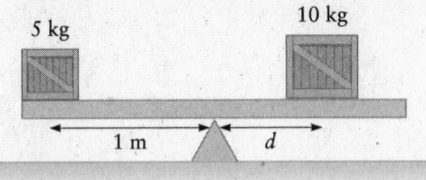

(a) 0.5 m

(b) 1.0 m

(c) 1.5 m

(d) 2.0 m

4. A solid cylinder ($I = mr^2/2$) has a string wrapped around it many times. The cylinder is released while holding the string so that the cylinder spins as it falls. What is the downward acceleration of the cylinder?

(a) 0

(b) 9.8 m/s^2

(c) 4.9 m/s^2

(d) 6.5 m/s^2

5. A point on the rim of a 0.30-m-radius wheel has a tangential speed of 4.0 m/s. What is the tangential speed of a point 0.20 m from the center of the wheel?

(a) 1.0 m/s

(b) 1.3 m/s

(c) 2.7 m/s

(d) 8.0 m/s

6. Consider a point on a bicycle wheel as the wheel turns about a fixed axis, neither slowing down nor speeding up. Compare the linear and angular accelerations of the point.

(a) Both are zero.

(b) Only the angular accelerating is zero.

(c) Only the linear acceleration is zero.

(d) Neither is zero.

7. A wheel is initially at rest and rotates with a constant angular acceleration of 2.5 rad/s². What is its angular speed at the instant it goes through an angular displacement of 5.0 rad?

   (a) 0.05 rad/s

   (b) 0.50 rad/s

   (c) 5.0 rad/s

   (d) 25 rad/s

8. Consider the following wheel with several forces applied at various points. Which force is producing the greatest torque?

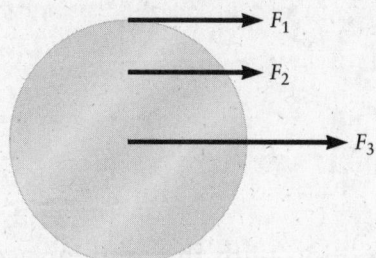

   (a) F_1

   (b) F_2

   (c) F_3

   (d) F_1 and F_2 produce the same torque.

9. Consider a hoop and a solid disk of the same mass and radius. Which object has the greatest moment of inertia?

   (a) The hoop.

   (b) The disk.

   (c) They have the same moment of inertia.

   (d) There is not enough information to tell.

10. Starting from rest a wheel undergoes a constant angular acceleration for a period of time T. At what time after the start of the rotation does the wheel reach and angular speed equal to its average angular speed for this interval?

   (a) 0.25 T

   (b) 0.50 T

   (c) 0.75 T

   (d) 1.0 T

SOLUTIONS

1. *MCAT strategies*

The force required to lift the entire beam off the ground would be 100 N so the answer is certainly less than 100 N. Answers (c) and (d) can be quickly eliminated.

No other answers can be easily eliminated so we proceed with our problem-solving strategy.

Recognize the principle

This is a situation where we can apply the condition for rotational equilibrium.

Sketch the problem

A quick sketch of the problem may help.

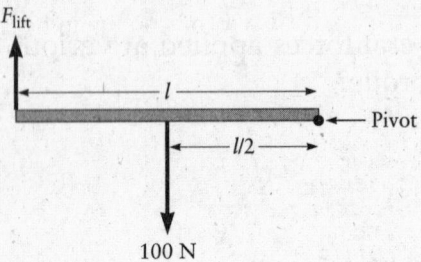

Identify the relationships and Solve

Applying Newton's second law for rotational motion with the angular acceleration zero,

$$\Sigma \tau = 0 \rightarrow F_{\text{lift}}\,(l) - 100(l/2) = 0 \rightarrow F_{\text{lift}} = 50 \text{ N}$$

So the correct answer is (b).

What does it mean?

Notice that the length of the beam was not required to solve this problem.

2. MCAT strategies

No answers can be easily eliminated so we proceed with our problem-solving strategy.

Recognize the principle

This is a situation where we can apply the condition for rotational equilibrium.

Sketch the problem

A quick sketch of the problem may help.

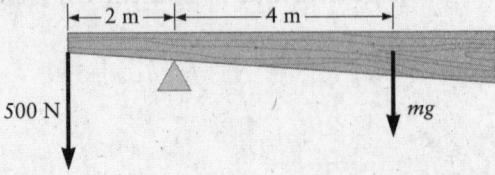

Identify the relationships and Solve

Applying Newton's second law for rotational motion with the angular acceleration zero,

$$\Sigma \tau = 0 \rightarrow (500 \text{ N})(2 \text{ m}) - mg(4 \text{ m}) = 0 \rightarrow mg = 250 \text{ N}$$

So the correct answer is (b).

What does it mean?

Notice that the length and shape of the beam is irrelevant as long as you know the location of its center of mass and how far it is from the pivot.

3. MCAT strategies

Since the 10 kg mass is greater than the 5 kg mass, the 10 kg mass must be located closer to the pivot than the 5 kg mass so the answer must be less than 1 m. The only answer less than 1 m is (a). So without any calculations, the correct answer must be (a).

4. MCAT strategies

The cylinder will certainly accelerate although not at 9.8 m/s² since it is not in freefall. So, answers (a) and (b) can be quickly eliminated. Of the two remaining answers it is difficult to determine which is correct without some analysis.

Recognize the principle

We can apply Newton's second laws for translational and rotational motion.

Sketch the problem

The quick sketch of the free-body diagram may help.

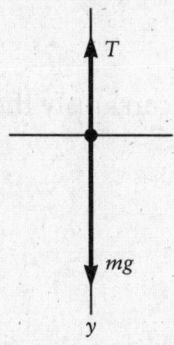

Identify the relationships

Applying Newton's second law:

$$\Sigma F_y = -T + mg = ma$$

Applying Newton's second law for rotational motion:

$$\Sigma \tau = Tr = I\alpha = \left(\frac{1}{2} mr^2\right)\alpha \rightarrow T = \frac{mr\alpha}{2}$$

Also, the angular acceleration and linear acceleration are related by,

$$\alpha = \frac{a}{r}$$

Solve

Combining these expressions and solving for a we find,

$$-T + mg = ma$$
$$-\frac{mr\alpha}{2} + mg = ma$$
$$-\frac{ma}{2} + mg = ma \rightarrow a = \frac{2g}{3}$$

So the correct answer is (d).

What does it mean?

Notice that the radius of the cylinder was not required.

5. MCAT strategies

This problem simply involves knowing the relationship between the radius, the tangential speed and the angular speed. Since the angular speed is the same for all points on the wheel, then we can write the following,

$$\omega = \frac{v_1}{r_1} = \frac{v_2}{r_2} \rightarrow v_2 = r_2 \frac{v_1}{r_1} = (0.2\ \text{m})\frac{(4\ \text{m/s})}{(0.3\ \text{m})} = \frac{0.8}{0.3}\ \text{m/s} = \frac{8}{3}\ \text{m/s}$$

So the correct answer is (c).

6. MCAT strategies

Since the wheel is neither slowing down nor speeding up, its angular acceleration is zero. Also, the tangential acceleration is zero. The change in direction of the tangential velocity is a centripetal acceleration not a tangential acceleration. So, the correct answer is (a).

7. MCAT strategies

No answers can be easily eliminated so we proceed with our problem-solving strategy.

Recognize the principle

Since the wheel rotates with a constant angular acceleration we can apply the kinematic relations for constant angular acceleration.

Sketch the problem

No sketch required.

Identify the relationships

Since the angular displacement, angular acceleration, and initial angular velocity are given in the problem, we use the following expression to solve for the final angular velocity:

$$\omega^2 = \omega_0^2 + 2\alpha(\theta - \theta_0)$$

Solve

Plugging in the given values we find,

$$\omega^2 = \omega_0^2 + 2\alpha(\theta - \theta_0) \rightarrow \omega = \sqrt{\omega_0^2 + 2\alpha(\theta - \theta_0)}$$

$$\omega = \sqrt{0 + 2(2.5\ \text{rad/s})(5.0\ \text{rad} - 0)} = \sqrt{25\ \text{rad}^2/\text{s}^2} = 5.0\ \text{rad/s}$$

So the correct answer is (c).

What does it mean?

Since the time interval was not provided in the problem, this method was the easiest; however, other kinematic expression could have been used by first solving for time and then using that time to find the angular velocity. Look for the quickest way to solve the problem.

8. *MCAT strategies*

Solving this problem simply requires us to know and understand the expression for torque, $\tau = Fr\sin\phi$. Force F_3 is applied at the wheel's center where $r = 0$, therefore it produces no torque. The other two forces are of the same magnitude and have the same angle $\phi = 90°$. Since F_1 is applied further from the wheel's center, it will produce the greater torque. So the correct answer is (a).

9. *MCAT strategies*

Solving this problem requires an understanding of the moment of inertia of an object. The moment of inertia depends not only on the mass of an object but how that mass is distributed about the axis of rotation. All else being equal, the object with more of its mass distributed further from the axis of rotation will have the larger moment of inertia. Since the hoop has more of its mass distributed further from the axis of rotation, it will have the larger moment of inertia. So the correct answer is (a).

10. *MCAT strategies*

If an object starts from rest and has a constant angular acceleration, then the average angular velocity for a given time interval is simply half of the angular velocity at the end of that time interval. Since the angular velocity is directly proportional to time, $\omega = \omega_0 + \alpha t$, the average angular velocity occurs half way through the time interval. For a time period of T, the average angular velocity occurs at $0.50\,T$. So the correct answer is (b).

9 Energy and Momentum of Rotational Motion

Part A. Summary of Key Concepts and Problem-Solving Strategies

KEY CONCEPTS

Rotational kinetic energy

The *rotational kinetic energy* of an object is

$$KE_{rot} = \frac{1}{2} I \omega^2$$

where I is the moment of inertia and ω is its angular speed. This energy is the kinetic energy associated with just the rotational motion of the object. If an object is also translating then we can write the total kinetic energy of an object as the sum of its rotational kinetic energy plus the kinetic energy associated with the translational motion:

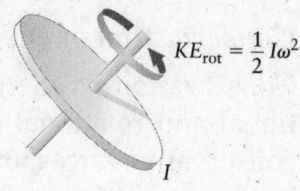

$$KE_{total} = KE_{translational} + KE_{rotational}$$

$$= \frac{1}{2} m v_{CM}^2 + \frac{1}{2} I \omega^2$$

If the rotation axis passes through the center of mass of the object, then v_{CM} in this expression is the speed of the center of mass of the object.

Work–energy theorem for rotational motion

Recall how we defined the work done by a force on an object when we limit our analysis to one dimension:

$$W = F\Delta x$$

We can define an analogous expression for work done in rotating an object. The work done by an applied torque is

$$W = \tau\theta$$

where θ is the angular displacement. Similarly, the work–energy theorem applied to rotational motion is

$$W = \tau\theta = \Delta KE = \frac{1}{2}I\omega_f^2 - \frac{1}{2}I\omega_i^2$$

Angular momentum

A rotating object has an ***angular momentum*** L. If the direction of the rotation axis is fixed, L is given by

$$L = I\omega$$

If the total external torque on a system is zero, the total angular momentum of the system is conserved.

Vector nature of angular momentum

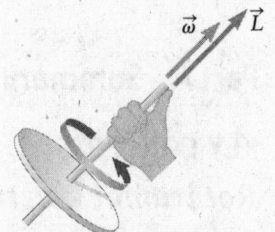

The direction of the angular velocity $\vec{\omega}$ is given by the ***right-hand rule***, and the angular momentum vector $\vec{L}$ is

$$\vec{L} = I\vec{\omega}$$

As is indicated in the figure, we curl our fingers in the direction of the angular velocity of the object. Our thumb then indicates the direction of the angular momentum.

APPLICATIONS

Conservation of energy

Conservation of energy principles can be applied to systems involving translational and rotational motion. When all the forces that do work on an object are conservative forces, the total mechanical energy is conserved. When the potential energy is due to gravity, as for an object that rolls up or down an incline, conservation of mechanical energy leads to the expression,

$$KE_i + PE_i = KE_f + PE_f$$

$$\frac{1}{2}mv_i^2 + \frac{1}{2}I\omega_i^2 + mgh_i = \frac{1}{2}mv_f^2 + \frac{1}{2}I\omega_f^2 + mgh_f$$

The kinetic energy here is the total kinetic energy, including the rotational contribution. Conservation of energy can be expanded to include other conservative forces such as springs, where a more general expression will be required.

Conservation of angular momentum

Angular momentum and its conservation play important roles in the motion of many objects, including figure skaters, cats, and motorcycles. As long as the system we define is isolated in the sense that the net external torque acting on the system is zero, then the angular momentum of the system will be conserved. This is a powerful tool we can use to analyze complicated situations. This is analogous to the translational momentum of an object being conserved if the net external force acting on the system is zero.

PROBLEM-SOLVING STRATEGIES

In this chapter, two very powerful concepts were discussed. The first is the principle of conservation of energy including rotational motion. In previous chapters we applied the principle of conservation of energy to systems which did not involve rotational motion and found that we could understand and predict the motion of very complicated systems. By generalizing our analysis to include rotational motion, we can tackle even more complicated systems. The second powerful concept introduced in this chapter is the principle of conservation of angular momentum. This principle explains why the rotational speed of a figure skater increases as she pulls her arms in closer to her body, Kepler's second law of planetary motion, and many other interesting phenomena. The following two schematics will help guide you through the problem-solving strategy used while tackling problems involving these two concepts.

Problem Solving: Applying the Principle of Conservation of Energy to Problems of Rotational Motion

Recognize the principle

The mechanical energy of an object is conserved only if all the forces that do work on the object are conservative forces. Only then can we apply the conservation of mechanical energy condition.

↓

Sketch the problem

Always use a sketch to collect your information concerning the initial and final states of the system.

↓

Identify the relationships

Find the initial and final kinetic and potential energies of the object. These may include several forms of potential energy (if springs and gravity are involved) and both translational and rotational kinetic energies. You will usually also need to find the moment of inertia of the object.

↓

Solve

Solve for the unknown quantities using the conservation of energy condition

$$KE_i + PE_i = KE_f + PE_f$$

The kinetic energy terms must account for both the translational and the rotational kinetic energies

$$KE_{total} = KE_{translational} + KE_{rotational}$$

$$= \frac{1}{2}mv_{CM}^2 + \frac{1}{2}I\omega^2$$

The potential energy may involve gravitational potential energy, $PE_{grav} = mgh$, and perhaps spring potential energy, $PE_{spring} = \frac{1}{2}kx^2$.

↓

What does it mean?

Always *consider what your answer means* and check that it makes sense.

Problem Solving: Applying the Principle of Conservation of Angular Momentum

---

Recognize the principle

If the total external torque on a system is zero, the angular momentum is conserved.

Sketch the problem

Always use a sketch to define the system and collect information concerning the initial and final states of the system.

Identify the relationships

- The system of interest and its initial and final states depend on the problem. The system might be a single object or a collection of objects.

- Express the initial and final angular velocities and moments of inertia.

- Sometimes it is also useful to gather information on the initial and final mechanical energies of the system.

Solve

Solve for the quantities of interest using the principle of conservation of angular momentum

$$L_i = I_i \omega_i = L_f = I_f \omega_f$$

What does it mean?

Always *consider what your answer means* and check that it makes sense.

Part B. Frequently Asked Questions

1.   *I don't understand the vector nature of angular quantities. How can a vector such as angular velocity, angular momentum, etc. not point in the direction the object is rotating?*

That's a good question and highlights a confusing point for many students. It is true that it makes more sense to define these angular quantities in terms of the direction the object is rotating, clockwise or counterclockwise. The problem is that these directions depend on how we're viewing the object. For instance, suppose that you are looking at a spinning wheel which is rotating clockwise from your perspective. Another observer standing on the other side of the wheel would see the wheel rotating counterclockwise. Both would be correct but disagree with each other. To get around this issue the right-hand rule was developed so that everyone agrees on the direction of rotation. If all observers use their right-hand fingers to indicate the direction the object is rotating from their perspective, then everyone's thumb will point in the same direction regardless of where they're standing. So, the direction of our thumb doesn't point to the direction of rotation, it helps us align our fingers in the direction the object is rotating so that all observers will agree.

2.   *Can I add linear momentum and angular momentum to calculate the total momentum of an object?*

No, definitely not. Linear momentum and angular momentum are two different quantities with different units. It makes no physical sense to add quantities with different units. For example, it is meaningless to add a length of 10 m to a time of 4 s. However, you can add or subtract quantities with the same units such as 10 m minus 4 m and obtain a meaningful result, 6 m.

3.   *If an object's angular momentum changes, does that mean that its angular velocity must have changed?*

No, not necessarily. Remember, the angular momentum of an object depends on both its angular velocity and its moment of inertia, $\vec{L} = I\vec{\omega}$. For the angular momentum to change either the moment of inertia or the angular velocity must change. The moment of inertia of an object depends on its mass and how that mass is distributed about the axis of rotation. So an object's moment of inertia can change if the object's mass or shape changes. The angular velocity of an object will change if its rate of rotation (angular speed) changes or the direction of the angular velocity changes. Any of these factors will affect the object's angular momentum.

4.   *Under what conditions is angular momentum conserved? What about rotational kinetic energy?*

If the net external torque acting on an object is zero, then the object's angular momentum will remain constant (i.e., it will be conserved). That's not to say that there cannot be torques acting on the object, only that the vector sum of these torques must be zero for angular momentum to be conserved. Note that this is analogous to our study of momentum conservation in Chapter 7 where the linear momentum of an object will remain constant if the net force acting on the object is zero. Also recall from Chapter 7 linear momentum is conserved during the collision of objects and in the case of an elastic collision the kinetic energy is also conserved. An analogous situation exists for rotational collision. Consider a disk being dropped onto a rotating disk as in the figure.

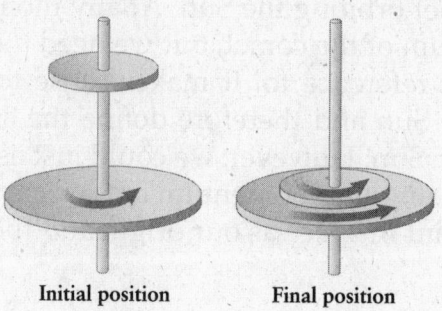

Initial position   Final position

This is a rotational collision and as long as this is an isolated system the total angular momentum of the two objects will be conserved. If the collision is elastic so that there are no non-conservative forces involved which could dissipate energy away in the form of light, heat, sound, etc., then the rotational kinetic energy will also be conserved.

5.   *When applying the conservation of angular momentum, how do I choose my system?*

When defining a system, we need to include the objects of interest and the forces and torques of interaction which are important. Also, the total external torque on the system should be zero so that we can apply the conservation of angular momentum. For example, consider the two colliding disks in the previous FAQ. The objects of interest are the two disks, and the interaction forces between the two disks must be included in the system since they produce internal torques which affect the rotation of the disks. Although the gravitational force from the Earth on the two disks is not zero, it is small compared to the interaction forces. So, we would define our system as the two disks and include their interaction forces.

6.   *Since it's almost impossible to completely isolate a system from external torques, is angular momentum ever really conserved?*

That's a good question. Certainly out in space far from the gravitational effects of stars and planets, the net external torque on our system is zero. In practice, however, this is seldom the case. Even air drag on a rotating object produces a small torque. Fortunately the forces (and therefore the torques) involved in the interactions within our system are usually very large compared to these external torques and so these external torques only change the total angular momentum of the system by a very small amount. For example, a spinning ice skater has a small force of friction between her skates and the ice and a small amount of air drag. Both these external forces produce small external torques on the skater; however, these torques are very small and can be neglected. For this reason, angular momentum conservation can still be applied and is considered to be a very useful and powerful tool in physics.

7. *Does the angular momentum of an object depend on how we define our coordinate system?*

Yes, absolutely. Though it's often obvious and convenient to define the origin of our coordinate system at the axis of rotating of the object of interest, it's not necessary to do so. For example, consider a comet orbiting the Sun. At any moment in time we can calculate the angular momentum of the comet, but we need to define what point that angular momentum is in reference to. It makes sense to place the origin of our coordinate system at the Sun and therefore define the angular momentum of the comet with respect to the Sun. However, we could just as easily place our origin at the Earth and define the angular momentum of the comet with respect to the Earth. We can choose any point in space as our origin and therefore have a different angular momentum.

8. *Is rotation necessary for angular momentum? That is, can an object have angular momentum but not be rotating?*

No, rotation is not necessary for angular momentum. This question is related to the previous FAQ in that angular momentum depends on where we choose our coordinate system. For example, consider an object moving through distant space far away from any star or planet. According to Newton's laws of motion, the object will travel in a straight line with a constant velocity as shown in the figure. The object is certainly not rotating, so does it have angular momentum? Well, we need to define a coordinate system and ask if the object has angular momentum with respect to a particular point (origin). As shown in the figure, the object has a momentum, $\vec{p} = m\vec{v}$, and is moving at a distance r from our origin. So, the object has an angular momentum with respect to the origin whose magnitude is given by the expression $L = mrv_{\perp}$.

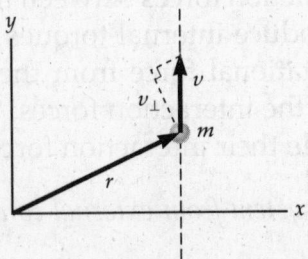

9. *How does a satellite use a gyroscope to orient itself in space?*

Typically, satellites use what's called a controlled momentum gyroscope to orient themselves in space. For example, on the Hubble space telescope several of these devices are used to align the telescope with a region of the sky of interest to astronomers. Essentially these devices work on the principle of conservation of angular momentum. The gyroscope consists of a spinning mass on several rotating axes. By sending a signal to the satellite, we can control the direction of the spinning mass, thereby adjusting the angular momentum vector of the spinning mass. Since the gyroscope and surrounding satellite are an isolated system, their total angular momentum is conserved. So, as the direction of the gyroscope is changed, the surrounding satellite rotates in response to this change such that the total angular momentum of the two remains constant. In this way the satellite can be turned to point in any direction desired.

10. *Why is it easier to balance a spinning basketball on my finger than when it's not spinning?*

This is the same reason why a rolling wheel is more stable than a stationary wheel as was discussed in your textbook. According to the right-hand rule, the angular momentum $\vec{L}$ of a spinning basketball on your finger is directed straight up or down, depending on which way it's spinning. If the external torque on the ball were exactly zero, $\vec{L}$ would remain in this direction forever and the ball would never slow down and would never fall over. However, in reality there is some torque acting on the ball due to friction with your finger, air drag, and eventually gravity as it starts to tip over. These torques eventually slow the ball down and it begins to tip over. Even so, this takes some time so the angular momentum of the spinning basketball allows you to balance it on your finger for several seconds. If it were not spinning, the basketball would simply fall over.

Part C. Selection of End-of-Chapter Answers and Solutions

QUESTIONS

Q9.5 Consider the translational and rotational kinetic energies of a wheel that rolls without slipping. Show that the ratio of these energies is independent of the size (the radius) of the wheel.

Answer

The rotational kinetic energy is given by the expression,

$$KE_{rot} = \frac{1}{2} I \omega^2$$

A wheel can be approximated as a disk, with momentum of inertia $I = \frac{1}{2} mR^2$. Since the wheel rolls without slipping, $\omega = v/R$. Combining these expressions we get,

$$KE_{rot} = \frac{1}{2}\left(\frac{1}{2}mR^2\right)\left(\frac{v}{R}\right)^2$$

$$KE_{rot} = \frac{1}{4} mv^2$$

The translational kinetic energy of the wheel is given by $KE_{trans} = \frac{1}{2} mv^2$, which does not depend on the radius of the wheel. The ratio of these energies is:

$$\frac{KE_{trans}}{KE_{rot}} = \frac{\frac{1}{2}mv^2}{\frac{1}{4}mv^2} = 2$$

A wheel that rolls without slipping always has twice as much translational kinetic energy as rotational kinetic energy, regardless of its size or mass.

Q9.17 (a) Why do most small helicopters have a rotor on their tail (Fig. Q9.17)? (b) Many large helicopters have two large rotors. Why do these helicopters not need a tail rotor? Do you think the two large propellers rotate in the same direction? Explain.

Figure Q9.17

Answer

(a) The engine in the helicopter must constantly provide torque on the rotor of the helicopter. As long as the helicopter is on the ground, the Earth can resist the equal and opposite torque that the rotor exerts on the engine (and therefore on the helicopter.) Once the helicopter is airborne, however, this counter-torque must be provided in another way. A rotor in the tail (a point far from the center of rotation) allows a small force to provide a counter torque.

(b) Larger helicopters with two rotors can provide the same counter-torque by carefully synchronizing the opposite rotations of two large rotors so that no tail rotor is needed.

PROBLEMS

P9.12 What is the rotational kinetic energy of the Earth as it spins about its axis?

Solution

Recognize the principle

We will need to determine the moment of inertia of the Earth about its axis of rotation. We can approximate the Earth as a solid sphere in order to determine its moment of inertia. Also, we will need the angular speed of the Earth.

Sketch the problem

No sketch needed.

Identify the relationships

The rotational kinetic energy is given by the expression,

$$KE_{rot} = \frac{1}{2}I\omega^2$$

The Earth makes one rotation every 24 h which is 86,400 s, so its rotational speed is:

$$\omega = \frac{1 \text{ rotation}}{\text{day}} = \frac{2\pi \text{ rad}}{86,400 \text{ s}} = 7.3 \times 10^{-5} \text{ rad/s}$$

If we approximate the Earth as a solid sphere, its moment of inertia is $I = \frac{2}{5}mr^2$. The mass and radius of the Earth can be found in Table 5.1 of the textbook.

Solve

Inserting our expression for moment of inertia into the rotational kinetic energy equation gives:

$$KE_{rot} = \frac{1}{2}\left(\frac{2}{5}mr^2\right)\omega^2 = \frac{1}{5}mr^2\omega^2$$

Inserting the values for angular speed, mass, and radius,

$$KE_{rot} = \frac{1}{5}(5.98 \times 10^{24}\,\text{kg})(6.37 \times 10^6\,\text{m})^2(7.3 \times 10^{-5}\,\text{rad/s})^2$$

$$KE_{rot} = \boxed{2.6 \times 10^{29}\,\text{J}}$$

What does it mean?

The energy stored in the Earth's rotation is far beyond our typical scales of energy use.

P9.31   A bucket filled with dirt of mass 20 kg is suspended by a rope that hangs over a pulley of mass 30 kg and radius 0.25 m (Fig. P9.31). Everything is initially at rest, but someone is careless and lets go of the pulley and the bucket then begins to move downward. What is the speed of the bucket when it has fallen a distance of 2.5 m?

Figure P9.31

Solution

Recognize the principle

Since the only force acting on the system is the force of gravity which is a conservative force, the mechanical energy is conserved in this system; the potential energy of the bucket is converted to the kinetic energy of the falling bucket and the rotational kinetic energy of the pulley.

Sketch the problem

No sketch needed.

Identify the relationships

The bucket only translates and the pulley only rotates, so our general conservation of energy equation can be written as:

$$KE_i + PE_i = KE_{trans,f} + KE_{rot,f} + PE_f$$

Defining the final potential energy as zero, the initial potential energy is $m_b gh$. Since the bucket and pulley are initially at rest there is no initial kinetic energy. The final kinetic energy will be the result of the translational kinetic energy of the bucket and the rotational kinetic energy of the pulley. So, we can write:

$$0 + m_b gh = \frac{1}{2}m_b v^2 + \frac{1}{2}I\omega^2 + 0$$

The moment of inertia of a pulley is $\frac{m_p R^2}{2}$ and the angular speed of the pulley will be the speed of the rope (and therefore the bucket) divided by R.

Solve

Solving for the speed we have,

$$0 + m_b g h = \frac{1}{2} m_b v^2 + \frac{1}{2} \frac{m_p R^2}{2} \left(\frac{v}{R}\right)^2 + 0$$

$$m_b g h = \frac{1}{2} m_b v^2 + \frac{1}{4} m_p v^2$$

$$v = \sqrt{\frac{m_b g h}{\frac{1}{2} m_b + \frac{1}{4} m_p}}$$

Inserting the numerical values,

$$v = \sqrt{\frac{(20 \text{ kg})(9.8 \text{ m/s}^2)(2.5 \text{ m})}{\frac{1}{2}(20 \text{ kg}) + \frac{1}{4}(30 \text{ kg})}}$$

$$\boxed{v = 5.3 \text{ m/s}}$$

What does it mean?

The bucket falls at a speed lower than it would if it were in free fall. The potential energy lost by the bucket is split between the rotating pulley and the falling bucket.

P9.43 Consider a person who is sitting on a frictionless rotating stool as in Figure P9.43. The person initially has his arms outstretched and is rotating with an angular velocity of 5.0 rad/s. He then pulls his arms close to his body. (a) Estimate his final angular velocity. (b) Estimate the kinetic energy before and after the person pulls his arms into his body.

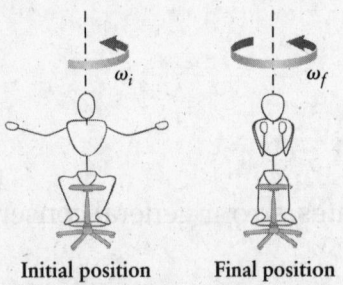

Initial position   Final position

Figure P9.43

Solution

Recognize the principle

The angular momentum of this system remains constant because there are no external torques. So, we can equate the initial and final angular momentum to determine the final angular speed of the person.

Sketch the problem

The sketch is provided in the problem.

Identify the relationships

(a) We can model the person as two separate objects. His legs, head, and torso have a fixed mass and radius and can be modeled as a solid cylinder. His arms from the tip of one hand to the tip of the other can be modeled as a bar of changeable length but constant mass spinning about its center. With these approximations, the moment of inertia of the person can be written as:

$$I_{total} = I_{body} + I_{arms} = \frac{1}{2} m_{body}\, r_{body}^2 + \frac{1}{12} m_{arms}\, \ell_{arms}^2$$

We know that angular momentum must remain constant since there are no external torques acting upon the person, so we can write,

$$L_{arms\ in} = L_{arms\ out}$$
$$I_{arms\ in}\, \omega_{arms\ in} = I_{arms\ out}\, \omega_{arms\ out}$$

(b) From the angular speeds and moments of inertia for each case in part (a), we can find the kinetic energy in each case using the relationship:

$$KE = \frac{1}{2} I \omega^2$$

Solve

(a) We can solve this equation for the final angular speed when his arms are out:

$$\omega_{arms\ in} = \frac{I_{arms\ out}\, \omega_{arms\ out}}{I_{arms\ in}}$$

A typical 70-kg person might have 60 kg in his head, legs, and torso and 10 kg in his arms. A reasonable estimate for the radius of the torso (considering we're averaging hips, shoulders, head, and feet as well as front to back) might be 25 cm. Twice this length (50 cm) serves as a good estimate for the "bar" associated with arm length when the arms are tucked in. When his arms are extended, 1.6 m is a good estimate.

We can use these values to find:

$$I_{arms\ out} = \frac{1}{2}(60\ \text{kg})(0.25\ \text{m})^2 + \frac{1}{12}(10\ \text{kg})(1.6\ \text{m})^2 = 4.0\ \text{kg} \cdot \text{m}^2$$

$$I_{arms\ in} = \frac{1}{2}(60\ \text{kg})(0.25\ \text{m})^2 + \frac{1}{12}(10\ \text{kg})(0.50\ \text{m})^2 = 2.1\ \text{kg} \cdot \text{m}^2$$

Then, including the given value for the arms-out angular speed gives:

$$\omega_{arms\ in} = \frac{(4.0\ \text{kg} \cdot \text{m}^2)(5.0\ \text{rad/s})}{(2.1\ \text{kg} \cdot \text{m}^2)} = \boxed{9.5\ \text{rad/s}}$$

(b) Inserting values in each case:

$$KE_{arms\ out} = \frac{1}{2}(4.0\ \text{kg} \cdot \text{m}^2)(5.0\ \text{rad/s})^2 = \boxed{50\ \text{J}}$$

$$KE_{arms\ in} = \frac{1}{2}(2.1\ \text{kg} \cdot \text{m}^2)(9.5\ \text{rad/s})^2 = \boxed{95\ \text{J}}$$

What does it mean?

Bringing your arms in while spinning on a platform almost doubles your spin rate! Also, the kinetic energy of the system is actually higher with arms pulled in. This means that it takes work to bring the masses closer to the center of rotation while the system is rotating.

P9.48 A meteor with a volume of 1.0 km³ strikes the Earth at the equator as shown in Figure P9.48, and all the fragments stick to the surface. (a) What is the change in the angular momentum of the Earth? (b) What is the change in the length of the day? Hint: Assume the average density of the meteor is 1.0×10^4 kg/m³.

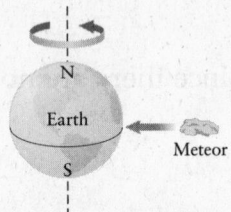

Figure P9.48

Solution

Recognize the principle

We recognize that this is a situation where the angular momentum of the system (the Earth–meteor system) is conserved during the collision process. This is because there is zero net external torque acting on the system. The change in the length of the day can be determined by comparing the initial and final angular speeds of the Earth.

Sketch the problem

The sketch is provided in the problem.

Identify the relationships

Conservation of angular momentum implies that the initial angular momentum is equal to the final angular momentum,

$$L_i = L_f \longrightarrow I_i \omega_i = I_f \omega_f$$

The initial moment of inertia of the system is of the Earth alone. We can approximate the Earth as a solid sphere, with moment of inertia given by the expression,

$$I_i = I_{Earth} = \frac{2}{5} m_{Earth} R^2_{Earth}$$

On the scale of the Earth's radius, the meteor can be considered to be a point mass and lands a distance of one Earth's radius from the center of rotation of the system. The moment of inertia of the point-mass-meteor is given by,

$$I_{meteor} = m_{meteor} R^2_{Earth}$$

Combining these two expressions we get the total final moment of inertia of the system,

$$I_f = I_{Earth} + I_{meteor} = \frac{2}{5}m_{Earth}R_{Earth}^2 + m_{meteor}R_{Earth}^2$$

We can find the mass of the meteor from the given density (ρ) and volume (V) since:

$$\rho = \frac{m}{V} \rightarrow m = \rho V$$

And we know the initial angular speed from the rotation speed of the Earth:

$$\omega_i = \frac{1\,\text{rotation}}{1\,\text{day}}\left(\frac{1\,\text{day}}{86{,}400\,\text{s}}\right)\left(\frac{2\pi\,\text{rad}}{1\,\text{rotation}}\right) = 7.3 \times 10^{-5}\,\text{rad/s}$$

The length of a day is the period of rotation of the Earth which we can get from the angular speed,

$$\text{period} = T = \frac{2\pi}{\omega}$$

Solve

(a) Since angular momentum is conserved, the angular momentum of the Earth remains constant. So, the change in angular momentum of the Earth is ⎢zero⎥.

(b) We first solve our conservation of angular momentum equation for the final angular speed of the system:

$$\omega_f = \frac{I_i \omega_i}{I_f}$$

Then inserting our expressions for the moments of inertia, we have:

$$\omega_f = \frac{\left(\frac{2}{5}m_{Earth}R_{Earth}^2\right)\omega_i}{\frac{2}{5}m_{Earth}R_{Earth}^2 + m_{meteor}R_{Earth}^2} = \left(\frac{\frac{2}{5}m_{Earth}}{\frac{2}{5}m_{Earth} + m_{meteor}}\right)\omega_i = \left(\frac{1}{1 + \frac{5\,m_{meteor}}{2\,m_{Earth}}}\right)\omega_i$$

Using our mass expression in terms of the density and volume, this becomes,

$$\omega_f = \left(\frac{1}{1 + \frac{5\rho V}{2m_{Earth}}}\right)\omega_i$$

Finally, inserting our values for the given density and volume of the meteor, the calculated initial angular speed, and the mass of the Earth, we have:

$$\omega_f = \left(\frac{1}{1 + \frac{5(1 \times 10^4\,\text{kg/m}^3)(1 \times 10^9\,\text{m}^3)}{2(5.98 \times 10^{24}\,\text{kg})}}\right)(7.3 \times 10^{-5}\,\text{rad/s})$$

$$\omega_f = \left(\frac{1}{1 + 4.2 \times 10^{-12}}\right)(7.3 \times 10^{-5}\,\text{rad/s}) \approx \omega_i$$

Therefore, ⎢$T_f \approx T_i$ so the length of the day does not change an appreciable amount.⎥

What does it mean?

The difference in the angular speed is less than 1 part in 1 billion, resulting in a change in the length of a day well less than 1 millionth of a second. Even a 1 km³ asteroid is just too small compared with the Earth to seriously affect its rotation.

P9.50 Halley's comet moves about the Sun in a highly elliptical orbit. At its closest approach, it is a distance of 8.9×10^{10} m from the Sun and has a speed of 54 km/s. When it is farthest from the Sun, the two are separated by 5.3×10^{12} m. Find the comet's speed at that point in its orbit.

Solution

Recognize the principle

The Sun–comet system is an isolated system with zero net external torque, so its angular momentum must be conserved.

Sketch the problem

The following sketch will help us organize the information.

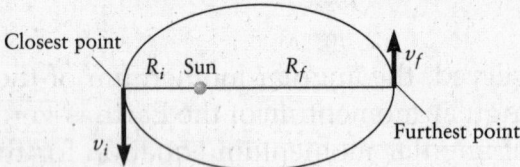

Identify the relationships

Conservation of angular momentum implies,

$$L_i = L_f \;\Rightarrow\; I_i\omega_i = I_f\omega_f$$

The comet is the system of interest, and can be modeled as a point mass revolving around the Sun. The comet has a moment of inertia about the Sun of,

$$I_{comet} = m_{comet}R^2$$

So, for the two points of interest we have,

$$I_i = m_{comet}R_i^2 \quad \text{and} \quad I_f = m_{comet}R_f^2$$

In both cases, the angular speed is related to the linear speed by the expression,

$$v = R\omega \longrightarrow \omega = \frac{v}{R}$$

Solve

Inserting both our moments of inertia and our expression for the angular speed in terms of the linear speed, we get,

$$m_{comet}R_i^2\!\left(\frac{v_i}{R_i}\right) = m_{comet}R_f^2\!\left(\frac{v_f}{R_f}\right)$$

$$R_iv_i = R_fv_f$$

Solving this equation for the final speed,

$$v_f = \frac{R_iv_i}{R_f}$$

Finally, we insert the numerical values and get,

$$v_f = \frac{(8.9 \times 10^{10}\ \text{m})(54\ \text{km/s})}{5.3 \times 10^{12}\ \text{m}}$$

$$v_f = \boxed{0.91\ \text{km/s}}$$

What does it mean?

Halley's comet moves more than 50 times faster at its closest approach to the Sun than it does at the far point of its orbit.

P9.56 A ball of mass 3.0 kg is tied to a string of length 2.0 m. The other end of the string is fastened to a ceiling, and the ball is set into circular motion as shown in Figure P9.56. If $\theta = 30°$, what is the magnitude of the angular momentum of the ball with respect to the vertical axis that passes through the center of the circle (shown dashed in the figure)?

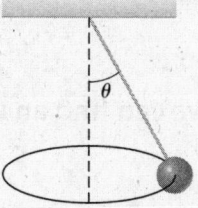

Figure P9.56

Solution

Recognize the principle

In order to determine the angular momentum of the ball we'll need to determine its moment of inertia and angular speed.

Sketch the problem

The sketch is provided in the problem.

Identify the relationships

The magnitude of the angular momentum is given by the expression,

$$L = I\omega$$

We can treat the ball as a point particle. The moment of inertia of a point particle is mR^2. We are given the mass, but need to determine the radius of the circular path the ball follows using the geometry of the problem,

$$r = l\sin\theta$$

where l is the length of the string. The angular speed is not given so we'll need to apply Newton's second law to the ball in both the vertical and horizontal directions. In the horizontal direction, we define in toward the center of the circular path as the positive x direction. The acceleration in this direction is centripetal acceleration, so applying Newton's second law we have,

$$\Sigma F_x = ma_x \longrightarrow T\sin\theta = \frac{mv^2}{r}$$

where $T\sin\theta$ is the component of the string's tension directed in toward the center of the circular path.

Defining up as the positive y direction and applying Newton's second law in this direction we have,

$$\sum F_y = ma_y$$
$$T\cos\theta - mg = 0 \longrightarrow T\cos\theta = mg$$

Once we solve this expression for the speed v we can use the following expression to determine the ball's angular speed:

$$v = r\omega \longrightarrow \omega = \frac{v}{R}$$

Solve

We first solve the vertical equation for T,

$$T = \frac{mg}{\cos\theta}$$

Then inserting this expression into the horizontal equation, we can find an expression for the linear speed:

$$\frac{mg}{\cos\theta}\sin\theta = \frac{mv^2}{r}$$

$$g\tan\theta = \frac{v^2}{r} \longrightarrow v = \sqrt{gr\tan\theta}$$

Our original equation of angular momentum can then be written,

$$L = I\omega = (mr^2)\left(\frac{v}{r}\right) = mrv = mr\sqrt{gr\tan\theta} = mr^{3/2}\sqrt{g\tan\theta}$$

And since $r = l\sin\theta$,

$$L = m(l\sin\theta)^{3/2}\sqrt{g\tan\theta}$$

Inserting the numerical values,

$$L = (3\text{ kg})[(2.0\text{ m})\sin 30°]^{3/2}\sqrt{[9.8\text{ m/s}^2\tan(30°)]}$$

$$L = \boxed{7.1\text{ kg}\cdot\text{m}^2/\text{s}}$$

What does it mean?

The ball can only swing at a 30° angle if it has the corresponding linear (and angular) speed. Therefore, the angle of the string is determined by the angular momentum of the ball!

P9.59 A child of mass 35 kg stands at the edge of a merry-go-round of mass 140 kg and radius 2.5 m, and both are initially at rest. The child then walks along the edge of the merry-go-round until she reaches a point opposite her starting point as measured on the ground (point B in Fig. P9.59). How far does the child walk as measured relative to the merry-go-round?

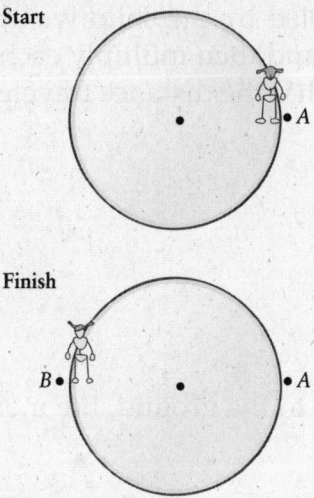

Figure P9.59

Solution

Recognize the principle

Since this is an isolated system with zero net external torque, the angular momentum is conserved in this system.

Sketch the problem

The sketch is provided in the problem.

Identify the relationships

Both the child and the merry-go-round are at rest initially, so the total angular momentum is initially zero. Since angular momentum is conserved, the total angular momentum of the girl plus the merry-go-round must remain at zero. This means that the angular momentum of the girl must have the same magnitude but opposite direction as the angular momentum of the merry-go-round. That is:

$$L_{girl} = L_{mgr}$$
$$I_{girl}\omega_{girl} = I_{mgr}\omega_{mgr}$$

We can therefore find a ratio of angular speeds with respect to the ground:

$$\frac{\omega_{girl}}{\omega_{mgr}} = \frac{I_{mgr}}{I_{girl}}$$

The girl can be modeled as a point mass moving in a circle with the same radius as the merry-go-round, with $I_{girl} = m_{girl}r^2$, while the merry-go-round can be modeled as a disk, with $I_{mgr} = \frac{1}{2}m_{mgr}r^2$.

Solve

Inserting our expressions for the moments of inertia we have,

$$\frac{\omega_{girl}}{\omega_{mgr}} = \frac{\frac{1}{2}m_{mgr}r^2}{m_{girl}r^2} = \frac{m_{mgr}}{2m_{girl}}$$

Since we are trying to determine the distance traveled by the child we multiply both angular speeds by the radius to get a velocity, and then multiply each velocity by a time to get a distance traveled. This means that the distance traveled with respect to the ground has the same ratio:

$$\frac{d_{girl}}{d_{mgr}} = \frac{m_{mgr}}{2m_{girl}}$$

Inserting the values of mass given:

$$\frac{d_{girl}}{d_{mgr}} = \frac{m_{mgr}}{2m_{girl}} = \frac{140 \text{ kg}}{2(35 \text{ kg})} = 2$$

Therefore, for every 2 m the girl walks with respect to the ground, the merry-go-round spins in the opposite direction 1 m.

What does it mean?

So, since the child walks around until she is opposite her starting point as measured on the ground, she has traveled a distance relative to the ground of,

$$\text{Half the circumference of a circle} = \frac{1}{2}(2\pi r) = \pi r = 2.5\pi = 7.85 \text{ m}$$

Therefore, the merry-go-round moves a distance of $\frac{1}{2}(7.85 \text{ m}) = 3.93$ m in the opposite direction relative to the ground. So, relative to the merry-go-round she walks a total distance of 7.85 m + 3.93 m = $\boxed{11.8 \text{ m}}$.

P9.64 Better Approximation! There also exists an item in nature that is extraordinarily spherical, where the approximation of a uniform sphere is astonishingly close. A pulsar is a rotating neutron star, the remnant of a supernova of a star between six and eight solar masses. A typical pulsar rotates at a rate of 600 rpm, has an average mass of twice that of the Sun, and is only 20 km in diameter, or about as big across as Washington D.C. (a) What is the rotational kinetic energy of an average pulsar? (b) What is the corresponding angular momentum?

Solution

Recognize the principle

Both the rotational kinetic energy and the angular momentum can be found from the star's moment of inertia and angular speed.

Sketch the problem

No sketch needed.

Identify the relationships

(a) The rotational kinetic energy is given by the expression

$$KE = \frac{1}{2}I\omega^2$$

We model the pulsar as a uniform sphere rotating about an axis through its center, so the moment of inertia is given by,

$$I = \frac{2}{5}MR^2$$

The angular speed can be expressed in radians as,

$$\omega = (2\pi \text{ rad/rotations})(600 \text{ rotations/min})\left(\frac{1 \text{ min}}{60 \text{ s}}\right) = 6.3 \times 10^1 \text{ rad/s}$$

(b) The magnitude of the angular momentum is given by the expression,

$$L = I\omega$$

Solve

(a) Inserting the expression for the moment of inertia into the kinetic energy expression gives,

$$KE = \frac{1}{2}I\omega^2 = \frac{1}{2}\left(\frac{2}{5}MR^2\right)\omega^2 = \frac{1}{5}MR^2\omega^2 = \frac{1}{5}(2M_{\text{Sun}})R^2\omega^2$$

Inserting the numerical values,

$$KE = \frac{1}{5}(4.0 \times 10^{30} \text{ kg})(2.0 \times 10^4 \text{ m})^2(6.3 \times 10^1 \text{ rad/s})^2 = \boxed{1.3 \times 10^{42} \text{ J}}$$

(b) Inserting the moment of inertia into the angular momentum expression gives,

$$L = \frac{2}{5}MR^2\omega = \frac{2}{5}(2M_{\text{Sun}})R^2\omega$$

Inserting the numerical values,

$$L = \frac{2}{5}(4.0 \times 10^{30} \text{ kg})(2.0 \times 10^4 \text{ m})^2(6.3 \times 10^1 \text{ rad/s}) = \boxed{4.0 \times 10^{40} \text{ N} \cdot \text{s}}$$

What does it mean?

Both the energy and angular momentum stored in a pulsar are far beyond anything comparable here on Earth.

Part D. Additional Worked Examples and Capstone Problems

The following worked examples and capstone problems provide you with additional practice with rotational motion, the connections between rotational and translational quantities, the principle of conservation of energy, and the principle of conservation of angular momentum. Working these problems will give you greater insight into these concepts. Although these problems do not incorporate all the material discussed in this chapter, they do highlight several of the key concepts. If you can successfully solve these problems then you should have confidence in your understanding of these key concepts, so use these problems as a test of your understanding of the chapter material.

WE 9.1 What Rolls the Fastest?

Consider the rolling motion of several different objects—a solid sphere, a cylinder, and a hoop—as they roll down an incline without slipping as shown in the figure. All three start from rest at the top of the incline and are released together. Which object will reach the bottom first?

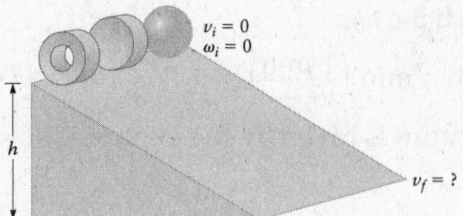

Solution

Recognize the principle

When an object rolls without slipping, the only force that does work on the object is gravity, which is a conservative force. Hence, mechanical energy is conserved, and we can apply the conservation of mechanical energy condition to find the speed of each object when it reaches the bottom of the incline. This final speed will depend on the moment of inertia. The object with the fastest final speed will be the one that gets to the bottom first because it will have had the largest velocity all along the incline.

Sketch the problem

The problem is sketched in the figure. The three objects have the same initial and final heights, and all start with $v_i = 0$.

Identify the relationships

We have already done some of this calculation in connection with the rolling ball in the textbook. As in that case,

$$KE_i + PE_i = KE_f + PE_f$$

The objects start from rest, so the initial speed is zero and $KE_i = 0$. Using this information plus the relation for rolling motion $\omega_f = v_f/R$ leads to the following,

$$mgh_i = \frac{1}{2}mv_f^2 + \frac{1}{2}I\omega_f^2 + mgh_f = \frac{1}{2}mv_f^2 + \frac{1}{2}I\left(\frac{v_f}{R}\right)^2 + mgh_f$$

The objects in the figure start from the top of the incline at a height h, so we can take $h_i = h$ and $h_f = 0$, and

$$mgh = \frac{1}{2}mv_f^2 + \frac{1}{2}I\left(\frac{v_f}{R}\right)^2 = v_f^2\left(\frac{m}{2} + \frac{I}{2R^2}\right)$$

Solve

Solving for v_f^2,

$$v_f^2 = \frac{mgh}{m/2 + (I/2mR^2)} \tag{1}$$

To compare the results for different objects (different moments of inertia, I), it is useful to divide both the top and bottom in Equation (1) by a factor of $m/2$, which leads to

$$v_f^2 = \frac{2gh}{1 + (I/mR^2)}$$

$$v_f = \sqrt{\frac{2gh}{1 + (I/mR^2)}} \tag{2}$$

Thus, as expected, each object's speed will depend on its moment of inertia. Because the moment of inertia is always proportional to the mass, the mass always cancels when we take the ratio I/m. Consulting Table 8.2 in your textbook we find the following: for a sphere, $I = 2mR^2/5$; for a cylinder, $I = mR^2/2$; and for a hoop (with no spokes), $I = mR^2$. Inserting these values into our result for v_f gives

$$\text{sphere: } v_f = \sqrt{\frac{2gh}{1 + [(2mR^2/5)/(mR^2)]}} = \sqrt{\frac{2gh}{1 + \frac{2}{5}}} \approx 1.20\sqrt{gh}$$

$$\text{cylinder: } v_f = \sqrt{\frac{2gh}{1 + [(mR^2/2)/(mR^2)]}} = \sqrt{\frac{2gh}{1 + \frac{1}{2}}} \approx 1.15\sqrt{gh}$$

$$\text{hoop: } v_f = \sqrt{\frac{2gh}{1 + [(mR^2)/(mR^2)]}} = \sqrt{\frac{2gh}{1 + 1}} = \sqrt{gh}$$

Hence, for these three shapes, the sphere will roll the fastest.

What does it mean?

The "fastest" rolling or rotating object will be the one with the smallest moment of inertia for given values of m and R. We can see that that is true by examining the denominator in the general expression for v_f (Eq. 2). In physical terms, a small moment of inertia is produced by placing as much of the mass as possible as close as possible to the rotation axis (see the definition of I in Chapter 8). In addition, the mass and the radius of the wheel both cancel from the final result for v_f, due to the general dependence of the moment of inertia on m and R. These considerations apply to the design of wheels and other rotating parts such as bearings.

WE 9.2 More Playground Physics: Kinetic Energy of a Merry-Go-Round

Let's revisit Example 9.5 (Fig. 9.12) in your textbook, in which a child steps onto a merry-go-round. Find the kinetic energy of the system (a) before and (b) after the child gets on the merry-go-round. Assume the mass of the child is 30 kg, the mass of the merry-go-round is 100 kg, and the radius of the merry-go-round is R_{mgr} = 2.0 m. Also take the initial angular velocity of the merry-go-round to be ω_i = 2.0 rad/s.

Solution

Recognize the principle

To find the initial and final kinetic energies, we need to know the initial and final angular velocities and the corresponding moments of inertia. They were found in Example 9.5 in your textbook, where ω_f was calculated using the principle of conservation of angular momentum. The kinetic energy of a rotating object is

$$KE = \frac{1}{2}I\omega^2$$

Sketch the problem

The problem is sketched in Figure 9.12 in your textbook.

Identify the relationships

Because the child is initially at rest, her initial kinetic energy is zero. All the (initial) kinetic energy is therefore associated with the merry-go-round, so

$$KE_i = \frac{1}{2} I_{mgr} \omega_i^2 \tag{1}$$

For the final kinetic energy, we have to account for the kinetic energy of both the merry-go-round and the child:

$$KE_f = \frac{1}{2} I_{mgr} \omega_f^2 + \frac{1}{2} I_c \omega_f^2 \tag{2}$$

where ω_f is the final angular velocity.

Solve

(a) Inserting the given value of ω_i into Equation (1) along with the moment of inertia of the merry-go-round from Example 9.5, we have

$$KE_i = \frac{1}{2} I_{mgr} \omega_i^2 = \frac{1}{2}\left(\frac{1}{2} m_{mgr} R_{mgr}^2\right) \omega_i^2 = \frac{1}{4}(100 \text{ kg})(2.0 \text{ m})^2 (2.0 \text{ rad/s})^2$$

$$KE_i = \boxed{400 \text{ J}}$$

(b) The final angular velocity of the merry-go-round was calculated in Example 9.5. There we found that the final (ω_f) and initial (ω_i) angular velocities are related by

$$\omega_f = \left(\frac{m_{mgr}}{m_{mgr} + 2m_c}\right)\omega_i$$

Inserting our values for the masses and for ω_i gives

$$\omega_f = \left[\frac{100 \text{ kg}}{(100 \text{ kg}) + 2(30 \text{ kg})}\right](2.0 \text{ rad/s}) = 1.3 \text{ rad/s}$$

The final kinetic energy is then (Eq. 2)

$$KE_f = \frac{1}{2} I_{mgr} \omega_f^2 + \frac{1}{2} I_c \omega_f^2 = \frac{1}{2}(I_{mgr} + I_c)\omega_f^2$$

$$KE_f = \frac{1}{2}\left(\frac{1}{2} m_{mgr} R^2 + m_c R^2\right)\omega_f^2$$

$$= \frac{1}{2}\left[\frac{1}{2}(100 \text{ kg})(2.0 \text{ m})^2 + (30 \text{ kg})(2.0 \text{ m})^2\right](1.3 \text{ rad/s})^2$$

$$KE_f = \boxed{270 \text{ J}}$$

What does it mean?

Comparing our results for the initial and final kinetic energies, we see that about half the kinetic energy is "lost." Where did this energy go? This example is again similar to the completely inelastic collisions studied in Chapter 7, in which two objects stick together and move with a common final velocity after the collision.

Here, the two objects—the merry-go-round and the child—stick together and move with a common final angular velocity. A portion of the initial kinetic energy is "lost" to the "sticking process." Typically, this energy would appear as heat energy when the child makes contact with the merry-go-round and there is a frictional force between her shoes and the merry-go-round.

CP 9.1 Conservation of Energy Including Rotational Motion

Consider the hypothetical child's toy shown in the figure. At the start of the track a ball ($m = 100$ g) is launched using a spring ($k = 10$ N/m). After rolling without slipping down the 3-m high track the ball travels upside down around a circular loop-the-loop, 1-m in radius. At the end of the track the ball bumps into an identical spring at point C. (a) If the initial spring is compressed 50 cm before launching the ball down the track, how much will the spring at the end of the track be compress by the collision with the ball? (b) Determine the speed of the ball at point B. (c) Is this speed sufficient to keep the ball on the track without falling off?

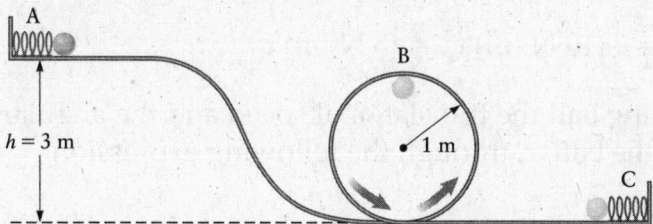

Solution

Recognize the principle

The main principle we can apply is conservation of energy. We'll need to take into account the potential energy of the spring, gravitational potential energy, and translational and rotational kinetic energies.

Sketch the problem

Refer to the figure in the problem.

Identify the relationships

We can begin by defining a few subscripts in our variables. Let the subscript "A" refer to the starting point, "B" be at point B and "C" for point C. Using this notation and applying the principle of conservation of energy we can write the following expression:

$$E_A = E_B = E_C$$

In words, we can describe this as follows: We start off with a certain amount of mechanical energy at point A and still have all this energy at point B and at point C since we are neglecting air resistance and any other non-conservative forces.

Now, if we define the origin of our vertical axis at the bottom of the hill, then we can define the gravitational potential energy in terms of the height above this point. Thinking about all the forms of energy at each point we can write the following expressions:

$$E_A = PE_{spring} + PE_{grav}$$

$$= \frac{1}{2}kx_A^2 + mgh$$

$$= \frac{1}{2}(10 \text{ N/m})(0.50 \text{ m})^2 + (0.10 \text{ kg})(9.8 \text{ m/s}^2)(3 \text{ m})$$

$$= 4.19 \text{ J}$$

$$E_B = KE_{tran, B} + KE_{rot, B} + PE_{B, grav}$$

$$= \frac{1}{2}mv_B^2 + \frac{1}{2}I\omega_B^2 + mgy_B$$

$$= \frac{1}{2}(0.10 \text{ kg})v_B^2 + \frac{1}{2}I\omega_B^2 + (0.10 \text{ kg})(9.8 \text{ m/s}^2)(2 \text{ m})$$

$$= (0.05 \text{ kg})v_B^2 + \frac{1}{2}I\omega_B^2 + 1.96 \text{ J}$$

$$E_C = PE_{spring} = \frac{1}{2}kx_C^2 = \frac{1}{2}(10 \text{ N/m})x_C^2 = (5 \text{ N/m})x_C^2$$

We also know that for a rolling ball the translational speed and the angular speed are related by the radius of the ball, r, through the following expression,

$$v = r\omega \longrightarrow \omega = \frac{v}{r}$$

and the moment of inertia for a solid sphere rotating about an axis through its center is,

$$I = \frac{2}{5}mr^2$$

Solve

Now we can solve for all of our unknowns by substituting these expressions into our general expression for energy conservation,

$$E_A = E_B = E_C$$

(a) Since the total energy at point C is equal to the total energy at point A we can equate the two and solve for the compression of the spring.

$$E_A = E_C$$

$$4.19 \text{ J} = (5 \text{ N/m})x_C^2$$

$$x_C = \sqrt{\frac{4.19 \text{ J}}{5 \text{ N/m}}} = \boxed{0.92 \text{ m}}$$

(b) To solve for the speed of the ball at point B we can equate the energy at point B to the energy at point A.

$$E_A = E_B$$

$$4.19 \text{ J} = (0.05 \text{ kg})v_B^2 + \frac{1}{2}I\omega_B^2 + 1.96 \text{ J}$$

$$4.19 \text{ J} = (0.05 \text{ kg}) v_B^2 + \frac{1}{2}\left(\frac{2}{5}(0.10)r^2\right)\left(\frac{v_B}{r}\right)^2 + 1.96 \text{ J}$$

$$v_B = \sqrt{\frac{4.19 \text{ J} - 1.96 \text{ J}}{0.05 \text{ kg} + \frac{1}{5}(0.10 \text{ kg})}} = \boxed{5.64 \text{ m/s}}$$

(c) We can determine the minimum speed required to keep the ball on the track by applying Newton's second law. Let's start by drawing the free-body diagram of the ball at point B.

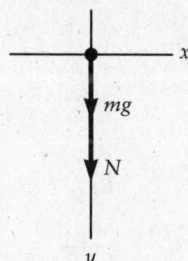

Applying Newton's second law in the y direction and recognizing that the acceleration of the ball at this point is centripetal acceleration we have,

$$\Sigma F = ma = ma_C$$

$$N + mg = m\left(\frac{v^2}{r}\right)$$

$$v = \left[\frac{r(N + mg)}{m}\right]^{1/2}$$

The minimum speed required for the ball to stay on the track is that for which $N = 0$ at the top. Substituting in the numerical values we find,

$$v_{\min} = \left[\frac{r(N + mg)}{m}\right]^{1/2} = \left[\frac{(1 \text{ m})(0 + (0.10 \text{ kg})(9.8 \text{ m/s}^2))}{0.10 \text{ kg}}\right]^{1/2} = 3.13 \text{ m/s}$$

Therefore, the ball is moving fast enough at point B to stay on the track.

What does it mean?

Note that the radius of the ball was not needed to solve the problem so the results are correct for any ball of the same mass. Although this seemed like a difficult problem, the key concept was energy conservation. By keeping track of the different forms of energy we were able to solve for various unknowns.

CP 9.2 Conservation of Angular Momentum

Consider the following rotational contraption in which a uniform disk ($M = 1.0$ kg, $R = 50$ cm) rotates without friction about an axis through its center. At the edge of the disk is fastened a container of 100-, 20-g ball bearings. As the disk rotates a laser beam strikes a sensor on the container and a spring shoots one ball bearing each rotation inward to a container placed half way to the center of the disk (neglect the mass of the containers). Also at the edge of the disk sits a block of mass

$m_{block} = 0.5$ kg. Static friction ($\mu_S = 0.4$) exists between the block and the disk and is responsible for the block staying in place. The disk is set in motion with an angular speed of 2.0 rad/s. (a) Determine the angular speed of the disk when the block slips off the disk. (b) How many times does the disk rotate before the block slides off?

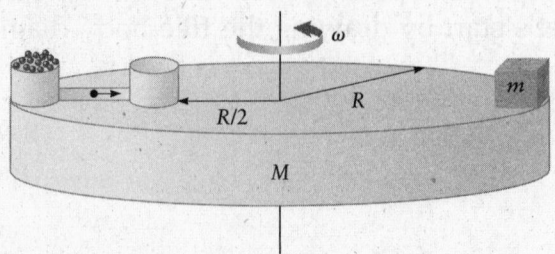

Solution

Recognize the principle

We can apply Newton's second law of motion to the block in order to determine the maximum rotational speed of the disk for which the block will not slide. Since the disk rotates without friction we can treat the entire contraption as an isolated system. Therefore, angular momentum is conserved. In order to calculate the angular momentum of the system we will need to know the moment of inertia of the disk and the masses on the disk.

Sketch the problem

In addition to using the figure provided in the problem, we can draw a free-body diagram for the block,

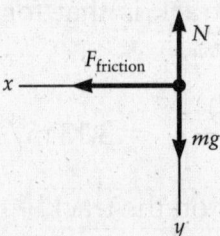

Identify the relationships

(a) Applying Newton's second law to the block in both the x and y directions we find,

$$\Sigma F_x = F_{friction} = m_{block}a_x$$
$$\Sigma F_y = m_{block}g - N = 0 \rightarrow N = m_{block}g$$

The acceleration in the x direction is centripetal acceleration and is given by the expression,

$$a_x = a_c = \frac{v^2}{R} = \frac{(R\omega)^2}{R} = R\omega^2$$

Combining these expression we solve for the angular speed ω,

$$\omega = \sqrt{\frac{a_x}{R}} = \sqrt{\frac{F_{friction}/m_{block}}{R}} = \sqrt{\frac{\mu_s N}{R m_{block}}} = \sqrt{\frac{\mu_s m_{block} g}{R m_{block}}} = \sqrt{\frac{\mu_s g}{R}}$$

(b) The angular momentum of a disk rotating about an axis through its center is given by the expression,

$$I_{disk} = \frac{1}{2}MR^2$$

and the rotational inertia of a mass, m, a distance r from an axis is given by,

$$I_{mass} = mr^2$$

Letting n represent the number of ball bearings that have rolled to the outer-most container we can write the following expression for the angular momentum of the system,

$$L = L_{disk} + L_{ball\ bearings\ at\ R/2} + L_{ball\ bearings\ at\ R} + L_{block}$$

$$= \left(\frac{1}{2}MR^2 + (100 - n)m_{ball\ bearing}R^2 + n\,m_{ball\ bearing}\left(\frac{R}{2}\right)^2 + m_{block}R^2\right)\omega$$

Once we find the angular speed when the block slides off the disk, we can determine the angular momentum of the system at this point in time and equate it to the initial angular momentum of the system, $L_i = L_f$. This will allow us to solve for the number of ball bearings, n, which have moved.

Solve

(a) Plugging in the values given in the problem we solve for the angular speed when the block just begins to slip,

$$\omega = \sqrt{\frac{\mu_s g}{R}} = \sqrt{\frac{(0.4)(9.8\ \mathrm{m/s^2})}{0.50\ \mathrm{m}}} = \boxed{2.8\ \mathrm{rad/s}}$$

(b) Conserving angular momentum we have,

$$L_i = L_f$$

$$\left(\frac{1}{2}MR^2 + (100)m_{ball\ bearing}R^2 + m_{block}R^2\right)\omega_i$$

$$= \left(\frac{1}{2}MR^2 + (100 - n)m_{ball\ bearing}R^2 + n\,m_{ball\ bearing}\left(\frac{R}{2}\right)^2 + m_{block}R^2\right)\omega_f$$

Rearranging this expression and solving for n we find,

$$n = \frac{2(M + 200\,m_{ball\ bearing} + 2m_{block})(\omega_f - \omega_i)}{3m_{ball\ bearing}\,\omega_f}$$

Substituting in the numerical values,

$$n = \frac{2(1.0\ \mathrm{kg} + 200(0.02\ \mathrm{kg}) + 1\ \mathrm{kg})(2.8\ \mathrm{rad/s} - 2.0\ \mathrm{rad/s})}{3(0.02\ \mathrm{kg})(2.0\ \mathrm{rad/s})}$$

$$= \boxed{80}$$

Therefore, the disk will rotate 80 times before the angular speed reaches the point where the block will slide off.

What does it mean?

Notice that the radius of the disk was not required for the answer. Again, at first glance this problem seemed difficult, however, by applying the Newton's second law and the conservation of angular momentum we were able to solve it quite easily.

Part E. MCAT Review Problems and Solutions

PROBLEMS

1. Consider the following three objects with the same mass and radius. Released at the same time from the top of an incline, they roll without slipping. Which one will reach the bottom of the incline first?

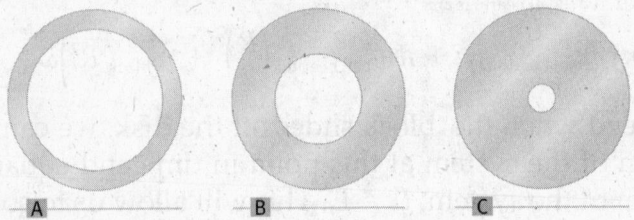

   (a) A
   (b) B
   (c) C
   (d) All at the same time.

2. A wheel with a moment of inertia of 0.02 kg·m² has a net torque acting on it of 0.20 N·m and is free to rotate about its axle. If it starts from rest, what rotational kinetic energy will it have 10.0 s later?
   (a) 0.04 J
   (b) 40 J
   (c) 100 J
   (d) 500 J

3. The total kinetic energy of a thrown, spinning baseball depends on:
   (a) Its linear speed but not its rotational speed.
   (b) Its rotational speed but not its linear speed.
   (c) Both its linear and rotational speeds.
   (d) Neither its linear nor rotational speeds.

4. Consider a ball with a mass of 7.0 kg, moment of inertia of 2.8×10^{-2} kg·m², and radius of 0.10 m rolling without slipping at a translational speed of 4.0 m/s. What is its total kinetic energy?

(a) 45 J

(b) 32 J

(c) 11 J

(d) 78 J

5. A gyroscope has a moment of inertia of 0.14 kg·m^2 and an angular speed of 15.0 rad/s. Friction in the bearings causes its speed to reduce to zero in 30 s. What is the magnitude of the average frictional torque on the gyroscope?

   (a) 3.3×10^{-2} N·s

   (b) 8.1×10^{-2} N·s

   (c) 14×10^{-2} N·s

   (d) 7.0×10^{-2} N·s

6. A cylinder with its mass concentrated toward the center has a moment of inertia of $0.1\,mr^2$. If this cylinder is rolling without slipping along a level surface with a translational speed v, what is the ratio of its rotational kinetic energy to its translational kinetic energy?

   (a) 1/10

   (b) 1/5

   (c) 1/2

   (d) 1/1

7. Consider a uniform solid sphere rolling without slipping down an incline of height 3.0 m after starting from rest. In order to calculate its speed at the bottom of the incline, one needs to know:

   (a) The mass of the sphere.

   (b) The radius of the sphere.

   (c) The mass and the radius of the sphere.

   (d) No additional information than what is given in the problem.

8. A figure skater with arms initially extended starts spinning on the ice at and angular speed of 3 rad/s. She then pulls her arms in close to her body. Which of the following results?

   (a) A smaller angular speed.

   (b) A greater angular speed.

   (c) A greater angular momentum.

   (d) A smaller angular momentum.

9. The Earth's gravity exerts no torque on a satellite orbiting the Earth in an elliptical orbit. Compare the motion of the satellite at the point nearest the Earth to the motion at the farthest from the Earth. At these two points:

   (a) The tangential velocities are the same.

   (b) The angular velocities are the same.

(c) The angular momenta are the same.

(d) The kinetic energies are the same.

10. An object of mass m and moment of inertia I has a rotational kinetic energy KE_{rot}. Its angular momentum is:

(a) $\dfrac{I}{2m}$

(b) $\sqrt{2I \cdot KE_{rot}}$

(c) $\sqrt{2m \cdot KE_{rot}}$

(d) None of the above.

SOLUTIONS

1. *MCAT strategies*

Rather than go through a full calculation to solve this problem, we can rely on our experience with rotating objects and our understanding of the moment of inertia. As was discussed in WE9.1, the speed of an object that rolls without slipping down an incline is related to the moment of inertia of that object. The "fastest" rolling or rotating object will be the one with the smallest moment of inertia for a given value of m and R. Since all three of these objects have the same radius and mass, the object which has more of its mass distributed closer the axis of rotation will have the smallest rotational inertia. Object C has the smallest rotational inertia and therefore reaches the bottom of the incline first. So, the correct answer is (c).

2. *MCAT strategies*

No answers can be easily eliminated so we proceed with our problem-solving strategy.

Recognize the principle

The rotational kinetic energy of an object is related to its rotational inertia and rotational speed. We can apply Newton's second law for rotational motion and to determine the angular acceleration of the wheel. Using this acceleration and our kinematic equations for rotational motion we can determine the angular speed of the object after 10 s and then calculate its rotational kinetic energy.

Sketch the problem

No sketch required.

Identify the relationships

The rotational kinetic energy is given by the expression

$$KE_{rot} = \frac{1}{2} I \omega^2$$

Applying Newton's second law for rotational motion we can determine the angular acceleration of the wheel,

$$\Sigma \tau = I \alpha \longrightarrow \alpha = \frac{\Sigma \tau}{I}$$

And using one of our kinematic expression for rotational motion we have,

$$\omega = \omega_i + \alpha t$$

Solve

Combining these expressions and noting that $\omega_i = 0$ we have,

$$KE_{rot} = \frac{1}{2} I\omega^2 = \frac{1}{2} I(\alpha t)^2 = \frac{1}{2} I\left(\frac{\Sigma \tau}{I} t\right)^2$$
$$= \frac{(\Sigma \tau)^2 t^2}{2I}$$

Now, substituting in the numerical values we find,

$$KE_{rot} = \frac{(\Sigma \tau)^2 t^2}{2I} = \frac{(0.2 \text{ N} \cdot \text{m})^2 (10 \text{ s})^2}{2(0.02 \text{ kg} \cdot \text{m}^2)} = 100 \text{ J}$$

So, the correct answer is (c).

What does it mean?

Checking the units of our final answer we see that we have manipulated our expressions correctly to obtain the correct units for energy.

3. MCAT strategies

To answer this question we simply need to understand the relationships of kinetic energy. The total kinetic energy of a translating and rotating object is given by the expression,

$$KE_{total} = KE_{translational} + KE_{rotational}$$
$$= \frac{1}{2} mv_{CM}^2 + \frac{1}{2} I\omega^2$$

So the total kinetic energy depends on the mass, translational speed, moment of inertia, and angular speed. So, the correct answer is (c). No further analysis is required.

4. MCAT strategies

No answers can be easily eliminated so we proceed with our problem-solving strategy.

Recognize the principle

The total kinetic energy is the sum of the translational and rotational kinetic energies. Also, the speed and angular speed are related by the radius of the ball.

Sketch the problem

No sketch required.

Identify the relationships

The total kinetic energy of a translating and rotating object is given by the expression,

$$KE_{total} = KE_{translational} + KE_{rotational}$$
$$= \frac{1}{2} mv_{CM}^2 + \frac{1}{2} I\omega^2$$

Also, the speed and angular speed for a ball rolling without slipping are related through the following expression,

$$\omega = \frac{v_{CM}}{r}$$

Solve

Combining these two expressions we have the following,

$$
\begin{aligned}
KE_{total} &= \frac{1}{2}mv_{CM}^2 + \frac{1}{2}I\omega^2 \\
&= \frac{1}{2}mv_{CM}^2 + \frac{1}{2}I\left(\frac{v_{CM}}{r}\right)^2 \\
&= \frac{1}{2}v_{CM}^2\left(m + \frac{I}{r^2}\right)
\end{aligned}
$$

Substituting in the numerical values we find,

$$KE_{total} = \frac{1}{2}v_{CM}^2\left(m + \frac{I}{r^2}\right) = \frac{1}{2}(4\text{ m/s}^2)^2\left(7.0\text{ kg} + \frac{2.8 \times 10^{-2}\text{ kg}\cdot\text{m}^2}{(0.10\text{ m})^2}\right)$$

$$= 8(7 + 2.8) = 8(9.8) = 78\text{ J}$$

So, the correct answer is (d).

What does it mean?

Again, checking the units of our final answer we see that we have manipulated our expressions correctly to obtain the correct units for energy.

5. MCAT strategies

No answers can be easily eliminated so we proceed with our problem-solving strategy.

Recognize the principle

We can apply Newton's second law for rotational motion to determine the torque on the gyroscope. Since we are not given the angular acceleration of the gyroscope we must apply one of our kinematic expressions for rotational motion to determine the angular acceleration from the information provided.

Sketch the problem

No sketch required.

Identify the relationships

Applying Newton's second law for rotational motion we have,

$$\Sigma\tau = I\alpha \longrightarrow \tau = I\alpha$$

The angular acceleration of the gyroscope can be determined using the following kinematic expression,

$$\omega = \omega_i + \alpha t \longrightarrow \alpha = \frac{\omega - \omega_i}{t}$$

Solve

Combining these expressions and noting that $\omega = 0$, $\omega_i = 15 \text{ rad/s}$, and $I = 0.14 \text{ kg} \cdot \text{m}^2$ we have,

$$\tau = I\alpha = I\left(\frac{\omega - \omega_i}{t}\right) = \frac{-I\omega_i}{t}$$

$$= \frac{-(0.14 \text{ kg} \cdot \text{m}^2)(15 \text{ m/s})}{30 \text{ s}} = -0.07 \text{ N} \cdot \text{m}$$

So the magnitude of the average torque is 0.07 N·m and the correct answer is (d).

What does it mean?

Checking the units of our final answer confirms that we have the units of torque.

6. MCAT strategies

To answer this question we simply need to know the relationships for translational and kinetic energies and how translational and angular speeds are related for a rolling object which is not slipping. The translational kinetic energy is given by $KE_{trans} = \frac{1}{2}mv^2$ and the rotational kinetic energy is given by $KE_{rot} = \frac{1}{2}I\omega^2$. The angular speed and translational speed are related by the radius, $\omega = v/r$. Combining these expressions into a ratio we find,

$$\frac{KE_{rot}}{KE_{trans}} = \frac{\frac{1}{2}I\omega^2}{\frac{1}{2}mv^2} = \frac{(0.1\ mr^2)(v/r)^2}{mv^2} = 0.1 = 1/10$$

So the correct answer is (a). No further analysis is required.

7. MCAT strategies

To answer this question we need to know the relationships which relate translational and rotational motion for a rolling object, our expressions for kinetic energies, and the expression for gravitational potential energy. If we were to actually solve this problem we would apply the principle of conservation of mechanical energy. However, the question is simply asking for what quantities are needed. To start, let's write down the expressions which are relevant,

$$KE_{rot} = \frac{1}{2}I\omega^2$$

$$KE_{trans} = \frac{1}{2}mv^2$$

$$PE_{grav} = mgh$$

Though we may not remember the expression for the moment of inertia of a sphere, we know that it is proportional to mr^2. Since we're not looking for a numerical answer let's just say that $I = mr^2$. Also, we know that $\omega = v/r$. Combining these expressions and applying the principle of conservation of mechanical energy we have,

$$PE_{grav} = KE_{trans} + KE_{rot}$$

$$mgh = \frac{1}{2}mv^2 + \frac{1}{2}I\omega^2$$

$$mgh = \frac{1}{2}mv^2 + \frac{1}{2}(mr^2)\left(\frac{v}{r}\right)^2$$

$$gh = \frac{1}{2}v^2 + \frac{1}{2}v^2$$

Our final expression only depends on the height of the ramp, h. So no additional information is required and the correct answer is (d).

8. *MCAT strategies*

To answer this question we simply need to know the principle of conservation of angular momentum. The angular momentum of a rotating object is given by the expression,

$$L = I\omega$$

As long as the net external torque on the figure skater is zero (which we can assume) then her angular momentum will remain constant. By pulling her arms in closer to her body, she reduces her moment of inertia. For her angular momentum to remain constant, her angular speed must increase. So, the correct answer is (b).

9. *MCAT strategies*

Since the net torque on the satellite is zero the principle of conservation of angular momentum applies. This means that the angular momentum of the satellite will remain constant throughout its orbit. So, the correct answer is (c). No additional analysis is required.

10. *MCAT strategies*

To answer this question we simply need to know the expression for the angular momentum and rotational kinetic energy of a rotating object. These expressions are given by,

$$KE_{rot} = \frac{1}{2}I\omega^2$$

$$L = I\omega$$

Rearranging the expression for kinetic energy and solving for ω we have,

$$\omega = \sqrt{\frac{2KE_{rot}}{I}}$$

Substitution this into our expression for angular momentum we find,

$$L = I\omega = I\sqrt{\frac{2KE_{rot}}{I}} = \sqrt{2I \cdot KE_{rot}}$$

So, the correct answer is (b). No further analysis is required.

10 Fluids

Part A. Summary of Key Concepts and Problem-Solving Strategies

KEY CONCEPTS

Pressure and density

The mechanics of fluids is described using the quantities *pressure* and *density*. Pressure is equal to the magnitude of the force per unit area,

$$P = \frac{F}{A}$$

Pressure can be associated with the force on the surface of a container, or it can involve a surface contained within the fluid. The pressure in the atmosphere at the Earth's surface is approximately

$$P_{atm} = 1.01 \times 10^5 \, \text{Pa}$$

The density of a substance is given by

$$\rho = \frac{M}{V}$$

An *incompressible* fluid is one for which the density is independent of pressure; most liquids are incompressible. The density of a gas varies with pressure, so gases are thus *compressible* fluids.

Dependence of pressure on depth

The pressure in a liquid varies with depth as

$$P = P_0 + \rho g h$$

The pressure increases as one moves deeper in a fluid because the pressure below must support the weight of the fluid above.

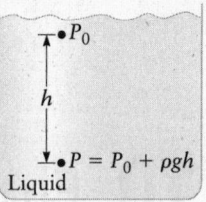

Real fluids

Fluids are composed of molecules, which helps us understand the origin of *viscosity* (i.e., friction), *surface tension*, and *capillary pressure*. Poiseuille's law relates the average flow speed to the viscosity of the fluid and the pressure difference across the ends of the pipe:

$$v_{\text{ave}} = \frac{P_1 - P_2}{8\eta L} r^2$$

where $P_1 - P_2$ is the pressure difference and η is the viscosity. The corresponding flow rate is

$$Q = v_{\text{ave}} A$$

Surface tension γ is the energy needed to increase the surface area of a fluid. The surface tension leads to an extra pressure when a fluid is confined to a narrow capillary. This capillary pressure is

$$P_{\text{cap}} = \frac{2\gamma}{r}$$

APPLICATIONS

Pascal's principle

If the pressure in a fluid at one location in a closed container is changed, this change is transmitted to all locations in the fluid. This principle is the basis of *hydraulics*. Hydraulic devices, such as a lift, are able to amplify forces. When they do so, however, they de-amplify the corresponding displacement so that the work done is conserved.

Archimedes's principle

When an object is submerged in a fluid, the fluid exerts an upward force on the object that is equal to the weight of the fluid that is displaced by the object. This upward force is called the **buoyant force**. The buoyant force on an object can be determined from the following expression:

$$F_B = \rho_L V g$$

where ρ_L is the density of the liquid, V is the volume of liquid displaced, and g is the acceleration due to gravity.

Archimedes's principle and buoyancy

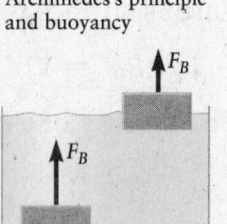

Buoyant force = F_B = weight of fluid *displaced* by object

Principle of continuity

When a fluid flows through a pipe (or similar geometry), the *equation of continuity* tells how the speed v of a fluid is related to the cross-sectional area A of the pipe:

$$v_1 A_1 = v_2 A_2$$

where the subscripts 1 and 2 denote two different sections of the pipe. The work-energy theorem leads to *Bernoulli's equation*, which relates the speed, pressure, and height at two different locations in an ideal fluid:

$$P_1 + \frac{1}{2}\rho v_1^2 + \rho g h_1 = P_2 + \frac{1}{2}\rho v_2^2 + \rho g h_2$$

According to Bernoulli's equation, when the flow speed increases, the pressure must decrease, and vice versa. This principle is one contributing factor to the lift force on an airplane wing.

PROBLEM-SOLVING STRATEGIES

In this chapter we introduced a new force called the buoyant force. The buoyant force is due to the fluid surrounding an object and is equal to the weight of the fluid displaced by the object. Problems involving the buoyant force often require us to apply Newton's second law, so the first of the problem-solving strategies outlines the thought process we should follow while tackling such problems. A second key concept introduced in this chapter was Bernoulli's principle. This principle is essentially the conservation of energy for fluids. We know from our previous applications of the conservation of energy that this is a very powerful tool for tackling complex problems. The second problem-solving strategy outlines the process of applying Bernoulli's principle.

Problem Solving: Applying Newton's 2nd Law in Problems Involving Fluids

---

Recognize the principle

Begin by recognizing the objects of interest and listing all the forces acting on each of them. Remember, the buoyant force is a force on an object due to the fluid it displaces and is directed upward.

Sketch the problem

- Start with a drawing that shows all the objects of interest in the problem along with all the forces acting on each.
- Draw the free-body diagram for each of these objects.

Identify the relationships

Apply Newton's second law for each of the objects of interest and in all directions which are important (usually this is only the vertical direction). The buoyant force on an object can be determined for the following expression:

$$F_B = \rho_L V g$$

where ρ_L is the density of the liquid, V is the volume of liquid displaced, and g is the acceleration due to gravity.

If the object is stationary or moving with a constant velocity then its acceleration is zero. Otherwise, you'll need to be given or asked to find its acceleration.

Solve

Solve for the unknowns. These may be forces, volumes, accelerations, or other quantities not provided.

What does it mean?

Always *consider what your answer means* and check that it makes sense.

Problem Solving: Bernoulli's Equation

Recognize the principle

Bernoulli's equation tells us that the total mechanical energy of a fluid is conserved as it flows from place to place. There is energy associated with the pressure of the fluid, with its motion (kinetic energy), and with its height (gravitational potential energy).

Sketch the problem

Start by defining a coordinate system, usually with the y axis as the vertical direction. Next, determine which points in the fluid are of interest and label them appropriately.

Identify the relationship

The main relationships we'll need to apply are Bernoulli's equation and the continuity equation. Bernoulli's equation is as follows:

$$P_1 + \frac{1}{2}\rho v_1^2 + \rho g h_1 = P_2 + \frac{1}{2}\rho v_2^2 + \rho g h_2$$

The continuity equation is a statement about the conservation of mass, i.e., what flows in must flow out,

$$A_1 v_1 = A_2 v_2$$

Solve

Solve these expression for the unknown quantities such as pressure, speed, height, or area.

What does it mean?

Always *consider what your answer means* and check that it makes sense.

Part B. Frequently Asked Questions

1. *What's the difference between force and pressure? Doesn't a greater force always mean a greater pressure?*

Force and pressure are different yet related quantities. Recall from earlier chapters that a force is simply a push or a pull and has the units of Newton's, N. Pressure on the other hand is a force applied over an area and has the units of a Newton per square meter, N/m^2. We abbreviate N/m^2 to a Pascal, Pa. So the expression which relates force and pressure is

$$P = \frac{F}{A}$$

While it is often the case that a greater force results in a greater pressure, this is only true if the area is held constant. In fact, it's possible to create a greater pressure with a smaller force if the area over which it is applied is sufficiently small.

2. *When calculating the buoyant force on an object, how do I know the volume of the displaced fluid?*

That's a good question and addresses a common difficulty many students have in applying Archimedes's principle. The expression for the buoyant force on an object is

$$F_B = \rho_L V g$$

where ρ_L is the density of the fluid (liquid), V is the volume of liquid displaced, and g is the acceleration due to gravity. The liquid which is displaced by the object is the amount of liquid which has been pushed away as a result of the object, or part of the object, being submerged in the fluid. So, to know how much liquid has been displaced we simply need to determine the volume of the object which is below the surface of the liquid.

3. *How does the size of the lake affect the size of the dam needed to hold it back?*

It is a common misconception that the larger the lake, the thicker the dam required to hold it back. The important factor when constructing a dam is the depth of the water at the dam, not the amount of water it is holding back. We've learned that the pressure in a fluid increases with depth so that the pressure at the bottom of a lake is much greater than the pressure near its surface. This is independent on the surrounding water. That is, the pressure at the bottom of a well of water 100 m deep is the same as at the bottom of a large lake 100 m deep. The following illustration shows two cross-sections of dams used to hold back different sized lakes:

Small lake   Large lake

Since the depth of the water at the dam is the same, the required thickness for the dams is the same as well.

4.   *Is the continuity equation valid for a compressible fluid?*

The equation of continuity we discussed in this chapter applies to fluids whose density remains constant, i.e., incompressible fluids. However, we can generalize this expression to include fluids whose density can vary. To do so, let's look at the expression more carefully. When a fluid flows through a pipe (or similar geometry), the equation of continuity tells how the speed v of a fluid is related to the cross-sectional area A of the pipe:

$$v_1 A_1 = v_2 A_2$$

where the subscripts 1 and 2 denote two different sections of the pipe. Essentially, it is a statement about the conservation of mass. That is, what flows into the pipe must also flow out. With that in mind we can generalize the expression as follows:

$$\text{mass in} = \text{mass out}$$
$$m_1 = m_2$$
$$\rho_1 V_1 = \rho_2 V_2$$
$$\rho_1 (l_1 A_1) = \rho_2 (l_2 A_2)$$
$$\rho_1 (v_1 t) A_1 = \rho_2 (v_2 t) A_2$$
$$\rho_1 v_1 A_1 = \rho_2 v_2 A_2$$

So, the more general equation of continuity which can be applied to compressible fluids is given by $\rho_1 v_1 A_1 = \rho_2 v_2 A_2$.

5.   *When applying Bernoulli's equation, does it matter where I define my coordinate system?*

No, it doesn't matter where you define your coordinate system. To see why this is the case, let's carefully consider Bernoulli's equation,

$$P_1 + \frac{1}{2}\rho v_1^2 + \rho g h_1 = P_2 + \frac{1}{2}\rho v_2^2 + \rho g h_2$$

The only terms in this equation which depend on the coordinate system are the terms involving h, the height of the point of interest above where you've defined zero on your vertical axis. Remember, these terms are really just the gravitational potential energy per volume associated with the fluid. As was the case when we dealt with gravitational potential energy in earlier chapters, the important factor is the difference in potential energy between the two points of interest. This difference is independent of where we've defined zero on our vertical axis.

6.   *As bubbles rise in a glass of soda they get further apart. Does this have something to do with pressure?*

Yes it does have to do with pressure, but several other factors are important as well. At any given time there are several forces acting on the bubble: the force of gravity (although this is negligibly small), the buoyant force from the surrounding fluid, and the drag force as it moves through the fluid. As the bubble rises the pressure from the surrounding soda becomes less. This is evident by examining our pressure at a depth expression

$$P = P_0 + \rho g h$$

As h becomes less, so does P. Since the pressure becomes less the radius of the bubble will increase. This expansion displaces more of the surrounding fluid and, therefore, increases the buoyant force on the bubble according to the expression,

$$F_B = \rho_L V g$$

So the buoyant force increases as the bubble rises. However, with an increase in radius also comes an increase in the magnitude of the drag force according to the expression

$$F_{\text{drag}} = C r v$$

where C is the drag coefficient, r is the radius of the bubble, and v is the speed of the bubble.

Since the buoyant force depends on the volume of the bubble, it increases proportionally to r^3. The drag force increases due to two factors. First, it is directly proportional to r which increases as the bubble rises. Second, the speed of the bubble increases as it rises. It turns out that these two contributing factors result in the drag force increasing at a slower rate than the buoyant force and so the bubble accelerates upward. This acceleration results in the bubbles getting further apart as they rise up through the fluid.

7. *What are ballast tanks used for on ships?*

Ballast tanks are large compartments on ships and submarines which can be filled with water. Filling these compartments with water increases the ballast force (weight) of the ship so that it will sit lower in the water, or in the case of a submarine, so that it can submerge below the ocean's surface. Large cargo ships fill their ballast tanks when they are carrying light loads so that they remain stable in the ocean. Fish have a similar mechanism for adjusting their buoyancy called a swim bladder or gas bladder which they fill or deflate with gas (air) to rise or sink in the water.

8. *Where did the expression "just the tip of the iceberg" come from?*

Though it's not clear when this expression was first used, it means we are only seeing a small fraction of something largely hidden. The reason why only a small portion of an iceberg is visible above the surface of the water has to do with small difference in density between ice and liquid water. The density of ice is 917 kg/m^3 whereas the density of sea water is around 1025 kg/m^3. So, ice is less dense than sea water, but only by a little bit. We can apply Archimedes's principle to determine what fraction of the iceberg's volume is below the surface of the water but there is an easy way to determine this based on the densities of the two substances. The ratio of the densities is equal to the ratio of the volumes,

$$\frac{\rho_{\text{ice}}}{\rho_{\text{water}}} = \frac{917}{1025} = 0.895 = \frac{V_{\text{ice below}}}{V_{\text{ice total}}}$$

This means that roughly 90% of the volume of an iceberg is below the surface of the water. So, we only see the "tip of the iceberg" above the surface of the water.

9. *When a truck passes me at high speeds going in the other direction my car gets pulled toward the truck. Why is that?*

This has to do with Bernoulli's principle. According to this principle, faster flowing fluids have less pressure. Let's look at the expression to see why this is true:

$$P_1 + \frac{1}{2}\rho v_1^2 + \rho g h_1 = P_2 + \frac{1}{2}\rho v_2^2 + \rho g h_2$$

Let's label the side of your car closest to the truck as point 1 and the opposite site to be point 2. The fluid in this case is the air surrounding your car. The height above the ground to midway up your car is the same on both sides of your car so the terms involving h_1 and h_2 cancel. As the truck passes you the speed of the airflow past your car is greater in the side closest to the truck, $v_1 > v_2$. As a result, the air pressure nearest the truck must be less than on the other side of your car, $P_1 < P_2$. This difference in pressure is called a pressure gradient and is responsible for the force you feel pushing your car toward the truck.

10. *What are the bends?*

Scuba diving involves breathing compressed air from a tank while under water. This compressed air is released from the tank (and enters your lungs) at a pressure equal to the surrounding water pressure. If this were not the case, the air would not be able to leave the tank and enter your lungs. This keeps your lungs inflated to their normal size which is quite different than holding your breath as you swim down below the water's surface. When holding your breath, the air inside your lungs compresses as you descend and your lungs shrink in size due to the increased pressure of the surrounding water. As you swim back to the surface your lungs return to their normal size as the air inside them expands. When scuba diving, the air from the tank enters your lungs at/near the pressure of the surrounding water, so your lungs remain at their normal size. When a diver stays under water at a significant depth (30 meters or more) for an appreciable amount of time, some of the nitrogen in the high-pressure air they are breathing dissolves into the water in their body. If they rise to the surface too quickly, this dissolved nitrogen can form bubbles in their body tissue and blood stream much like what happens when you open a warm bottle of soda. This condition is quite painful and can be fatal. To avoid the affect of rapidly decompressing, divers must ascend slowly and spend time waiting at intermediate depths along the way.

Part C. Selection of End-of-Chapter Answers and Solutions

QUESTIONS

Q10.4   Figure Q10.4 is a photograph of a graduated cylinder filled with four fluids. Starting with mercury on the bottom and going up, we have salt water, water, and vegetable oil. In addition, a solid object rests at the interface between each liquid. At the bottom is a steel ball bearing, next an egg, followed by a block of wood, and a table-tennis ball on top. (a) Rank these eight substances in terms of their densities from the highest to the lowest. (b) Which items have a specific gravity greater than 1? Less than 1?

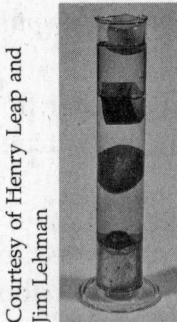

Figure Q10.4

Answer

(a) According to Archemedes's principle, if an object floats with part of its volume above the surface of a liquid, then it must have a smaller overall density than the liquid. Furthermore, if a liquid floats on top of another liquid then it must have a lower density. Based on this, the densities will be ranked from greatest to least in the order they are stacked, counting from the bottom up. (Mercury, steel, salt water, egg, water, wood, vegetable oil, and ping pong ball, which has an "average" density very near to that of air.)

(b) Water has a specific gravity of 1, so all items above water have a specific gravity of less than 1 (wood, vegetable oil, and ping pong ball), and the items below the water level have a specific gravity greater than 1. (Mercury, steel, salt water, egg)

Q10.16 Figure Q10.16 shows a popular demonstration involving a moving fluid. Here an air "jet" is aimed upward and is used to levitate a small object such as a table-tennis ball. The ball is drawn *to the center of the jet*, where the velocity is greatest. This behavior is found when the air jet is directed vertically (Fig. Q10.16 left) and also when the jet is directed at an angle (Fig. Q10.16 right). Use Bernoulli's equation to explain this behavior.

Figure Q10.16

Answer

Bernoulli's equation states that faster moving air exerts a lower pressure. The ball is drawn to the center of the jet because when it moves away from the center of the air jet, the side closer to the center of the jet experiences faster moving air, and therefore less pressure, than the side further from the center. For instance, if it is to the right side of the jet, the pressure on the right side of the ball is higher than the pressure on the left side of the ball, which causes it to move to the left. It should be noted that the ball will return to the middle of the jet if it is perturbed slightly, but if the perturbation is too great, the ball finds itself in relatively still air on both sides, and the pressure difference is no longer great enough to return it to the center of the jet.

PROBLEMS

P10.12   A suction cup works by virtue of a vacuum that is created within the cup. When the cup is pressed against a flat surface, most of the air is forced out, leaving a region of very low pressure. If a suction cup of area 1.0 cm^2 attached to a ceiling is able to support an object with a mass as large as 0.70 kg, what is the pressure inside the suction cup?

Solution

Recognize the principle

The pressure is the force divided by the area.

Sketch the problem

No sketch needed.

Identify the relationships

The suction cup experiences an upward force due to atmospheric pressure, and a smaller downward force due to reduced air pressure inside (which is what we'd like to find). The ceiling actually provides a normal force when the supported weight is below the maximum, but we're looking at the maximum weight case, so $N = 0$. The weight force (mg) plus the force due to the downward pressure ($P_{reduced}$) equals the force due to the upward pressure (P_{atm}). Since, according to our equation for pressure,

$$P = \frac{F}{A} \Rightarrow F = PA$$

our sum of forces condition is:

$$\Sigma F = P_{atm} A - P_{reduced} A - mg = 0$$

Solve

We can solve this expression for the reduced pressure:

$$P_{reduced} = P_{atm} - \frac{mg}{A}$$

Then inserting values,

$$P_{\text{reduced}} = -\frac{(0.7 \text{ kg})(9.8 \text{ m/s}^2)}{(1.0 \text{ cm}^2)}\left(\frac{1.0 \text{ cm}}{0.01 \text{m}}\right)^2 + 1.01 \times 10^5 \text{ Pa} = \boxed{32000 \text{ Pa}}$$

What does it mean?

The pressure under the suction cup is less than 1/3 of atmospheric pressure.

P10.27   A U-tube contains two fluids with densities $\rho_1 = 1000 \text{ kg/m}^3$ and $\rho_2 = 600 \text{ kg/m}^3$ as sketched in Figure P10.27. What is the difference d in the heights of the top surfaces of the two fluids?

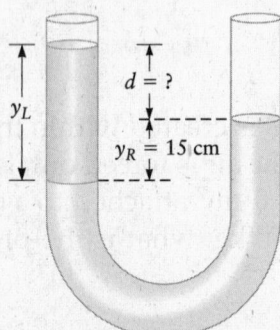

Figure P10.27

Solution

Recognize the principle

Use the concepts of pressure as a function of depth and density.

Sketch the problem

Refer to Figure P10.27.

Identify the relationships

This problem is similar to Example 10.3. Let the depth from the point where the liquids touch to the surface on the right-hand side be y_R:

$$P = P_0 + \rho_1 g y_R$$

Let y_L be the depth of liquid 2 from the surface to the interface:

$$P = P_0 + \rho_2 g y_L$$

At the point of contact of the two liquids we have

$$P_1 = P_0 + \rho_2 g y_R$$

from liquid 2 and

$$P_2 = P_0 + \rho_1 g y_L$$

from liquid 1. The pressure at this point must be equal, that is $P_1 = P_2$, therefore

$$P_0 + \rho_2 g y_L = P_0 + \rho_1 g y_R$$

Solve

Solving for d:

$$\rho_2 g y_L = \rho_1 g y_R$$

$$y_L = \frac{\rho_1 y_R}{\rho_2}$$

The height is the difference $d = y_L - y_R$:

$$d = y_R\left(\frac{\rho_1}{\rho_2} - 1\right)$$

$$d = (0.15 \text{ m})\left(\frac{1000 \text{ kg/m}^3}{600 \text{ kg/m}^3} - 1\right) = 0.10 \text{ m} = \boxed{10 \text{ cm}}$$

What does it mean?

The height of the fluids differ by 10 cm.

P10.38   A block of mass 20 g sits at rest on a plate that is at the top of the fluid on one side of a U-tube, as shown in Figure P10.38. The U-tube contains two different fluids with densities $\rho_1 = 1000 \text{ kg/m}^3$ and $\rho_2 = 600 \text{ kg/m}^3$ and has a cross-sectional area of $A = 5.0 \times 10^{-4} \text{ m}^2$. If the surfaces are offset by an amount h as shown, (a) which side of the U-tube contains the fluid with the greater density? (b) Find h. (Assume h is positive.)

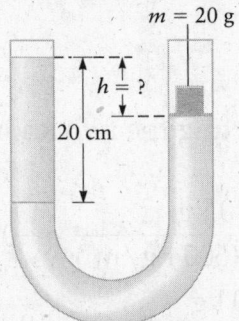

Figure P10.38

Solution

Recognize the principle

The liquids move until the pressure at the interface is equal from both sides.

Sketch the problem

Refer to Figure P10.38.

Identify the relationships

To find the height difference we equate the pressure in liquid 1 to the pressure in liquid 2 at the point of the interface using the expression,

$$P = P_0 + \rho g h$$

Since both liquids are open to the atmosphere at the top, $P_0 = P_{atm}$. The left-hand liquid is a typical column, so:

$$P_L = P_{atm} + \rho_L g(0.20 \text{ m})$$

The "extra" pressure generated by the block is just the weight of the block per unit area, or:

$$P_{block} = \frac{F}{A} = \frac{mg}{A}$$

For the liquid on the right-hand side, the pressure would increase as we went down the right side, reaching a maximum at the bottom of the U-tube, then decrease as we worked our way up the other side. The net pressure at the interface then depends on the height difference between the interface and the top of the right liquid. Including the weight of the block acting on the cross-sectional area of the tube, we can write this as:

$$P_R = P_{atm} + \rho_R g[(0.20 \text{ m}) - h] + \frac{mg}{A}$$

Equating these two

$$P_{atm} + \rho_L g(0.20 \text{ m}) = P_{atm} + \rho_R g[(0.20 \text{ m}) - h] + \frac{mg}{A}$$

Solve

(a) and (b) We can solve this expression for h,

$$\rho_L(0.20 \text{ m}) = \rho_R(0.20 \text{ m}) - \rho_R h + \frac{m}{A}$$

$$-\rho_L(0.20 \text{ m}) + \rho_R(0.20 \text{ m}) + \frac{m}{A} = \rho_R h$$

$$h = 0.20 \text{ m} - \frac{\rho_L(0.20 \text{ m})}{\rho_R} + \frac{m}{A\rho_R}$$

We cannot tell which liquid is more dense, but by inserting the two known densities, we see that only one value for h is reasonable:

$$h = 0.20 \text{ m} - \frac{(1000 \text{ kg/m}^3)(0.20 \text{ m})}{600 \text{ kg/m}^3} + \frac{0.020 \text{ kg}}{5 \times 10^{-4} \text{ m}^2(600 \text{ kg/m}^3)} = -0.067 \text{ m}$$

$$h = 0.20 \text{ m} - \frac{(600 \text{ kg/m}^3)(0.20 \text{ m})}{1000 \text{ kg/m}^3} + \frac{0.020 \text{ kg}}{5 \times 10^{-4} \text{ m}^2(1000 \text{ kg/m}^3)} = \boxed{0.12 \text{ m}}$$

Since the height of the column must be positive, only the second result makes sense, and the $\boxed{\text{liquid on the right side of the column must be denser.}}$

What does it mean?

The less dense liquid on the left rises 12 cm higher than the more dense liquid on the right.

P10.48 You want to build a raft that can hold you (80 kg) plus some supplies (40 kg) for a long trip. You decide to make the raft out of logs that are each 20 cm in diameter and 3.0 m long. If the density of each log is 600 kg/m^3, how many logs do you need?

Solution

Recognize the principle

The buoyant force on the logs is equal to the weight of the water displaced.

Sketch the problem

We begin by drawing the free-body diagram for this situation.

Identify the relationships

We can find the number of logs needed by solving for the buoyant force needed to support not only the logs but also the weight of you and your supplies. If V is the total volume of the raft (which is just barely submerged), the buoyant force is

$$F_B = \rho_{\text{water}} V g$$

Each cylindrical log has a volume of

$$V_{\text{log}} = \pi r^2 l = \pi (0.10 \text{ m})^2 (3 \text{ m}) = 0.0942 \text{ m}^3$$

The mass of the raft can be written as $m_{\text{raft}} = \rho_{\text{log}} V$, and the total weight is:

$$F_{\text{grav}} = (m_{\text{you}} + m_{\text{supplies}} + \rho_{\text{log}} V) g$$

Solve

To just float at the surface, the magnitudes of the weight and buoyant force must be equal:

$$\rho_{\text{water}} V g = (m_{\text{you}} + m_{\text{supplies}} + \rho_{\text{log}} V) g$$

Then solving this expression for the total volume:

$$\rho_{\text{water}} V - \rho_{\text{log}} V = m_{\text{you}} + m_{\text{supplies}}$$

$$V = \frac{m_{\text{you}} + m_{\text{supplies}}}{\rho_{\text{water}} - \rho_{\text{log}}}$$

Inserting values,

$$V = \frac{80 \text{ kg} + 40 \text{ kg}}{1000 \text{ kg/m}^3 - 600 \text{ kg/m}^3} = 0.30 \text{ m}^3$$

Then dividing this total volume by the volume of each log should gives the number of logs (n):

$$n = \frac{V}{V_{\text{log}}} = \frac{0.30 \text{ m}^3}{0.0942 \text{ m}^3} = 3.18$$

What does it mean?

Since this was a minimum to just keep the raft slightly submerged, you can't have any fewer than this, so rounding won't work. This implies you should build your raft from 4 logs .

P10.59 The blood flow rate through the aorta is typically 100 cm³/s, and a typical adult has about 5.0 liters of blood. (a) How long does it take for all your blood to pass through the aorta? (b) If your aorta has a diameter of 2.0 cm, what is the speed of blood as it flows through the aorta?

Solution

Recognize the principle

Apply the concept of flow rate.

Sketch the problem

No sketch is required.

Identify the relationships

The flow rate of a fluid is given by the expression,

$$Q = vA$$

which has units of volume per unit time. Therefore, we can also write the flow rate as the volume per unit time,

$$Q = \frac{V}{\Delta t}$$

Note that one liter is equivalent to 1000 cm³.

Solve

(a) Solving for the elapsed time and inserting values we find,

$$\Delta t = \frac{V}{Q} = \frac{5000 \text{ cm}^3}{100 \text{ cm}^3/\text{s}} = \boxed{50 \text{ s}}$$

(b) The flow rate equation will give the speed of the blood give the cross section of the aorta, where the area is $A = \pi\left(\frac{d}{2}\right)^2$,

$$v = \frac{Q}{A} = \frac{Q}{\pi\left(\frac{d}{2}\right)^2} = \frac{4Q}{\pi d^2} = \frac{4(100 \text{ cm}^3/\text{s})}{\pi(2.0 \text{ cm})^2} = \boxed{32 \text{ cm/s}}$$

What does it mean?

The heart pumps 1 liter of blood every 10 s!

P10.65 A Frisbee is observed to fly nearly horizontally, which implies that the lift force must be approximately equal to the weight of the Frisbee. If the air speed over the top of the Frisbee is 9.0 m/s, what is the flow speed across the bottom? Assume the Frisbee has a mass of 0.15 kg.

Solution

Recognize the principle

Apply Bernoulli's equation and the concept of lift.

Sketch the problem

No sketch is required.

Identify the relationships

The lift force is the difference in pressure between the top and bottom multiplied by the area,

$$F_{lift} = (P_{bot} - P_{top})A$$

and the lift force is approximately equal to the weight of the Frisbee, mg. The pressure difference is also given by

$$P_{bottom} - P_{top} = \frac{1}{2}\rho(v_{top}^2 - v_{bot}^2) + \rho g(h_{top} - h_{bot}) = \frac{F_{lift}}{A}$$

which can be solved to find the speed at the bottom

$$v_{bot} = \sqrt{v_{top}^2 - \frac{2}{\rho}\left[\frac{F_{lift}}{A} - \rho g(h_{top} - h_{bot})\right]}$$

We need the area and weight of a Frisbee. The mass and diameter of an Ultimate Frisbee is 0.15 kg and 270 mm, respectively. This gives a weight of

$$mg = (0.15 \text{ kg})(9.8 \text{ m/s}^2) = 1.47 \text{ N}$$

and an area of

$$A = \pi\left(\frac{d}{2}\right)^2 = \pi\left(\frac{0.27 \text{ m}}{2}\right)^2 = 0.06 \text{ m}^2$$

We also need the thickness of the Frisbee which we will estimate at

$$h_{top} - h_{bot} \approx 0.03 \text{ m}$$

Solve

The velocity across the bottom of the Frisbee while in flight is

$$v_{bot} = \sqrt{v_{top}^2 - \frac{2}{\rho}\left[\frac{F_{lift}}{A} - \rho g(h_{top} - h_{bot})\right]}$$

Inserting our supplied and estimated values gives

$$v_{bot} = \sqrt{(9.0 \text{ m/s})^2 - \frac{2}{(1.29 \text{ kg/m}^3)}\left[\left(\frac{1.47 \text{ N}}{0.06 \text{ m}^3}\right) - (1.29 \text{ kg/m}^3)(9.8 \text{ m/s}^2)(0.03 \text{ m})\right]}$$

$$\approx \boxed{7 \text{ m/s}}$$

What does it mean?

The minimum velocity across the top to allow for lift equal to the weight of the Frisbee is the value obtained by letting the velocity across the bottom equal zero, which provides a minimum overall velocity for the Frisbee at 7 m/s for the estimated specifications. Also, notice that including the thickness of the Frisbee in our calculations contributed little to the answer and could have been neglected.

P10.72 What is the capillary pressure for water in a vertical tube of diameter 0.10 mm?

Solution

Recognize the principle

The capillary pressure depends on the surface tension of the material and the radius of the capillary.

Sketch the problem

No sketch needed.

Identify the relationships

The pressure required is the capillary pressure as given by equation 10.35:

$$P_{cap} = \frac{2\gamma}{r}$$

The surface tension of water is given as $\gamma_{water} = 7.3 \times 10^{-2}$ J/m^2 in table 10.4, and the radius $r = \frac{d}{2} = .05$ mm $= 5 \times 10^{-5}$ m.

Solve

Inserting these values into equation 10.35,

$$P_{cap} = \frac{2(7.3 \times 10^{-2} \text{ J/m}^2)}{(5 \times 10^{-5} \text{ m})} = \boxed{2920 \text{ Pa}}$$

What does it mean?

This is a fairly small pressure, which allows capillary tubes to be used with small amounts of liquid near atmospheric pressures.

P10.91   The density of air decreases with height above the Earth, and when the density is too low, people get sick or even become unconscious. That is why the cabins of commercial airplanes are "pressurized" during flight. Cabin pressure, however, is lower than the pressure at sea level. Cabins typically have a pressure equal to the air pressure at an altitude of about 8000 feet above the Earth's surface. This reduced pressure is one reason some people feel fatigued after a long flight. (a) Use Equation 10.11 to calculate this cabin pressure. (b) Estimate the total net force on the walls of a Boeing 737 flying at 35,000 feet. Approximate the airplane as a cylinder of length 30 m and diameter 4 m. *Hint:* You will need to estimate the air pressure outside the plane, using Figure 10.15. (c) What would the force on the walls be if the cabin pressure equals the pressure at the Earth's surface?

Solution

Recognize the principle

Use the concepts of pressure as a function of height for a compressible gas.

Sketch the problem

A sketch is not required for this solution.

Identify the relationships

The pressure at an altitude from 8000 ft can be obtained from,

$$P = P_{atm} \exp\left(\frac{-y}{y_0}\right)$$

where $P_{atm} = 1.01 \times 10^5$ Pa, and $y_0 = 1.0 \times 10^4$ m is described in the text as a constant with reference to sea level which applies to low regions of the atmosphere. An estimate of the magnitude of the total force on the walls of the jet's pressurize fuselage can be obtained from $F = \Delta PA$.

Solve

(a) The pressure inside the aircraft is taken to be the pressure at an altitude of 8000 ft = 2438 m,

$$P_{cabin} = (1.01 \times 10^5 \text{ Pa}) \exp\left(\frac{-2438 \text{ m}}{1.0 \times 10^4 \text{ m}}\right) = \boxed{7.9 \times 10^4 \text{ Pa}}$$

(b) The pressure outside the aircraft is that at an altitude of 35,000 ft = 10700 m,

$$P_{out} = (1.01 \times 10^5 \text{ Pa}) \exp\left(\frac{-10700 \text{ m}}{1.0 \times 10^4 \text{ m}}\right) = 3.5 \times 10^4 \text{ Pa}$$

which gives a difference in pressure of,

$$\Delta P = P_{cabin} - P_{out} = 7.9 \times 10^4 \text{ Pa} - 3.5 \times 10^4 \text{ Pa} = 4.4 \times 10^4 \text{ Pa}$$

The problem asks us to approximate the area of the interior of the aircraft walls to be that of a cylinder. The area is then,

$$A = 2A_{end} + A_{tube} = 2\pi\left(\frac{d}{2}\right)^2 + 2\pi\left(\frac{d}{2}\right)L = \pi d\left(\frac{d}{2} + L\right)$$

$$A = \pi(4.0 \text{ m})\left(\frac{4.0 \text{ m}}{2} + 30 \text{ m}\right) = 4.0 \times 10^2 \text{ m}$$

So the magnitude of total force on the walls of the aircraft pushing outward is,

$$F = \Delta PA = (4.4 \times 10^4 \text{ Pa})(4.0 \times 10^2 \text{ m}) = \boxed{1.8 \times 10^7 \text{ N}}$$

(c) If the cabin pressure were instead $P_{atm} = 1.01 \times 10^5$ Pa, the difference in pressure at the cruising altitude would be,

$$\Delta P = P_{cabin} - P_{out} = 1.01 \times 10^5 \text{ Pa} - 3.5 \times 10^4 \text{ Pa} = 6.6 \times 10^4 \text{ Pa}$$

which would generate a magnitude of the force on the walls of the aircraft of,

$$F = \Delta PA = (6.6 \times 10^4 \text{ Pa})(4.0 \times 10^2 \text{ m}) = \boxed{2.6 \times 10^7 \text{ N}}$$

What does it mean?

Maintaining atmospheric pressure in the aircraft cabin would put an additional stress of 8 million Newtons of force at cruising altitude.

Part D. Additional Worked Examples and Capstone Problems

The following worked example and capstone problems provide you with additional practice with calculating pressure at a depth, applying Archimedes's principle, and applying Bernoulli's principle. Working these problems will give you greater insight into these concepts. Though these three problems do not incorporate all the material discussed in this chapter, they do highlight several of the key concepts and draw on material from previous chapters. If you can successfully solve these problems then you should have confidence in your understanding of these key concepts, so use these problems as a test of your understanding of the chapter material.

WE 10.1 Designing a Water System

You are given the job of designing the water system for a new skyscraper. This building will be 500 m tall, and you want to be sure that the plumbing works properly. If you want to pump water from ground level to the top of the building using a single pump at the bottom as in the figure, what will be the pressure in the pipes at the bottom of the building?

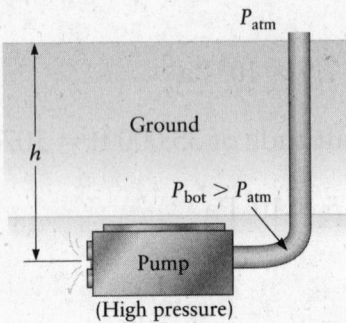

Solution

Recognize the principle

We can use the relation between pressure and depth for an incompressible fluid, $P = P_0 + \rho g h$. The pump at the bottom of a building operates in the same way as the pump at the bottom of the well in the figure, generating a pressure P_{bot} that is greater than atmospheric pressure. The water in the pipes near the pump will thus be at this pressure, whereas the water on the top floor must have a pressure slightly greater than P_{atm} so that water will flow out.

Sketch the problem

The provided figure describes the problem.

Identify the relationships

Applying our expression for pressure at a depth and taking the reference point at the top of the building, we have $P_0 = P_{atm}$ and

$$P_{bot} = P_{atm} + \rho g h$$

where h is the height of the building and ρ is the density of water.

Solve

Inserting the given values of h and ρ gives

$$P_{bot} = P_{atm} + \rho g h$$
$$= 1.01 \times 10^5 \, \text{Pa} + (1000 \, \text{kg/m}^3)(9.8 \, \text{m/s}^2)(500 \, \text{m})$$
$$P_{bot} = \boxed{5.0 \times 10^6 \, \text{Pa}}$$

What does it mean?

This pressure is rather large; it is about 50 times atmospheric pressure. The water pressure in the pipes of a typical house is only about one tenth of this value. Although it is possible to make pumps that can reach this pressure, it is greater than normal plumbing can withstand; that is, the pipes would leak or burst. Even if the pipes were specially built to hold this high pressure, the velocity of water coming from the faucets and showers on the lower floors would be uncomfortably high. So that the pressure at any one location is not too high, the water system for a tall building has pumps located on many different floors.

CP 10.1 Archimedes's Principle: Another Application of Newton's 2nd Law

Suppose you want to support a 1.0-kg block of aluminum in a beaker of water such that half its volume is above the water's surface. You do this by connecting a helium-filled balloon to the block. The rubber balloon (when empty) plus the string has a mass of 2.0 g. (a) Draw a free-body diagram for both the balloon and the block of aluminum, being sure to include all the forces involved. (b) Apply Newton's second law to solve for the tension in the string and the volume of helium required for the system to remain stable.

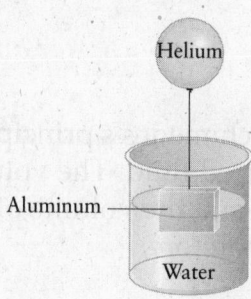

Solution

Recognize the principle

We'll need to apply several concepts to solve this problem. To apply Newton's second law we'll need to determine all the forces acting on both the balloon and the aluminum block. On the balloon these forces include the force of gravity on the balloon and the helium within the balloon, tension, and the buoyant force from the surrounding air. The aluminum block has the force of gravity, tension, and the buoyant force from the water.

Sketch the problem

(a) We begin by drawing the free-body diagram for both the balloon and the block of aluminum.

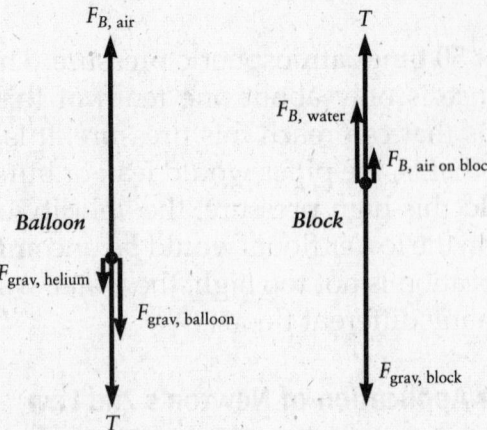

Identify the relationships

We can apply Newton's second law to each of the objects of interest, recognizing that the acceleration is zero. Defining up as positive we have,

Balloon

$$\Sigma F = F_{B,\,air} - F_{grav,\,balloon} - F_{grav,\,helium} - T = 0$$

Block

$$\Sigma F = F_{B,\,water} + F_{B,\,air\,on\,block} + T - F_{grav,\,block} = 0$$

Also, we have expressions for the buoyant forces since Archimedes's principle tells us that buoyant force is equal to the weight of the displaced fluid. The volume of air displaced by the balloon is the volume of the balloon and the volume of water displaced by the block is half the volume of the block. Therefore,

$$F_{B,\,air} = \rho_{air} V_{balloon} g$$
$$F_{B,\,water} = \rho_{water}\left(\frac{1}{2}V_{block}\right)g$$
$$F_{B,\,air\,on\,block} = \rho_{air}\left(\frac{1}{2}V_{block}\right)g$$

Finally, we can substitute for the volume of the block since we know the mass of the block and can look up the density of aluminum,

$$V_{block} = \frac{m_{block}}{\rho_{aluminum}}$$

and the mass of the helium is equal to the density of helium times the volume of the balloon,

$$m_{helium} = \rho_{helium} V_{balloon}$$

Solve

Combining these expressions and rearranging we find the following:

$$F_{B,\,air} - F_{grav,\,balloon} - F_{grav,\,helium} - T = 0$$
$$\rho_{air} V_{balloon}\, g - m_{balloon}\, g - m_{helium}\, g - T = 0$$
$$\rho_{air} V_{balloon}\, g - m_{balloon}\, g - \rho_{helium} V_{balloon}\, g - T = 0$$
$$\rightarrow V_{balloon} = \frac{m_{balloon}\, g + T}{(\rho_{air} - \rho_{helium})g}$$

and,

$$F_{B,\,water} + F_{B,\,air\;on\;block} + T - F_{grave,\,block} = 0$$

$$\rho_{water}\left(\frac{1}{2}V_{block}\right)g + \rho_{air}\left(\frac{1}{2}V_{block}\right)g + T - m_{block}\,g = 0$$

$$(\rho_{water} + \rho_{air})\left(\frac{1}{2}\frac{m_{block}}{\rho_{aluminum}}\right)g + T - m_{block}\,g = 0$$

$$\rightarrow T = m_{block}\,g\left(1 - \frac{(\rho_{water} + \rho_{air})}{2\rho_{aluminum}}\right)$$

Substituting in the numerical values we have,

$$T = (1.0\text{ kg})(9.8\text{ m/s}^2)\left(1 - \frac{(1000 + 1.29)\text{ kg/m}^3}{2(2{,}700\text{ kg/m}^3)}\right)$$

$$= \boxed{7.98\text{ N}}$$

Substituting this value for the tension into our expression for the volume of the balloon we have,

$$V_{balloon} = \frac{(0.001\text{ kg})(9.8\text{ m/s}^2) + (7.98\text{ N})}{(1.29\text{ kg/m}^3 - 0.18\text{ kg/m}^3)(9.8\text{ m/s}^2)}$$

$$= 0.73\text{ m}^3$$

Since the volume of the helium is the volume of the balloon,

$$V_{helium} = \boxed{0.73\text{ m}^3}$$

What does it mean?

If we assume that the balloon is spherical, this balloon would have a diameter of just over 1 meter!

CP 10.2 Bernoulli's Equation: Conservation of Energy in Fluids

A small town has a large water tank from which water can be drained in emergencies, periods of high demand, and times of severe drought. The tank is open at the top and water is drained through a hose with an inside diameter of 5.0 cm. At the end of the hose is a nozzle whose inside diameter is 2.0 cm. The water level in the tank is maintained at 10 m above the nozzle by pumping water from a nearby river. (a) Calculate the absolute pressure inside the hose at the nozzle when the nozzle is closed. (b) The nozzle is then opened and water begins to flow. How many cubic meters of water flow from the nozzle in 1 hour? (c) Calculate the absolute pressure in the water in the hose just before the nozzle.

Solution

Recognize the principle

This problem involves the application of several key concepts from this chapter. We'll need to use our expression relating pressure to depth when the nozzle is closed. Once the nozzle is opened we can apply Bernoulli's equation and the continuity equation.

Sketch the problem

No sketch is required.

Identify the relationship

(a) We begin by applying the expression for pressure as a function of depth, remembering that the absolute pressure includes the pressure due to the surrounding atmosphere,

$$P = P_0 + \rho g h$$
$$= P_{atm} + \rho g h$$

(b) Once the nozzle is opened the water begins to flow; therefore, we must apply Bernoulli's equation which is the conservation of energy expression for flowing fluids,

$$P_1 + \frac{1}{2}\rho v_1^2 + \rho g h_1 = P_2 + \frac{1}{2}\rho v_2^2 + \rho g h_2$$

where we'll take point 1 to be at the surface of the water tank and point 2 to be just as the water exits the nozzle. Since the surface of the tank is open to the atmosphere, $P_1 = P_{atm}$. Similarly, as the water exits the nozzle is at atmospheric pressure, $P_2 = P_{atm}$. Since the water level in the tank is maintained, $v_1 = 0$. Combining these expressions we have,

$$P_2 = P_1 + \frac{1}{2}\rho(v_1^2 - v_2^2) + \rho g(h_1 - h_2)$$
$$\cancel{P}_{atm} = \cancel{P}_{atm} + \frac{1}{2}\rho(\cancel{v_1^2} - v_2^2) + \rho g(h_1 - h_2)$$
$$v_2 = \sqrt{2g(h_1 - h_2)}$$

This is the speed of the water as it exits the nozzle. Knowing this speed we can determine the volume of water which exits the nozzle in a time, *t*, simply by working with the dimension:

$$\text{Total Volume} = Avt = A_{nozzle}v_2 t$$

(c) Since we have the speed of the water in the nozzle from part (b) we can apply the continuity equation to determine the speed of the water in the hose,

$$A_{hose}v_{hose} = A_{nozzle}v_{nozzle} \quad \rightarrow \quad v_{hose} = \frac{A_{nozzle}v_{nozzle}}{A_{hose}}$$

and from this we can apply Bernoulli's equation to determine the pressure in the water in the hose,

$$P_{hose} = P_1 + \frac{1}{2}\rho(v_1^2 - v_{hose}^2) + \rho g(h_1 - h_2)$$
$$= P_{atm} + \frac{1}{2}\rho(\cancel{v_1^2} - v_{hose}^2) + \rho g(h_1 - h_2)$$
$$= P_{atm} - \frac{1}{2}\rho v_{hose}^2 + \rho g(h_1 - h_2)$$

where we have again used $P_1 = P_{atm}$ and $v_1 = 0$.

Solve

Substituting in the numerical values of $P_{atm} = 1.01 \times 10^5$ Pa, $h = h_1 - h_2 = 10$ m, $\rho = 1000$ kg/m³, $A_{nozzle} = \pi\left(\frac{0.02 \text{ m}}{2}\right)^2 = 3.14 \times 10^{-4}$ m³ and $A_{hose} = \pi\left(\frac{0.05 \text{ m}}{2}\right)^2 = 1.96 \times 10^{-3}$ m³, we can solve for the unknowns.

(a) $\quad P = P_{atm} + \rho g h = (1.01 \times 10^5 \text{ Pa}) + (1000 \text{ kg/m}^3)(9.8 \text{ m/s}^2)(10 \text{ m})$

$\qquad = \boxed{1.99 \times 10^5 \text{ Pa}}$

(b) $\quad$ Total Volume $= Avt = A_{nozzle}v_2 t$

$\qquad = A_{nozzle}\sqrt{2g(h_1 - h_2)}\, t$

$\qquad = (3.14 \times 10^{-4} \text{ m}^2)\left(\sqrt{2(9.8 \text{ m/s}^2)(10 \text{ m})}\right)(3600 \text{ s})$

$\qquad = \boxed{1.58 \times 10^1 \text{ m}^3}$

(c) $\quad P_{hose} = P_{atm} - \frac{1}{2}\rho v_{hose}^2 + \rho g(h_1 - h_2) = P_{atm} - \frac{1}{2}\rho\left(\frac{A_{nozzle}v_{nozzle}}{A_{hose}}\right)^2 + \rho g(h_1 - h_2)$

$\qquad = (1.01 \times 10^5 \text{ Pa}) - \frac{1}{2}(1000 \text{ kg/m}^3)\left(\frac{(3.14 \times 10^{-4} \text{ m}^2)(14 \text{ m/s})}{(1.96 \times 10^{-3} \text{ m}^2)}\right)^2$

$\qquad\quad + (1000 \text{ kg/m}^3)(9.8 \text{ m/s}^2)(10 \text{ m})$

$\qquad = \boxed{1.96 \times 10^5 \text{ Pa}}$

What does it mean?

Notice that the pressure in the hose is slightly less when the water is flowing as in part (c) than when it is static as in part (a). This is consistent with Bernoulli's equation which predicts that faster flowing fluids have less pressure.

Part E. MCAT Review Problems and Solutions

PROBLEMS

1. A scuba diver is 10 m below the surface of a lake. There is no current, the air above the lake has a pressure of 1 atmosphere, and the density of the water is 1000 kg/m³. What is the absolute pressure experienced by the diver?

   (a) 1.1 atm

   (b) 1.99×10^5 Pa

   (c) 11 atm

   (d) 1.01×10^5 Pa

2. The aorta of a 70-kg man has a cross-sectional area of 3.0 cm² and carries blood with a speed of 30 cm/s. What is the average volume flow rate of the blood through the aorta?

   (a) 10 cm/s

   (b) 33 cm³/s

(c) $10 \text{ cm}^2/\text{s}$

(d) $90 \text{ cm}^3/\text{s}$

3. At 20°C the density of water is 1 g/cm^3. What is the density of an object that has a weight of 1.0 N in air and an apparent weight of 0.25 N when fully submerged in water?

(a) 0.25 g/cm^3

(b) 0.75 g/cm^3

(c) 1.3 g/cm^3

(d) 4.0 g/cm^3

4. Two insoluble objects appear to lose the same weight when submerged in alcohol. Which statement is true?

(a) Both objects have the same mass in air.

(b) Both objects have the same volume.

(c) Both objects have the same density.

(d) Both objects have the same weight in air.

5. The bottom of each foot of an 80-kg person has an area of 400 cm^2. What is the effect on wearing snowshoes with an area of 0.40 m^2?

(a) The pressure exerted on the snow becomes 10 times greater.

(b) The pressure exerted on the snow becomes 1/10 as great.

(c) The pressure exerted on the snow remains the same.

(d) The magnitude of the force exerted on the snow is 1/10 as great.

6. Two rectangular water tanks are filled to the same depth with water. Tank A has a bottom surface area of 2 m^2 and tank B has a bottom surface area of 4 m^2. Which statement about the magnitude of the forces and pressures at the bottom of the tanks is correct?

(a) Since P_A is less than P_B, F_A is less than F_B.

(b) Since P_A is less than P_B, F_A is equal to F_B.

(c) Since P_A is equal to P_B, F_A is less than F_B.

(d) Since P_A is equal to P_B, F_A is greater than F_B.

7. In a hydraulic lift, the surface of the input piston is 10 cm^2 and that of the output piston is 3000 cm^2. What is the work done if a 100-N force is applied to the input piston raising the output piston by 2.0 m?

(a) 120 kJ

(b) 30 kJ

(c) 40 kJ

(d) 60 kJ

8. Consider the following objects at rest in the same container of water. Rank the densities of the objects.

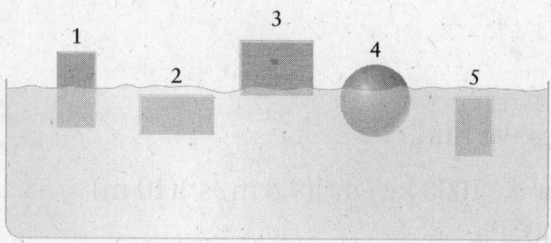

(a) $5 > 2 > 4 > 1 > 3$

(b) $(2 = 5) > (1 = 3 = 4)$

(c) $(2 = 5) > 4 > 1 > 3$

(d) $5 > 2 > (1 = 3 = 4)$

9. According to Bernoulli's equation, the faster a fluid flows,

(a) the greater the pressure within the fluid.

(b) the greater the density of the fluid.

(c) the lesser the pressure within the fluid.

(d) the lesser the density within the fluid.

10. Water flows at 6 m/s through a pipe. The pipe goes up 1 m and the pressure of the water in the pipe remains the same. What is the speed of the water flow in the higher portion of the pipe?

(a) 2 m/s

(b) 4 m/s

(c) 6 m/s

(d) 8 m/s

SOLUTIONS

1. *MCAT strategies*

We know that the pressure within a fluid increases as a function of depth, so the answer must be greater than 1 atm = 1.01×10^5 Pa. Since all of the answers are greater than this value we cannot eliminate any of them from our choice. Answer (c) seems large, but we should continue with our analysis to be sure of the correct answer.

Recognize the principle

The main principle is that the pressure in a fluid increases with depth.

Sketch the problem

No sketch required.

Identify the relationships

The absolute pressure at a depth, h, beneath the surface of the lake is given by,

$$P = P_0 + \rho g h$$

Solve

Substituting in the numerical values we find,

$$P = P_0 + \rho g h = 1.01 \times 10^5 \, \text{Pa} + (1000 \, \text{kg/m}^3)(9.8 \, \text{m/s}^2)(10 \, \text{m})$$
$$= 1.01 \times 10^5 \, \text{Pa} + 0.98 \times 10^5 \, \text{Pa}$$
$$= 1.99 \times 10^5 \, \text{Pa}$$

So the correct answer is (b).

What does it mean?

If we simply remember that the pressure increases by about 1 atm for every 10 m we go below the surface of water, then we can answer questions like this without doing any calculations!

2. MCAT strategies

Volume flow rate is the volume per second flowing past a point. Therefore the correct answer should have the units of cm^3/s. Answers (a) and (c) do not have these units and can be eliminated. Volume flow rate is the product of the cross-sectional area and the speed giving us $(3.0 \, \text{cm}^2)(30 \, \text{cm/s}) = 90 \, \text{cm}^3/\text{s}$ as the correct answer. So the correct answer is (d).

3. MCAT strategies

Since the object still has an apparent weight in water (i.e., it sinks!), its density must be greater than the density of water. Therefore answers (a) and (b) can be eliminated. No additional answers can be easily eliminated so we proceed with our problem-solving strategy.

Recognize the principle

We'll need to apply Archimedes's principle.

Sketch the problem

No sketch required.

Identify the relationships

The difference in the object's weight and apparent weight is due to the buoyant force from the surrounding water. So the buoyant force is equal to $1 \, \text{N} - 0.25 \, \text{N} = 0.75 \, \text{N}$. Buoyant force is given by the following expression,

$$F_B = \rho_L V g$$

where $\rho_L = 1 \, \text{g/cm}^3$, V is the volume of water displaced by the object, and $g = 9.8 \, \text{m/s}^2$. We can solve this expression for the volume displaced. Recognizing that the volume displaced is the volume of the object we can then determine the density of the object from its mass (which we'll get from its weight) and its volume.

Solve

We'll first solve for the volume of liquid displaced by the object,

$$F_B = \rho_L V g \rightarrow V = \frac{F_B}{\rho_L g} = \frac{0.75 \text{ N}}{(1000 \text{ kg/m}^3)(9.8 \text{ m/s}^2)} \approx 7.7 \times 10^{-5} \text{ m}^3$$

The mass of the object is determine from its weight in air,

$$m = \frac{1.0 \text{ N}}{9.8 \text{ m/s}^2} \approx 0.1 \text{ kg}$$

Therefore, the density of the object is given by,

$$\rho = \frac{m}{V} = \frac{0.1 \text{ kg}}{7.7 \times 10^{-5} \text{ m}^3} \approx 1.3 \text{ g/cm}^3$$

So the correct answer is (c).

What does it mean?

Note that we need to be a little careful with our units, working in N and kg.

4. *MCAT strategies*

This question is testing our understanding of Archimedes's principle. The buoyant force on an object due to the surrounding fluid is related to the density of the fluid and the volume of the displaced fluid. Also, the apparent weight loss for an object is equal to the buoyant force. So, if the apparent weight loss for the two objects is the same, then they experience the same buoyant force and therefore displace the same amount of fluid. This means that the objects must have the same volume in order to displace the same amount of fluid when submerged. So the correct answer is (b). Remember, buoyant force has nothing to do with properties of the object such as its mass, weight in air, or density!

5. *MCAT strategies*

This question is simply testing our understanding of the relationship among the magnitude of the force, area, and pressure. Since $P = F/A$, and the area over which your weight is distributed increases when you put on snowshoes, then the pressure will decrease. The magnitude of the force does not change since it is due to your weight which remains constant. The only answer which states a decrease in pressure is (b). So the correct answer is (b).

6. *MCAT strategies*

Once again this question is testing our understanding of the relationship among the magnitude of the force, area, and pressure. Also, we need to know that the pressure in a fluid is proportional to depth. Since the tanks are the same depth, the pressure will be the same at the bottom of each tank. This eliminates answers (a) and (b). Now, since the area of the bottom of tank A is smaller than tank B, the magnitude of the force $F = PA$ will be greater on the bottom of tank A. So the correct answer is (d).

7. *MCAT strategies*

No answers can be easily eliminated so we proceed with our problem-solving strategy.

Recognize the principle

According to Pascal's principle, the pressure in the fluid at each piston is the same. As a result the piston with the smaller area experiences the larger force. However, the work done on each of the pistons is the same.

Sketch the problem

No sketch required.

Identify the relationships

When the force is in the direction of the displacement, the work done moving an object is given by,

$$W = Fd$$

Also, pressure is related to the magnitude of the force and area by the following expression,

$$P = F/A$$

Using the appropriate subscripts to describe the quantities at the two pistons we have,

$$W_{out} = F_{out}d_{out}$$

But since the pressure at the two pistons is the same,

$$P_{in} = P_{out}$$

$$\frac{F_{in}}{A_{in}} = \frac{F_{out}}{A_{out}} \quad \rightarrow \quad F_{out} = \frac{F_{in}A_{out}}{A_{in}}$$

Combining these expressions we have,

$$W_{out} = F_{out}d_{out} = \frac{F_{in}A_{out}d_{out}}{A_{in}}$$

Solve

Inserting the numerical values we find,

$$W_{out} = \frac{F_{in}A_{out}d_{out}}{A_{in}} = \frac{(100 \text{ N})(3000 \text{ cm}^2)(2.0 \text{ m})}{(10 \text{ cm}^2)} = 60{,}000 \text{ J} = 60 \text{ kJ}$$

So the correct answer is (d).

What does it mean?

Notice that we didn't need to convert our areas to m^2 since the units canceled in the expression. This saved us a little time.

8. MCAT strategies

We can tell a lot about the density of an object by how it floats (or sinks) in a liquid. If half of the volume of the object floats above the surface of the liquid, then the object is half the density of the liquid. If only 10% of the object's volume is above the surface, then it's 90% the density of the fluid. If an object "floats" beneath the surface of the liquid then its density is the same as the liquid's density. With this in mind we can easily rank the densities of the objects. Object 3 floats highest in the water and is therefore the least dense. Next is object 1 and then object 4. Both object 2 and 5 are fully submerged and therefore have the same density as water. So the correct answer is (c).

9. MCAT strategies

One consequence of Bernoulli's equation is that as the speed of the fluid increases, the pressure within the fluid decreases. Also, since Bernoulli's equation only applies to ideal fluids whose density remains constant, answers (b) and (c) cannot be correct. So the correct answer is (c).

10. MCAT strategies

No answers can be easily eliminated so we proceed with our problem-solving strategy.

Recognize the principle

This problem requires us to apply Bernoulli's equation.

Sketch the problem

No sketch required.

Identify the relationships

We begin by writing Bernoulli's equation for two points within the pipe (one at the lower point and the second at the higher point),

$$P_1 + \frac{1}{2}\rho v_1^2 + \rho g h_1 = P_2 + \frac{1}{2}\rho v_2^2 + \rho g h_2$$

Since the pressures are said to be the same we have,

$$P_1 = P_2$$

and the difference in height is given by,

$$h_2 - h_1 = 1 \text{ m}$$

Combining these expressions and solving for v_2 we have,

$$v_2 = \sqrt{v_1^2 - 2g\,(h_2 - h_1)} = \sqrt{v_1^2 - 2g}$$

Solve

Substituting in the numerical values we have,

$$v_2 = \sqrt{v_1^2 - 2\,mg} = \sqrt{(6 \text{ m/s})^2 - 2 \text{ m}(9.8 \text{ m/s}^2)} = \sqrt{36 \text{ m}^2/\text{s}^2 - 19.6 \text{ m}^2/\text{s}^2}$$

$$\approx \sqrt{16 \text{ m}^2/\text{s}^2} = 4 \text{ m/s}$$

So the correct answer is (b).

What does it mean?

Since the pressure did not change though the pipe increased in height, the water must have decreased in speed. So, we could have eliminated answers (c) and (d) without having done any calculations.

11 Harmonic Motion and Elasticity

CONTENTS

Part A. Summary of Key Concepts and Problem-Solving Strategies

KEY CONCEPTS

Frequency and period

Any motion that repeats with time is **harmonic**. The repeat time is called the **period**, and the **frequency** (measured in hertz) is equal to the number of cycles that are completed each second. Frequency and period are related by the expression

$$f = \frac{1}{T}$$

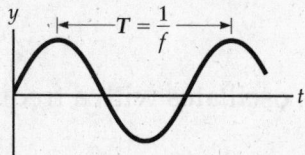

Simple harmonic motion

The position and velocity of a simple harmonic oscillator vary with time according to

$$y = A \sin(2\pi f t)$$
$$v = 2\pi f A \cos(2\pi f t)$$

The total mechanical energy of a simple harmonic oscillator is conserved. The KE and PE oscillate with time, but their sum is constant. Applying the principle of conservation of mechanical energy to objects undergoing simple harmonic motion allows us to simplify the analysis of complex problems.

Stress, strain, and Hooke's law

Elastic objects stretch or compress in response to applied forces. For a compressive or tensile force, the deformation is given by

$$\frac{\Delta L}{L_0} = \frac{1}{Y}\left(\frac{F}{A}\right)$$

where Y is an elastic constant called the **Young's modulus**. The ratio F/A is called the **stress**, and $\Delta L/L_0$ is called the **strain**. Other types of deformations (such as shear stresses) are described by other elastic constants. For a shear stress, the deformation is given by

$$\frac{\Delta x}{L_0} = \frac{1}{S}\left(\frac{F}{A}\right)$$

where S is an elastic constant called the **shear modulus**. Similarly, for a bulk compression, the deformation is given by

$$\frac{\Delta V}{V_0} = \frac{-P}{B}$$

where B is an elastic constant called the **bulk modulus**.

Damping and resonance

Most real harmonic oscillators are affected by friction, causing the oscillations to be damped. There are three classes of damped oscillators: **underdamped** (weak damping), **overdamped** (large damping), and **critically damped**. When a damped harmonic oscillator is subject to a time-dependent driving force, the amplitude will be largest when the driving frequency is close to the natural frequency of the oscillator. This is called **resonance**.

APPLICATIONS

Mass connected to a spring

A mass connected to a spring with spring constant, k, oscillates with a frequency

$$f = \frac{1}{2\pi}\sqrt{\frac{k}{m}}$$

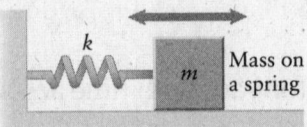

Simple pendulum

A *simple pendulum* consists of a mass attached to a massless string of length L. Its oscillation frequency is

$$f = \frac{1}{2\pi}\sqrt{\frac{g}{L}}$$

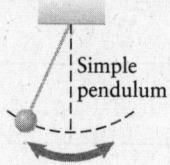

Simple
pendulum

Torsional oscillator

An object with a moment of inertia I suspended by a torsion fiber is a *torsional oscillator*. Its frequency is

$$f = \frac{1}{2\pi}\sqrt{\frac{\kappa}{I}}$$

where κ is the torsion constant of the fiber.

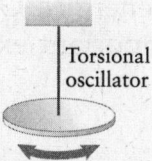

Torsional
oscillator

Properties of simple harmonic oscillators

All simple harmonic oscillators have the following properties:

- The frequency is independent of amplitude.
- The force or torque that causes the oscillatory motion is a restoring force proportional to the displacement of the oscillator from its equilibrium position. Such a restoring force is described by Hooke's law.

PROBLEM-SOLVING STRATEGIES

Once again, in this chapter we saw the importance of the principle of conservation of mechanic energy. This principle provides us with a powerful tool for analyzing complex mechanical systems including objects undergoing simple harmonic motion. The following schematic outlines the thought process we should follow when applying this concept.

Problem Solving: Applying the Principle of Conservation of Energy to Harmonic Motion

Recognize the principle

Start by finding the object (or system of objects) whose mechanical energy is conserved. This may include masses on springs, bobs on strings or other objects which are undergoing harmonic motion. Also, forms of energy may include kinetic, gravitational potential, and spring potential energy.

Sketch the problem

In your sketch, show the initial and final states of the object. This sketch should contain a coordinate system, including an origin, with which to measure the potential energy. Also, you should label the forces involved and the speeds of the object at the points of interest.

Identify the relationships

Find expressions for the initial and final kinetic and potential energies. One or more of these energies may involve unknown quantities. Important expressions include:

$$KE = \frac{1}{2} mv^2, PE_{\text{spring}} = \frac{1}{2} kx^2, \text{ and } PE_{\text{grav}} = mgh$$

Remember, the total energy of a harmonic oscillator is equal to the maximum potential energy, which is the potential energy of the oscillator when it is at its maximum displacement from equilibrium. If there are nonconservative forces involved such as friction, determine the work done by these forces. Use the appropriate expression for energy conservation:

With conservative forces → $KE_i + PE_i = KE_f + PE_f$

With nonconservative forces → $KE_i + PE_i + W_{\text{noncon}} = KE_f + PE_f$

Solve

Substitute in the known quantities and solve for the unknown quantities.

What does it mean?

Always *consider what your answer means* and check that it makes sense.

Part B. Frequently Asked Questions

1.   *For a mass on a spring or a simple pendulum, the mass is moving the fastest as it goes through the equilibrium point. Doesn't this imply that the acceleration is greatest at that point as well?*

No. In fact, the acceleration of the mass at the equilibrium point is zero! To understand this let's look carefully at the force responsible for the oscillation. In the case of the mass on the spring it is the spring force which acts on the mass. This force is always directed toward the equilibrium point, is proportional to the displacement of the mass from equilibrium, and therefore is zero at the equilibrium point. Newton's second law tells us that the acceleration of the mass is directly proportional to the force. So, the acceleration of the mass is also zero at the equilibrium point. We can apply this same reasoning to a simple pendulum and arrive at the same conclusion that the tangential acceleration is zero at the equilibrium point.

2.   *How does mass affect the period of an object undergoing simple harmonic motion?*

To answer this questions let's look carefully at our expression for the period of oscillation for both a mass on a spring and a simple pendulum. For a mass on a spring the period is given by the expression,

$$T = \frac{1}{f} = 2\pi \sqrt{\frac{m}{k}}$$

where m is the mass and k is the spring constant. So, we see that the period of oscillation is proportional to the square root of the mass. This means that if we double the mass on the spring, then the period will increase by a factor of $\sqrt{2}$. Similarly, if we quadruple the mass, the period will double.

For a simple pendulum the period is given by the expression,

$$T = \frac{1}{f} = 2\pi \sqrt{\frac{L}{g}}$$

where g is the acceleration due to gravity and L is the length of the pendulum. Since the mass of the pendulum does not appear in this expression, the period is independent of the mass! This means that two pendulums with different masses will have the same period as long as they are the same length.

3.   *How does amplitude of oscillation affect the period of an object undergoing simple harmonic motion?*

Once again we can look carefully at our expressions for the period of oscillation as in the previous FAQ. Since the amplitude of oscillation does not appear in either of these expressions we can conclude that the period is independent of the amplitude.

4.   *For an object undergoing simple harmonic motion, where is its velocity zero? Where is it a maximum?*

To answer this question let's consider a mass on a spring oscillating back and forth in simple harmonic motion. As the mass reaches a point where its displacement is a maximum (i.e., when it is farthest away from the equilibrium point), it momentarily stops as it changes direction. This is where the speed is zero. As the mass accelerates back toward the equilibrium point its speed increases until it reaches maximum at its equilibrium point. On the other hand, the force on the spring (and therefore the acceleration of the mass) is a maximum when the mass is at its maximum displacement and is zero at the equilibrium point. The following diagram helps to illustrate this:

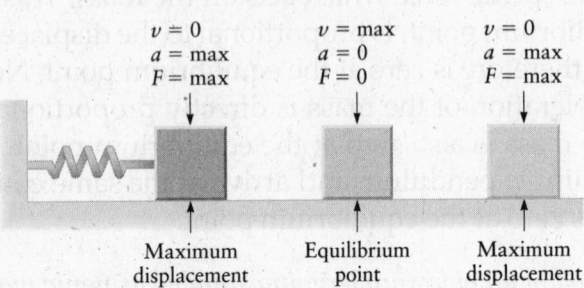

$v = 0$
$a = \text{max}$
$F = \text{max}$

$v = \text{max}$
$a = 0$
$F = 0$

$v = 0$
$a = \text{max}$
$F = \text{max}$

Maximum displacement   Equilibrium point   Maximum displacement

5. *How can I find the total energy of a simple harmonic oscillator?*

When studying the motion of a simple harmonic oscillator, there are two forms of energy that are important. First, the object is moving so there is kinetic energy, which is dependent on the speed of the object. Since the speed of an oscillator is a maximum when it is at its equilibrium point as we discussed in the previous FAQ, then the kinetic energy is a maximum at this point as well. Also, the kinetic energy is zero when the speed is zero at the points of maximum displacement. The second form of energy is potential energy. In the case of the simple pendulum, it's gravitational potential energy and for a mass on a spring it's spring potential energy. For either case the potential energy is a maximum when the object is at its maximum displacement from equilibrium and is zero at the equilibrium point. So, with this in mind we can easily determine the total energy of a simple harmonic oscillator by knowing either its maximum speed or its maximum displacement. Since the kinetic energy is zero at the points of maximum displacement, all of the energy is in the form of potential energy. If we know the maximum displacement (i.e., the amplitude of oscillation), then the total energy is simply the potential energy at this point given by the expressions,

$$E = PE_{\text{spring, max}} = \frac{1}{2}kA^2 \text{ (mass on a spring)}$$

$$E = PE_{\text{grav, max}} = mgh_{\text{max}} \text{ (simple pendulum)}$$

If we know the maximum speed of the oscillator, then the total energy is the kinetic energy when the speed is a maximum, given by the expression,

$$E = KE_{\text{max}} = \frac{1}{2}mv_{\text{max}}^2$$

6. *Is mechanical energy conserved for an object undergoing a damped oscillation?*

No. When an object is undergoing a damped oscillation, there is a nonconservative force such as friction or air drag which is responsible for the damping. These forces dissipate energy in the form of heat energy, thus reducing the total mechanical energy of the system. So, the mechanical energy is not conserved for an object undergoing a damped oscillation. The result of damping is that the object eventually comes to rest, meaning that all of its mechanical energy has been dissipated.

7. *How will hanging a mass on a spring vertically affect the frequency of the oscillation?*

It won't! The frequency of oscillation of a mass on a spring is given by the expression,

$$f = \frac{1}{2\pi} \sqrt{\frac{k}{m}}$$

So, the frequency only depends on the spring constant k and the mass m. If the mass-spring system is hanging vertically, the oscillation will have a different equilibrium point but the frequency of the oscillation will be the same.

8. *Can I apply the kinematic equations we learned about in Chapter 3 to a simple harmonic oscillator?*

No. The kinematic equations we derived in Chapter 3 apply for the case when the acceleration of the object remains constant. For object's undergoing simple harmonic motion, the force on the object is proportional to its displacement from equilibrium. This is given by Hooke's law. This means that as the object moves farthest from the equilibrium point, the restoring force increases proportionally. So the force is not a constant. From Newton's second law, the acceleration is proportional to the force, therefore, the acceleration changes as well. Since the acceleration does not remain constant we cannot apply the kinematic expressions derived in Chapter 3.

9. *What's the difference between angular and linear frequency?*

The term angular frequency is sometimes used in place of angular velocity, ω. When discussing the connection between simple harmonic motion and angular motion we introduce the expression,

$$f = \frac{\omega}{2\pi}$$

where f is the linear frequency (or just frequency), and ω is the angular frequency (or angular velocity). Frequency is the number of oscillations per second an oscillator undergoes while the angular frequency is the number of radians per second an object rotates through. Since one complete oscillation of a simple harmonic oscillator was compared to one complete rotation of an object, then f and ω are related by 2π.

10. *How do I know when to apply Young's, shear, or bulk moduli?*

It all depends on how you are stressing the object. In other words, you need to consider how the force is being applied to the object. If the force is applied in such a way as to stretch or compress the object, then this is a tensile or compressive force and Young's modulus would apply. If the force is applied *across* the face of the object, then this is a situation where the shear modulus would be used. Finally, if the entire volume of the object is compressed or expanded due to an increase or decrease in external pressure, then the bulk modulus must be used. These three situations are illustrated in the following figures:

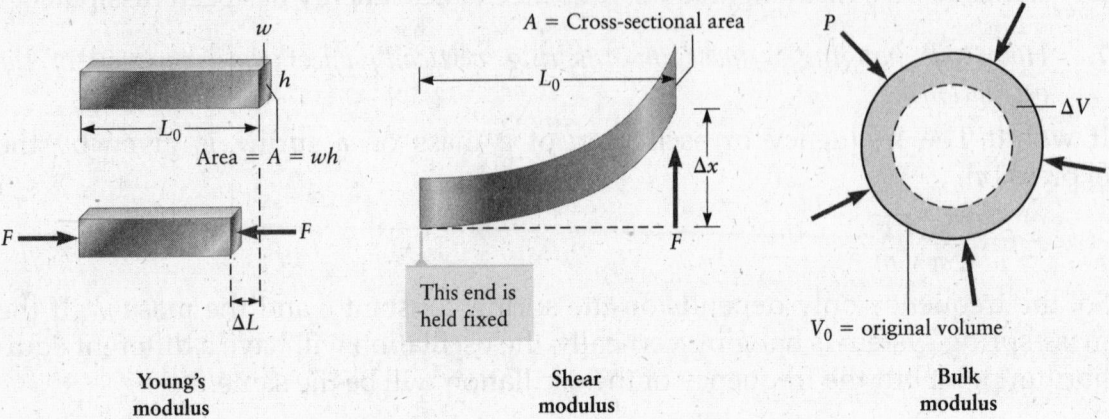

Young's modulus   Shear modulus   Bulk modulus

Part C. Selection of End-of-Chapter Answers and Solutions

QUESTIONS

Q11.4 Consider a simple pendulum that is used as a clock. (a) What should the length be to make one oscillation (one "tick" of the clock) every second when it is at is at sea level? (b) Will this clock "speed up" or "slow down" when it is taken to the top of Mount Everest? (c) How long will this clock take to make 60 ticks on Mount Everest?

Answer

(a) For a pendulum, an oscillation consists of a full swing "there and back." One oscillation is the period of the pendulum. The period of a simple pendulum of length L with acceleration g on Earth is,

$$T = 2\pi\sqrt{\frac{L}{g}}$$

Solving for the length of the pendulum with a period of 1 s we find,

$$L = \frac{T^2 g}{4\pi^2} = \frac{(1.0 \text{ s})^2 (9.8 \text{ m/s}^2)}{4\pi^2} = 0.248 \text{ m} \approx \boxed{0.25 \text{ m}}$$

(b) When the clock is taken to the top of Mount Everest g is SMALLER than at sea level. This will make the period LARGER, and so $\boxed{\text{the clock will slow down}}$.

(c) Using our expression for the period of a pendulum, we can write the following ratio:

$$\frac{T_{ME}}{T_{SL}} = \frac{2\pi\sqrt{\dfrac{L}{g_{ME}}}}{2\pi\sqrt{\dfrac{L}{g_{SL}}}} = \sqrt{\frac{g_{SL}}{g_{ME}}}$$

where the subscripts ME and SL refer to Mount Everest and sea level, respectively. Recall from Chapter 5 that the acceleration of gravity is given by the following expression,

$$g = \frac{GM_{Earth}}{(r_E + h)^2}$$

where h is the height above sea level. Combining these expressions we can set up the following ratio:

$$\frac{T_{ME}}{T_{SL}} = \sqrt{\frac{\dfrac{GM_E}{r_E^2}}{\dfrac{GM_E}{(r_E + h)^2}}} = \sqrt{\frac{(r_E + h)^2}{r_E^2}} = \frac{r_E + h}{r_E}$$

The height of Mount Everest is 8850 m above sea level, therefore $h = 8550$ m. Substituting in the numerical values we have,

$$\frac{T_{ME}}{T_{SL}} = \frac{r_E + h}{r_E} = 1 + \frac{h}{r_E} = \left(1 + \frac{8550 \text{ m}}{6.37 \times 10^6 \text{ m}}\right) s = (1 + 0.00134)\,s$$

So the period of our clock will be 0.00134 s longer on Mount Everest than at sea level. Therefore, 60 ticks on Mount Everest will take,

$$60\,(1 + 0.00134)s = (60 + 0.080)s = \boxed{60.08 \text{ s}}$$

Q11.13 Use energy considerations to derive the oscillation frequency for a mass-on-a-spring oscillator. *Hint*: The maximum potential energy stored in the spring must be equal to the maximum kinetic energy of the mass.

Answer

The kinetic energy of the mass is given by the expression,

$$KE = 2\pi^2 mf^2 A^2 \cos^2(2\pi ft)$$

This has a maximum when $\cos^2(2\pi ft) = 1$. So,

$$KE_{max} = 2\pi^2 mf^2 A^2$$

The potential energy stored in the spring is given by the expression,

$$PE = \frac{kA^2}{2}\sin^2(2\pi ft)$$

The maximum potential energy occurs when $\sin^2(2\pi ft) = 1$. So,

$$PE_{max} = \frac{kA^2}{2}$$

Using the hint given in the question, we set the maximum kinetic energy equal to the maximum potential energy

$$2\pi^2 mf^2 A^2 = \frac{kA^2}{2}$$

Now we can solve for the frequency of oscillation

$$f^2 = \frac{k}{4\pi^2 m} \Rightarrow f = \sqrt{\frac{k}{4\pi^2 m}} = \frac{1}{2\pi}\sqrt{\frac{k}{m}}$$

$$f = \frac{1}{2\pi}\sqrt{\frac{k}{m}}$$

PROBLEMS

P11.10 A simple harmonic oscillator has a frequency of 300 Hz and an amplitude of 0.10 m. What is the maximum velocity of the oscillator?

Solution

Recognize the principle

We can find the maximum velocity from the amplitude and frequency using the general form of the simple harmonic oscillator.

Sketch the problem

No sketch needed.

Identify the relationships

We begin by writing our expression for the velocity of an oscillator in its general form:

$$v = 2\pi f A \cos(2\pi f t)$$

The maximum velocity of a simple harmonic oscillator occurs when the sine or cosine portion of the expression has its maximum value of 1.

Solve

This gives

$$v_{max} = 2\pi f A = 2\pi(300 \text{ Hz})(0.10 \text{ m}) = \boxed{190 \text{ m/s}}$$

What does it mean?

In order to oscillate 300 times per second with an amplitude = 10 cm, a simple harmonic oscillator must reach a maximum speed of 190 m/s. This places the oscillator under tremendous force and acceleration!

P11.16 A mass $m = 4.5$ kg is attached to a vertical spring with $k = 200$ N/m and is set into motion. (a) What is the frequency of the oscillation? (b) If the amplitude of the oscillation is 3.5 cm, what is the maximum value of the velocity? (c) How long does it take the mass to move from $y = 1.5$ cm to $y = 2.5$ cm? (d) If the mass is oscillating with a maximum speed of 45 m/s, what is the amplitude? (e) If the spring constant is increased by a factor of two and the maximum kinetic energy of the mass is the same, by what factor does the amplitude change?

Solution

Recognize the principle

(a) We can find the frequency of an oscillation from the given mass and spring constant.

(b)–(d) The displacement and velocity of a harmonic oscillators are given by the expressions,

$$y(t) = A\sin(2\pi ft) \quad \text{and} \quad v(t) = A(2\pi f)\cos(2\pi ft)$$

Sketch the problem
No sketch needed.

Identify the relationships
(a) Using our expression for the frequency in terms of the spring constant and mass,

$$f = \frac{1}{2\pi}\sqrt{\frac{k}{m}}$$

(b) The maximum velocity occurs when the sine function evaluates to 1. So,

$$v_{max} = 2\pi fA$$

(c) We can use the general form of the displacement to find the time it takes, t_1, to reach 1.5 cm and the time it takes, t_2, to reach 2.5 cm. We can then determine the difference in these two times.

(d) The maximum velocity (when the cosine function in the expression for velocity evaluates to 1) gives,

$$v_{max} = 2\pi fA$$

Solve
(a) Inserting the numerical values we find,

$$f = \frac{1}{2\pi}\sqrt{\frac{200\ \text{N/m}}{4.5\ \text{kg}}} = 1.06\ \text{Hz} = \boxed{1.1\ \text{Hz}}$$

(b) Inserting the numerical values for frequency (with an extra significant figure) and amplitude gives,

$$v_{max} = 2\pi(1.06\ \text{Hz})(0.035\ \text{m}) = 23\ \text{cm/s} = \boxed{0.23\ \text{m/s}}$$

(c) Inserting the amplitude and frequency values into our general form for the displacement and solving for t gives,

$$y = 3.5\ \text{cm}[\sin(2\pi)(1.06\ \text{Hz})(t)]$$

$$\frac{y}{3.5\ \text{cm}} = \sin[2\pi(1.06\ \text{Hz})(t)]$$

$$2\pi(1.06\ \text{Hz})(t) = \sin^{-1}\left(\frac{y}{3.5\ \text{cm}}\right)$$

$$t = \frac{\sin^{-1}\left(\dfrac{y}{3.5\ \text{cm}}\right)}{2\pi(1.06\ \text{Hz})}$$

We can then insert the displacement and determine each time, while being sure to measure our angles in radians we have

$$t_1 = \frac{\sin^{-1}\left(\dfrac{1.5\ \text{cm}}{3.5\ \text{cm}}\right)}{2\pi(1.06\ \text{Hz})} = 0.067\ \text{s}$$

$$t_2 = \frac{\sin^{-1}\left(\frac{2.5\ \text{cm}}{3.5\ \text{cm}}\right)}{(2\pi(1.06\ \text{Hz}))} = 0.119\ \text{s}$$

Then, the time difference, and, therefore, the time required, is,

$$\Delta t = t_2 - t_1 = 0.119\ \text{s} - 0.067\ \text{s} = \boxed{0.052\ \text{s}}$$

(d) We can solve our expression for the maximum velocity for the amplitude,

$$A = \frac{v_{max}}{2\pi f}$$

Inserting the given maximum velocity and the frequency from part (a) yields,

$$A = \frac{(45\ \text{m/s})}{2\pi(1.06\ \text{Hz})} = \boxed{6.8\ \text{m}}$$

(e) We can combine the expressions for maximum velocity and the frequency, and solve for the amplitude,

$$v_{max} = 2\pi f A \qquad f = \frac{1}{2\pi}\sqrt{\frac{k}{m}}$$

$$A = v_{max}\sqrt{\frac{m}{k}}$$

If we call the original amplitude $A_0 = v_{max}\sqrt{\frac{m}{k}}$ then the new amplitude is

$$A_{new} = v_{max}\sqrt{\frac{m}{2k}} = \frac{A_0}{\sqrt{2}} = \boxed{0.71 A_0}$$

What does it mean?

The new amplitude is smaller—about 71% of the original amplitude. A larger k means a stiffer spring and hence a smaller amplitude.

P11.18 Estimate the spring constant for a trampoline. Assume a person is standing on the trampoline and oscillating up and down without leaving the trampoline. *Hint*: Begin by estimating the mass of the oscillator and the period of the motion.

Solution

Recognize the principle

We can find the spring constant from a given mass and period using the expression which relates these two variables.

Sketch the problem

No sketch needed.

Identify the relationships

The period of a mass on a spring undergoing simple harmonic motion is given by the expression,

$$T = \frac{1}{f} = 2\pi\sqrt{\frac{m}{k}}$$

We estimate the mass of a child to be about 50 kg, and estimate the period of oscillation of a trampoline to be about 1 s.

Solve

We can solve this equation for the spring constant.

$$\left(\frac{T}{2\pi}\right)^2 = \frac{m}{k}$$

$$k = m\left(\frac{2\pi}{T}\right)^2$$

Then, inserting our estimated values we have,

$$k = (50 \text{ kg})\left(\frac{2\pi}{1\text{ s}}\right)^2 \approx \boxed{2000 \text{ N/m}}$$

What does it mean?

We should be very careful with this type of exercise, since the trampoline only provides a restorative force while the person is in contact. A person jumping on a trampoline is not truly moving in simple harmonic motion since they are accelerated downward by the gravitational force (which does not vary with the distance from equilibrium) during the part of the motion above the trampoline level.

P11.37   A particle is attached to a spring with $k = 50$ N/m is undergoing simple harmonic motion, and its position is described by the equation $x = (5.7 \text{ m})\cos(7.5t)$, with t measured in seconds. (a) What is the mass of the particle? (b) What is the period of the motion? (c) What is the maximum speed of the particle? (d) What is the maximum potential energy? (e) What is the total energy?

Solution

Recognize the principle

From the general expression for the displacement of a simple harmonic oscillator, we can find the amplitude and frequency. These will let us calculate the period, mass, and energies.

Sketch the problem

No sketch needed.

Identify the relationships

(a) We know that for a simple harmonic oscillator (mass on a spring),

$$f = \frac{1}{2\pi}\sqrt{\frac{k}{m}}$$

The general expression for the displacement of a harmonic oscillator is given by,

$$x = A\sin(2\pi f t)$$

(b) The period of the motion is related to the frequency by the expression, $T = \frac{1}{f}$, and we know that,

$$2\pi f = 7.5 \text{ rad/s} \rightarrow f = \frac{7.5 \text{ rad/s}}{2\pi}$$

(c) The maximum speed for an oscillator can be found from the amplitude and frequency:

$$v_{\text{max}} = 2\pi f A$$

(d) The maximum potential energy is when the spring is stretched or compressed to its maximum. This occurs when the oscillator has its maximum displacement. Therefore,

$$PE_{max} = \frac{1}{2}kA^2$$

(e) When the mass is at its maximum displacement, all the energy is in the form of spring potential energy and the mass is momentarily at rest. This means that the kinetic energy is momentarily zero. So, $KE = 0$.

Solve

(a) We can see from the general expression that,

$$A = 5.7 \text{ m} \quad \text{and} \quad 2\pi f = 7.5 \text{ rad/s}$$

Rearranging our equation for the frequency and solving for the mass gives,

$$m = \frac{k}{(2\pi f)^2}$$

Inserting the numerical values,

$$m = \frac{50 \text{ N/m}}{(7.5 \text{ rad/s})^2} = \boxed{0.89 \text{ kg}}$$

(b) Combining these two expressions yields,

$$T = \frac{1}{\left(\frac{7.5}{2\pi}\right)} = \boxed{0.84 \text{ s}}$$

(c) We can insert our numerical expressions for $2\pi f$ and A into our expression for the maximum velocity and find,

$$v_{max} = (7.5 \text{ rad/s})(5.7 \text{ m}) = \boxed{43 \text{ m/s}}$$

(d) Inserting the numerical values for the spring constant and amplitude gives,

$$PE_{max} = \frac{1}{2}(50 \text{ N/m})(5.7 \text{ m})^2 = \boxed{810 \text{ J}}$$

(e) Therefore, $PE_{max} = \text{Energy}_{total} = \boxed{810 \text{ J}}$

What does it mean?

Given the general expression for the displacement of a harmonic oscillator and the spring constant, we can determine the mass, the period, the maximum velocity of the oscillator, and the maximum potential and total energies.

P11.57 An aluminum sphere has a radius of 45 cm on the Earth. It is then taken to a distant planet where the atmospheric pressure P is much larger than on the Earth. If the sphere has a radius of 43 cm on that planet, what is P?

Solution

Recognize the principle

The volume change of an object due to a pressure change can be calculated using the bulk modulus.

Sketch the problem

No sketch needed.

Identify the relationships

Since we are talking about a volume deformation we use our expression involving the bulk modulus,

$$P = -B\frac{\Delta V}{V_0}$$

Since the object is a sphere, the volume is given by,

$$V = \frac{4}{3}\pi r^3$$

The bulk modulus for aluminum can be found to be 7.1×10^{10} Pa.

Solve

The volume change ΔV is given by,

$$\Delta V = V_{\text{planet}} - V_{\text{Earth}}$$

$$\Delta V = \frac{4}{3}\pi r^3_{\text{planet}} - \frac{4}{3}\pi r^3_{\text{Earth}} = \frac{4}{3}\pi\left(r^3_{\text{planet}} - r^3_{\text{Earth}}\right)$$

where r_{planet} and r_{Earth} represent the radius of the sphere on the distant planet and on the Earth, respectively.

So, the pressure on the distant planet is,

$$P = -B\frac{\Delta V}{V_0} = -B\frac{\frac{4}{3}\pi\left(r^3_{\text{planet}} - r^3_{\text{Earth}}\right)}{\frac{4}{3}\pi r^3_{\text{Earth}}} = -B\frac{\left(r^3_{\text{planet}} - r^3_{\text{Earth}}\right)}{r^3_{\text{Earth}}}$$

Inserting the numerical values gives,

$$P = -(7.1 \times 10^{10}\text{ Pa})\frac{(43 \times 10^{-2}\text{ m})^3 - (45 \times 10^{-2}\text{ m})^3}{(45 \times 10^{-2}\text{ m})^3} = 9.1 \times 10^9\text{ Pa}$$

$$\boxed{P = 9.1 \times 10^9\text{ Pa}}$$

What does it mean?

The pressure on the distant planet is more than 1000 times higher than on the Earth!

P11.60 A damped harmonic oscillator is displaced from equilibrium and then released. The oscillator displacement as a function of time is shown in Figure P11.60. Is this oscillator underdamped or overdamped?

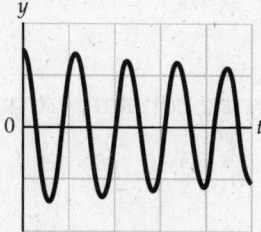

Figure P11.60

Solution

Recognize the principle

An oscillator is underdamped if it makes more than one oscillation before coming to rest at equilibrium, and overdamped if it just returns to the equilibrium point without oscillating.

Sketch the problem

The sketch is provided in the question.

Identify the relationships

According to the figure, this oscillator passes through the equilibrium point ($y = 0$) several times.

Solve

This oscillator is underdamped.

What does it mean?

If the oscillator was overdamped, it would reach $y = 0$ only once.

P11.64 Get new shocks! A car with worn-out (underdamped) shock absorbers is observed driving over a dip in the road. The car bounces up and down a total of three times in a period of 5 seconds after hitting the dip. If an average automobile shock absorber spring has a spring constant of 6.0 kN/m, what is the approximate mass of the observed car? *Note*: The body of a car is supported on four shock absorbers.

Solution

Recognize the principle

We can determine the oscillation frequency from the description of the oscillations. From this frequency, we can determine the mass of the car from the given spring constants.

Sketch the problem

No sketch needed.

Identify the relationships

From the information provided in the question, the frequency of the oscillation is

$$f = \frac{3 \text{ oscillations}}{5 \text{ s}} = 0.6 \text{ Hz}$$

The car body is supported on the four springs of the four shock absorbers. As was discussed in the textbook, multiple springs in this configuration can be summed to find an effective spring constant. In this case,

$$k_{eff} = 4k = 4(6.0 \text{ kN/m}) = 24 \text{ kN/m}$$

The frequency of oscillation is related to the mass and spring constant through the expression,

$$f = \frac{1}{2\pi}\sqrt{\frac{k}{m}}$$

Solve

Solving this expression for the mass gives,

$$\sqrt{\frac{k}{m}} = 2\pi f$$

$$m = \frac{k}{(2\pi f)^2}$$

Then inserting the numerical values we find,

$$m = \frac{24 \text{ kN/m}}{(2\pi(0.6 \text{ Hz}))^2} = \boxed{1.7 \times 10^3 \text{ kg}}$$

What does it mean?

The mass of the car is about 1700 kg, or 3740 lbs.

P11.73 Two playground swings are side by side. The children using the swings notice that one of the swings (swing 1) has a period of exactly 4.5 s. They also find that when the first swing has completed 10 oscillations, the other swing (swing 2) has completed 11. (a) Is swing 2 longer or shorter than swing 1? (b) Find the length of swing 2.

Solution

Recognize the principle

The period of a pendulum is determined only by its length and the acceleration of gravity.

Sketch the problem

No sketch needed.

Identify the relationships

We know that swing 1 completes 10 oscillations in the time swing 2 completes 11 oscillations. So, we can write a relationship between the periods of these two swings

$$10T_1 = 11T_2 \;\rightarrow\; T_1 = \frac{11}{10}T_2$$

This tells us that swing 1 has a larger period. We also have the expression for the period of a pendulum (swing),

$$T = 2\pi\sqrt{\frac{L}{g}}$$

Solve

(a) Since swing 1 has a longer period and only the length can change, we know: $\boxed{\text{Swing 2 must be shorter than swing 1.}}$

(b) To find the length of swing 2 we go back to the ratio for the periods:

$$T_1 = \frac{11}{10}T_2$$

We insert the expression for the period of swing 2 in terms of its length L_2 and solve for the length,

$$T_1 = \frac{11}{10}2\pi\sqrt{\frac{L_2}{g}}$$

$$L_2 = \left(\frac{10T_1}{11(2\pi)}\right)^2(g)$$

Then, inserting numerical values we find,

$$L_2 = \left(\frac{10(4.5\text{ s})}{11(2\pi)}\right)^2 (9.8\text{ m/s}^2) = \boxed{4.2\text{ m}}$$

What does it mean?

Swing 2 hangs 4.2 m (about 13 feet), and is close to a meter shorter than swing 1.

Part D. Additional Worked Examples and Capstone Problems

The following worked example provides you with further practice in analyzing a system undergoing simple harmonic motion. The capstone problems combine several of the key concepts from this chapter with concepts from previous chapters, including the physics of fluids and the conservation of momentum. Although these problems do not incorporate all the material discussed in this chapter, they do highlight several of the key concepts. If you can successfully solve these problems then you should have confidence in your understanding of these key concepts, so use these problems as a test of your understanding of the chapter material.

WE 11.1 Acceleration of a Simple Harmonic Oscillator

The speakers connected to your computer or stereo system (see figure below) contain a flexible membrane in the speaker "cone" that produces sound when it vibrates. If the speaker is playing a single-frequency tone, it undergoes simple harmonic motion. Suppose the speaker cone is oscillating with an amplitude of $A = 1.0$ mm at $f = 500$ Hz. Find (a) the maximum speed of the speaker cone and (b) the maximum acceleration.

Speaker cone

Solution

Recognize the principle

The displacement and velocity of a simple harmonic oscillator are given by the expressions,

$$y = A\ \sin(2\pi ft)$$
$$v = 2\pi fA\ \cos(2\pi ft)$$

so we can use the given values of f and A to find the speed. The speaker cone can be modeled as a mass on a spring, and we can get the acceleration from Newton's second law.

Sketch the problem

The figure provided shows the problem. The speaker cone oscillates along the x direction.

Identify the relationships

(a) The velocity of a simple harmonic oscillator is given by

$$v = 2\pi f A \cos(2\pi f t)$$

The cosine factor has a maximum value of 1, so the largest speed is just

$$v_{max} = 2\pi f A$$

(b) Using Newton's second law for a mass m attached to a spring k leads to

$$F = ma = -kx$$
$$a = -\frac{k}{m}x$$

The maximum acceleration (largest value of a) thus occurs when the displacement has its "largest negative" value, which is $-A$; hence,

$$a_{max} = -\frac{k}{m}(-A) = \frac{k}{m}A$$

The values of k and m are not given, but we can derive their ratio because the frequency of the oscillator is $f = (1/2\pi)\sqrt{k/m}$.

Solve

(a) Substituting in the numerical values we find

$$v_{max} = 2\pi f A = 2\pi(500\ \text{Hz})(1.0 \times 10^{-3}\ \text{m}) = \boxed{3.1\ \text{m/s}}$$

(b) Let's first solve for k/m:

$$f = \frac{1}{2\pi}\sqrt{\frac{k}{m}}$$

$$\sqrt{\frac{k}{m}} = 2\pi f$$

$$\frac{k}{m} = 4\pi^2 f^2$$

Substituting this expression into our expression for acceleration and substituting in the numerical values gives

$$a_{max} = \frac{k}{m}A = [4\pi^2(500\ \text{Hz})^2](1.0 \times 10^{-3}\ \text{m}) = \boxed{9900\ \text{m/s}^2}$$

What does it mean?

The maximum speed and acceleration are both quite large. That's why you can feel the vibration if you touch the case of a loudspeaker. In fact, a_{max} is about 1000 times greater than the acceleration due to gravity!

CP 11.1 Simple Harmonic Motion, Energy Conservation, and Collisions

A mass of 10 kg is attached to a spring of spring constant $k = 50$ N/m. The mass is free to oscillate on a frictionless, horizontal surface. The mass is displaced from equilibrium a distance of 20 cm and released. (a) Determine the frequency of oscillation. (b) Determine the total energy of the system. (c) Determine the maximum speed obtained by the mass. (d) Suppose that during the oscillation the mass broke away from the spring at the instant the mass was moving to the right and passing the equilibrium position. Comment on the subsequent motion of the mass. (e) Repeat part (d) if the mass breaks away from the spring at the far right position (20 cm to the right of equilibrium). (f) Suppose a blob of clay (5 kg) is dropped vertically on the oscillating mass as the mass passes through the equilibrium point and sticks to the mass. Determine the new amplitude of oscillation.

Solution

Recognize the principle

For parts (a)–(c) of this problem we can apply the principle of conservation of mechanical energy. The total energy of the system, which is the sum of the kinetic energy of the mass and the spring potential energy, will remain constant throughout the oscillations. For parts (d) and (e) we need to think about the motion of a mass when there are no external forces. Newton's laws will help in this analysis. Finally, for part (f) we will need to apply the principle of conservation of momentum since this is a perfectly inelastic collision.

Sketch the problem

It is helpful to begin by sketching the initial configuration of the system, being sure to label all the given information.

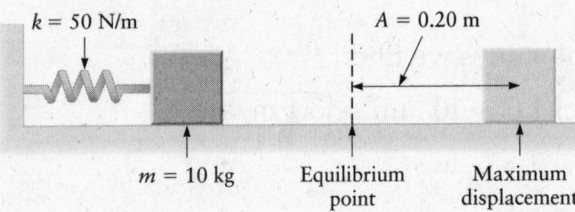

$k = 50$ N/m $A = 0.20$ m

$m = 10$ kg   Equilibrium   Maximum
point   displacement

Identify the relationships

(a) The frequency of a mass on a spring undergoing simple harmonic motion is given by the expression,

$$f = \frac{1}{2\pi}\sqrt{\frac{k}{m}}$$

(b) We know that the mass momentarily comes to rest when it is at its maximum displacement from equilibrium. This means that all of the energy of the system is in the form of spring potential energy. So, we can write the following expression,

$$E = PE_{\text{spring, max}} = \frac{1}{2}kA^2$$

(c) We know that as the mass passes through the equilibrium point, the spring potential energy momentarily is zero so the kinetic energy is a maximum. We can use this information to solve for the speed of the mass at the equilibrium point which will be the maximum speed of the mass,

$$E = KE_{\text{max}} = \frac{1}{2}mv_{\text{max}}^2 \rightarrow v_{\text{max}} = \sqrt{\frac{2E}{m}}$$

(d) and (e) In both of these cases the spring breaks away from the mass and so there is no net force acting on the mass. Since Newton's first law tells that an object will remain at rest or remain in motion when there is zero net external force, we only need to determine the motion of the mass at the instant the mass breaks away in order to determine its subsequent motion.

(f) When the blob of clay is dropped vertically on the mass, a collision occurs in which linear momentum is conserved. Since the blob has no horizontal velocity, we can write the following expression describing the momentum conservation,

$$P_i = P_f$$
$$mv_{i,max} = (m + m_{blob})\, v_{f,max}$$

where $v_{i,max}$ is the speed of the mass before the collision which we determined in part (c), and $v_{f,max}$ is the speed of the mass + blob immediately after they collide and stick together. Solving this expression for the final speed we get,

$$v_{f,max} = \frac{mv_{i,max}}{(m + m_{blob})}$$

Finally, we can use this speed and the combined mass to determine the new amplitude of oscillation by applying the principle of conservation of energy,

$$E = PE_{spring,max} = KE_{max}$$

Therefore,

$$\frac{1}{2} kA^2_{new} = \frac{1}{2}(m + m_{blob})\, v^2_{f,max}$$

Solving for the new amplitude and substituting in for $v_{f,max}$ we have,

$$A_{new} = \sqrt{\frac{1}{k(m + m_{blob})}}\, mv_{i,max}$$

Solve

(a) Substituting in the numerical values we have,

$$f = \frac{1}{2\pi}\sqrt{\frac{k}{m}} = \frac{1}{2\pi}\sqrt{\frac{50\,\text{N/m}}{10\,\text{kg}}} = \boxed{0.36\,\text{Hz}}$$

(b) Substituting in the numerical values, the total energy is,

$$E = \frac{1}{2} kA^2 = \frac{1}{2}(50\,\text{N/m})\,(0.20\,\text{m})^2 = \boxed{1.0\,\text{J}}$$

(c) Substituting in the numerical values, the maximum speed is,

$$v_{max} = \sqrt{\frac{2E}{m}} = \sqrt{\frac{2(1.0\,\text{J})}{10\,\text{kg}}} = \boxed{0.45\,\text{m/s}}$$

(d) If the spring were to break away at the moment the mass was traveling to the right through the equilibrium point, the mass would have its maximum speed of 0.45 m/s. Since no forces would be acting on the mass it would continue to move to the right at a speed of 0.45 m/s.

(e) If the spring were to break at the moment the mass was at its maximum displacement, the mass would have a velocity of zero. Since no forces would be acting on the mass it would remain at rest at this point.

(f) Substituting in the numerical values we have,

$$A_{\text{new}} = \sqrt{\frac{1}{k(m + m_{\text{blob}})}} \, mv_{i,\text{max}}$$

$$= \sqrt{\frac{1}{50 \text{ N/m} (10 \text{ kg} + 5 \text{ kg})}} \, (10 \text{ kg}) \, (0.45 \text{ m/s})$$

$$= \boxed{0.16 \text{ m}}$$

What does it mean?

Since the total mass of the oscillator increased without adding more energy to the system (i.e., the velocity of the clay blob was zero before the collision), the amplitude must decrease. The frequency of oscillation would also change!

CP11.2 Deep-Sea Submersible: Combining the Elasticity of a Solid with the Physics of Fluids

Suppose a spherically shaped deep-sea submersible of radius 1.5 m and mass of 15,000 kg (including its passengers and instrumentation) is lowered to a depth of 1500 m in the ocean. The submersible is made of steel and is lowered down using a steel cable 10 cm in diameter. (a) Determine the change in volume of the submersible due to the pressure of the ocean water. (b) What is the tension in the steel cable when the submersible is deployed? (c) Determine the change in length of the steel cable from when it is suspending the submersible above the ocean's surface to when it is lowered beneath the ocean's surface.

Solution

Recognize the principle

There are several key principles we'll need to apply to tackle this problem. As the submersible is lowered beneath the surface of the water, the pressure surrounding the submersible increases with depth. We can calculate the change in pressure using the expression for the pressure-depth expression we discussed in Chapter 10. The change in pressure will compress the submersible in such a way as to change its volume, so we'll need the bulk modulus for steel. Since the submersible displaces sea water, there will be a buoyant force which we'll need to calculate in order to determine the tension in the cable once the submersible is submerged. We can compare this tension with that when the submersible is above the water to get the change in tension. We can use this change in tension to determine the change in length of the cable.

Sketch the problem

To answer part (a) of the problem we'll need to determine the pressure on the submersible when it is 1500 m beneath the surface of the ocean. A simple sketch will help us set things up.

For part (b) of this problem we'll need to apply Newton's second law so a free-body diagram for the submersible, both above and below the ocean's surface, will be helpful.

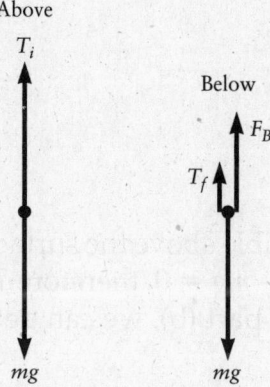

Identify the relationships

(a) First we need to determine the change in pressure on the submersible as it is lowered to a depth of 1500 m. From Chapter 10, the pressure increases with depth according to the expression,

$$P = P_0 + \rho g h$$

where ρ is the density of seawater, $g = 9.8 \text{ m/s}^2$, and h is the height of water above the submersible. Since the pressure above the ocean's surface is P_0, the change in pressure on the submersible as it is lowered is simply $\rho g h$.

Next we can use this change in pressure to determine the change in volume of the submersible according to the expression,

$$P = -B \frac{\Delta V}{V_0}$$

where B is the bulk modulus of steel. Since the submersible is a sphere, its volume is given by the expression $V = \frac{4}{3} \pi r^3$. Combining all these expressions we can solve for the change in volume of the submersible,

$$\Delta V = -\frac{PV_0}{B} = -\frac{(\rho g h)\left(\frac{4}{3} \pi r_i^2\right)}{B} = -\frac{4\pi \rho g h r_i^2}{3B}$$

(b) From our free-body diagram for when the submersible is below the surface of the ocean, we can apply Newton's second law and obtain the following expression,

$$\Sigma F = ma$$
$$T_f + F_B - mg = 0$$

From Chapter 10, the buoyant force on the submersible is given by the following expression,

$$F_B = \rho g V$$

where V is the volume of seawater displaced by the submersible. Since the volume displaced is the volume of the submersible, we can substitute $V = V_f = V_i - \Delta V$

into this expression and combine this with our expression from applying Newton's second law to determine the tension in the cable,

$$T_f = mg - F_B$$
$$= mg - \rho g V$$
$$= mg - \rho g(V_i - \Delta V)$$
$$= g\left[m - \rho\left(\frac{4}{3}\pi r_i^3 - \Delta V\right)\right]$$

(c) Finally, from our free-body diagram of the submersible above the surface of the ocean and applying Newton's second law we have, $T_i - mg = 0$, therefore $T_i = mg$. Combining this with our expression for the tension in part (b), we can determine the change in tension in the cable,

$$\Delta T = T_f - T_i$$
$$= g\left[m - \rho\left(\frac{4}{3}\pi r_i^3 - \Delta V\right)\right] - mg$$
$$= -\rho g\left(\frac{4}{3}\pi r_i^3 - \Delta V\right)$$

Since this is a tensile stress on the cable we can apply our expression involving Young's modulus for steel, and since the force in this expression is the change in tension, we have

$$\Delta L = \frac{FL_0}{YA} = \frac{\Delta T L_0}{YA}$$

The initial length of the cable is $L_0 = 1500$ m. The cross-sectional area of the cable (which we will assume remains constant) is given by its diameter, d, according to the expression $A = \pi d^2/4$. Combining this with the expression for ΔT, we have,

$$\Delta L = \frac{\Delta T L_0}{YA} = \frac{-\rho g\left(\frac{4}{3}\pi r_i^3 - \Delta V\right)L_0}{Y\left(\frac{\pi d^2}{4}\right)}$$

Solve

(a) Substituting in the numerical values of $\rho = 1025\,\text{kg/m}^3$, $g = 9.8\,\text{m/s}^2$, $h = 1500$ m, $r_i = 1.5$ m, and $B = 1.4 \times 10^{11}$ Pa, we have,

$$\Delta V = -\frac{4\pi\rho g h r_i^3}{3B} = \frac{4\pi(1025\,\text{kg/m}^3)(9.8\,\text{m/s}^2)(1500\,\text{m})(1.5\,\text{m})^3}{3(1.4 \times 10^{11}\,\text{Pa})}$$
$$= \boxed{-1.52 \times 10^{-3}\,\text{m}^3}$$

(b) Substituting the numerical values into our expression for the tension in the cable when it is fully deployed we have,

$$T_f = g\left[m - \rho\left(\frac{4}{3}\pi r_i^3 - \Delta V\right)\right]$$
$$= 9.8\,\text{m/s}^2\left[15{,}000\,\text{kg} - (1025\,\text{kg/m}^3)\left(\frac{4}{3}\pi(1.5\,\text{m})^3 - (-1.52 \times 10^{-3}\,\text{m}^3)\right)\right]$$
$$= \boxed{4.98 \times 10^3\,\text{N}}$$

(c) Substituting in all the numerical values into our expression for the change in length of the cable we have,

$$\Delta L = \frac{-\rho g \left(\frac{4}{3} \pi r_i^3 - \Delta V \right) L_0}{Y \left(\frac{\pi d^2}{4} \right)}$$

$$= \frac{-(1025 \text{ kg/m}^3)(9.8 \text{ m/s}^2) \left[\frac{4}{3} \pi (1.5 \text{ m})^3 - (-1.52 \times 10^{-3} \text{ m}^3) \right] (1500 \text{ m})}{(2 \times 10^{11}) \left(\frac{\pi (0.10 \text{ m})^2}{4} \right)}$$

$$= \boxed{-1.4 \times 10^{-1} \text{ m}}$$

What does it mean?

Several comments can be made about our results. Notice that the change in volume we found in part (a) is negative. This is expected since the submersible is subjected to an increase in pressure as it is lowered beneath the surface of the ocean. Our change in length is also negative, but for a different reason. As the submersible is lowered into the ocean it experiences an upward buoyant force which reduces the tension in the cable. So, the tension is less when the submersible is under the water.

Part E. MCAT Review Problems and Solutions

PROBLEMS

1. The Young's modulus for steel is 2.0×10^{11} Pa. What is the stress experienced by a steel rod that is 100 cm long and 20 mm in diameter when it is stretched by a force of 6.2×10^3 N?

   (a) 2.0×10^7 N/m^2
   (b) 12.6×10^{11} N/m^2
   (c) 3.2×10^7 N/m^2
   (d) 4.0×10^7 N/m^2

2. A steel rod 100 cm in length experiences a stress of 4.0×10^8 N/m^2 when it is stretched by a force of 10 N. The Young's modulus of steel is 2.0×10^{11} Pa. What is the strain on the rod?

   (a) 5.0×10^3 m
   (b) 5.0×10^3
   (c) 2.0×10^{-3} m
   (d) 2.0×10^{-3}

3. A simple pendulum has a period of 4.63 s at a particular point on the Earth's surface where the acceleration due to gravity is 9.82 m/s^2. The pendulum is moved to a different point and the period increases to 4.64 s. What is the value of g at this new location?

   (a) 9.78 m/s^2
   (b) 9.82 m/s^2

(c) $9.86 \, \text{m/s}^2$

(d) Cannot be determined with the information provided.

4. A simple pendulum with a length L has a period of 2 s. In order for the pendulum to have a period of 4 s we must

(a) halve the length.

(b) quarter the length.

(c) double the length.

(d) quadruple the length.

5. If a pendulum is 12 m long and has a frequency of 0.25 Hz, what will be the period of a second pendulum at the same location if its length is 3.0 m?

(a) 2.0 s

(b) 3.0 s

(c) 4.0 s

(d) 6.0 s

6. A pendulum clock is losing time. How should the pendulum be adjusted?

(a) The weight of the bob should be decreased so it can move faster.

(b) The length of the pendulum should be decreased.

(c) The amplitude of the swing should be reduced so the path covered is shorter.

(d) None of the above.

7. A 20-kg mass is attached to a wall by a spring. A 5.0-N force is applied horizontally to the mass so that the mass is displaced 1.0 m from its equilibrium position along a horizontal frictionless surface. Once released, with approximately what period will the mass oscillate?

(a) 2.0 s

(b) 6.0 s

(c) 13.0 s

(d) 16.0 s

8. Suppose a 0.3-kg mass on a spring that has been compressed 0.10 m has spring potential energy of 1.0 J. How much further must the spring be compressed to triple the spring potential energy?

(a) 0.30 m

(b) 0.20 m

(c) 0.17 m

(d) 0.07 m

9. Consider the following expression for the position as a function of time of a mass on a spring undergoing simple harmonic motion, $x = (0.5)\sin(4.0\pi t)$, where x is in meters and t is in seconds. What is the frequency of oscillation?

(a) 0.5 Hz

(b) 4.0 Hz

(c) 2.0 Hz

(d) 4.0π Hz

10. When the shock absorbers on an automobile wear out and lose their ability to damp out oscillations, the resulting oscillation is

(a) underdamped.

(b) critically damped.

(c) overdamped.

(d) hyperdamped.

SOLUTIONS

1. *MCAT strategies*

No answers can be easily eliminated so we proceed with our problem-solving strategy.

Sketch the problem

No sketch required.

Recognize the principle

To solve this problem we must understand the expression relating the stress, strain, and Young's modulus for a material.

Identify the relationships

The expression relating these quantities is,

$$\frac{F}{A} = Y\frac{\Delta L}{L_0}$$

In this expression, F/A is the stress, Y is the Young's modulus, and $\Delta L/L_0$ is the strain. So, much of the information provided in the question is not necessary for solving the problem. We are simply asked for the stress,

$$\text{stress} = \frac{F}{A} = \frac{F}{\pi r^2}$$

Solve

Substituting in for $F = 6.2 \times 10^3$ N and $r = 0.01$ m we have,

$$\text{stress} = \frac{F}{\pi r^2} = \frac{6.2 \times 10^3 \text{ N}}{\pi(0.01 \text{ m})^2} \approx \frac{6.2 \times 10^3 \text{ N}}{(3.1)(0.01 \text{ m})^2}$$

$$\approx 2 \times 10^7 \text{ N/m}^2$$

So, the correct answer is (a).

What does it mean?

Notice that we approximated π to be 3.1 to simplify the mathematics. Remember, calculators are not allowed on the MCAT exam so you need to be able to make approximations when necessary.

2. *MCAT strategies*

Since the strain is given by $\Delta L / L_0$, it is dimensionless. Therefore, the correct answer should be dimensionless, thus eliminating answers (a) and (c). Of the two remaining answers it's difficult to know which is correct without doing the calculation so we proceed with our problem-solving strategy.

Recognize the principle

We can rearrange our expression involving Young's modulus to solve for the strain on the rod.

Sketch the problem

No sketch required.

Identify the relationships

Rearranging our expression involving Young's modulus we have,

$$\text{Strain} = \frac{\Delta L}{L_0} = \frac{\text{Stress}}{Y}$$

Solve

Substituting in for the stress = $4.0 \times 10^8 \text{ N/m}^2$ and $Y = 2.0 \times 10^{11}$ Pa we have,

$$\text{Strain} = \frac{\text{Stress}}{Y} = \frac{4.0 \times 10^8 \text{ N/m}^2}{2.0 \times 10^{11} \text{ Pa}} = \frac{4.0 \times 10^8 \text{ N/m}^2}{2.0 \times 10^{11} \text{ N/m}^2} = 2.0 \times 10^{-3}$$

So, the correct answer is (d).

What does it mean?

Note once again that some of the information provided in the question was not needed to solve the problem. Don't be tempted by this type of distraction technique. They are commonly used on the MCATs.

3. *MCAT strategies*

To solve this problem we simply need to understand how the period of a simple pendulum depends on the value of g. The period is inversely proportional to the square root of g,

$$T \propto \frac{1}{\sqrt{g}}$$

If the value of g is decreased, then the period will increase, whereas if g is increased, then the period will decrease. In this problem the period is said to increase from 4.63 s to 4.64 s so the value of g must decrease. The only answer which decreases the value of g is (a). Also, it's clear that we could have used ratios to determine the answer precisely so answer (d) is incorrect. So, the correct answer is (a).

4. *MCAT strategies*

To solve this problem we need to understand how the period of a simple pendulum depends on the length of the pendulum. This relationship is given by the expression,

$$T \propto \sqrt{L}$$

In order to increase the period from 2 s to 4 s (i.e., double the period), we must quadruple the length of the pendulum. So, the correct answer is (d).

5. *MCAT strategies*

Since we are asked for the period of oscillation, let's begin by determining the initial period from the frequency,

$$T = \frac{1}{f} = \frac{1}{0.25 \text{ Hz}} = 4 \text{ s}$$

By decreasing the length of the pendulum from 12 m to 3 m we are reducing it to ¼ of its initial length. Since the period of a pendulum is proportional to the square root of its length, this will reduce the period to half its initial value. So, the new period will be 2 s and thus the correct answer is (a).

6. *MCAT strategies*

Once again, we simply need to understand what factors affect the period of a simple pendulum. The period of a simple pendulum is given by,

$$T = 2\pi \sqrt{\frac{L}{g}}$$

From this expression we see that the period does not depend on the mass of the bob or the amplitude of oscillation so answers (a) and (c) can be quickly eliminated. The period does depend on the length of the pendulum, so shortening the pendulum will decrease the period causing the clock to run faster. So, the correct answer is (b).

7. *MCAT strategies*

No answers can be easily eliminated so we proceed with our problem-solving strategy.

Recognize the principle

We'll need to apply our expression for the period of a mass on a spring and also apply Hooke's law describing how the force from a spring is related to how much the spring is stretched or compressed.

Sketch the problem

No sketch required.

Identify the relationships

Let's start by applying Hooke's law. The magnitude of the force required to compress a spring by a distance x is given by,

$$F = kx$$

where k is the spring constant. Rearranging this expression we can solve for the spring constant,

$$k = \frac{F}{x}$$

Now, the period of oscillation of a mass on a spring is given by the expression,

$$T = 2\pi \sqrt{\frac{m}{k}}$$

Combing these two expressions we have,

$$T = 2\pi\sqrt{\frac{m}{k}} = 2\pi\sqrt{\frac{mx}{F}}$$

Solve

Substituting in the numerical values given in the question we have,

$$T = 2\pi\sqrt{\frac{mx}{F}} = 2\pi\sqrt{\frac{(20\text{ kg})(1.0\text{ m})}{5.0\text{ N}}} = 2\pi\sqrt{4}\text{ s}$$

$$\approx 12\text{ s}$$

So the closest answer is (c).

What does it mean?

Notice that we did not log any number in until we have simplified the expression. Also, we approximated π to be 3. By waiting until the end to crunch the numbers and approximating π we simplified the mathematics required to solve the problem.

8. *MCAT strategies*

To solve this problem we need to know how the spring potential energy depends on the amount you compress or stretch the spring. The spring potential energy is given by the expression,

$$PE_{\text{spring}} = \frac{1}{2}kx^2$$

So, if we want the spring potential energy to triple, we'll need to compress the spring by a factor of $\sqrt{3}$ time the initial amount. Since $\sqrt{3} \approx 1.7$, this means that the spring must be compressed by 0.7 m more than the original 1 m. So, the correct answer is (d).

9. *MCAT strategies*

To solve this problem we simply need to compare our general expression describing the position of a mass on a spring undergoing simple harmonic motion with the expression provided in the question. Our general expression is,

$$x = A\sin(2\pi ft)$$

Comparing this with that provided in the question we see that the frequency must be 2 Hz. So, the correct answer is (c).

10. *MCAT strategies*

This question is simply testing our understanding of damped, underdamped, and overdamped oscillations. There is no such thing as a hyperdamped oscillation so answer (d) can be eliminated. After hitting a bump in the road an automobile with worn out shock absorbers would oscillate up and down for a long time before finally coming to rest. This is an example of an underdamped oscillation. So, the correct answer is (a).

12 Waves

> ## CONTENTS

Part A. Summary of Key Concepts and Problem-Solving Strategies

KEY CONCEPTS

Waves

A *wave* is a disturbance that transports energy from one place to another without transporting matter. Waves travel through a medium; for mechanical waves, the medium is a material substance. For *transverse waves*, the displacement of the medium is perpendicular to the direction of the velocity of the wave. With *longitudinal waves*, the displacement is parallel to the direction of the wave velocity. Fluids (liquids and gases) cannot support transverse waves but they can support longitudinal waves. Solids can support both transverse and longitudinal waves.

Frequency and period

A wave can consist of a "single" disturbance—a wave pulse—or the disturbance can have a repetitive form. A periodic wave has a characteristic repeat time called the *period* T. The *frequency* of a periodic wave is related to the period by $f = 1/T$.

Displacement of a transverse wave

The displacement in the y direction associated with a periodic transverse wave is described by

$$y = A \sin\left(2\pi ft - \frac{2\pi x}{\lambda}\right)$$

373

Here A is the amplitude of the wave, f is the frequency, and λ is the wavelength. The wave represented by this expression is traveling in the positive x direction.

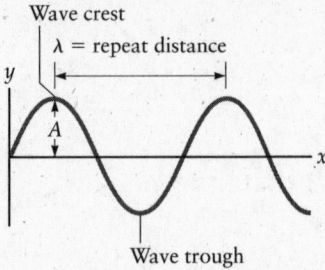

The frequency, wavelength, and speed of a wave are related by the expression,

$$v = f\lambda$$

Superposition

According to the ***principle of superposition***, when two waves are present together, the total displacement of the medium is the sum of the displacements of the individual waves. One result of this principle is that two waves can "pass through" each other.

Wave fronts and rays

A ***wave front*** is a surface that describes how the energy carried by a wave travels through space. ***Rays*** indicate the direction of energy flow and are perpendicular to the wave fronts.

Reflection and refraction

Waves can be reflected when they reach the boundary of the medium in which they travel. ***Refraction*** is the change in direction of a wave due to a change in velocity. The wave velocity can change at a boundary or within the medium.

APPLICATIONS

Examples of waves

There are many examples of waves. The speed of a wave depends on the properties of the medium.

- ***Waves on a string*** are transverse waves and are employed in many musical instruments. The speed of a wave on a string is dependent on the tension in the string F_T and the mass per unit length of the string μ according to the expression,

$$v = \sqrt{\frac{F_T}{\mu}}$$

- *Sound* is a longitudinal wave that can exist in solids, liquids, and gases. We'll visit this again in Chapter 13.

- *Light* is an electromagnetic wave. Because they are not mechanical waves, light and other electromagnetic waves can travel through a vacuum as well as through a material medium such as air.

Interference and standing waves

When two waves pass through the same location at the same time, they are said to *interfere*. One example of interference is the phenomenon of *standing waves*. A point along a standing wave where the displacement of the medium is always zero is called a *node*, and the points where the displacement is largest are *antinodes*. Standing waves are used to produce musical tones in stringed instruments such as guitars and pianos. The frequency of the standing wave with the lowest frequency (longest wavelength) is called the *fundamental frequency*. For a string held rigidly at both ends, the standing waves form a series of *harmonics* with frequencies $f_n = nf_1$ with $n = 1, 2, 3 \ldots$ (standing waves on a string with fixed ends)

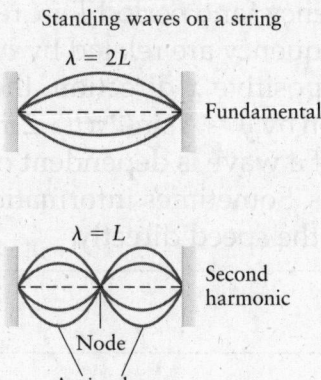

Standing waves on a string

$\lambda = 2L$ Fundamental

$\lambda = L$ Second harmonic

Node
Antinodes

Seismic waves

Seismic waves provide information about the Earth's inner structure.

PROBLEM-SOLVING STRATEGIES

In this chapter we were introduced to the concept of a wave, how to describe a wave, and the basic properties of a wave. Speed, frequency, and wavelength were defined and related by a basic expression. Furthermore, how waves reflect and interact with each other was discussed in the context of interference. The following two problem-solving schematics should help you tackle many of the end-of-chapter problems in this chapter.

Problem Solving: Understanding and Describing Waves

Recognize the principle

Waves are described by their speed, wavelength, frequency, period, the medium through which they propagate, and their nature (transverse or longitudinal). Some of this information may be provided while some may need to be determined.

Sketch the problem

Often it is helpful to sketch the wave involved in the problem, being sure to label all key parameters describing the wave.

Identify the relationships

Several key relationships are important. The frequency f and period T are related by $f = 1/T$ and the speed v, wavelength λ and frequency are related by $v = \lambda f$. For a periodic transverse wave traveling in the positive x direction, the displacement of the wave along the y direction is given by $y = A\sin(2\pi ft - 2\pi x/\lambda)$. Here A is the amplitude of the wave. The speed of a wave is dependent on the properties of the medium through which it travels. Sometimes information on the medium is provided, allowing us to calculate the speed directly.

Solve

Solve for the unknown quantities using the information provided in the problem and the key relationships. Problems may involve the application of concepts from previous chapters such as Newton's laws.

What does it mean?

Always consider what your answer means and check that it makes sense.

Problem Solving: Interference of Waves

Recognize the principle

When two or more waves pass through the same location at the same time they interfere, resulting in a combined wave. By applying the principle of superposition we can determine the combined displacement of the medium. If the displacements of the waves add at a particular point then we call this constructive interference, whereas if they subtract we call this destructive interference.

Sketch the problem

Begin by sketching the situation described in the problem, being sure to label all the key parameters. For example, for the collision of sound waves it's important to label the distance from each source (speaker) to where the sound waves collide. It's also important to note the speed and frequency of the sound.

Identify the relationships

For maximum constructive interference to occur the crest of one wave must overlap with the crest of the other wave. For destructive interference the crest of one wave must overlap with the trough of the other wave. If the waves are produced by different sources (such as speakers in the case of sound) then knowing the distance from the sources to where the interference occurs allows us to determine if the crests align or not. Consider the following example where sound waves are produced by two speakers driven by the same frequency generator. If the distances from each source differ by 1, 2, 3, . . . wavelengths then the crests will align, leading to constructive interference,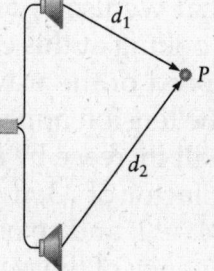

$$d_2 - d_1 = n\lambda; \quad n = 0, 1, 2, 3 \ldots$$

If the distance from each source differ by 1/2, 3/2, 5/2, . . . wavelengths then the crests will align with the troughs, leading to destructive interference,

$$d_2 - d_1 = (n + \tfrac{1}{2})\lambda; \quad n = 0, 1, 2, 3 \ldots$$

Solve

Solve for the unknown quantities which may include the location of constructive or destructive interference, the frequency, wavelength, or speed of the waves.

What does it mean?

Always *consider what your answer means* and check that it makes sense.

Part B. Frequently Asked Questions

1. *I'm confused by the relationship among the speed, wavelength, and frequency of a wave. I thought that speed is a distance divided by a time. If so, how can the speed of a wave equal the wavelength times the frequency?*

To answer this question let's start with the expression relating speed, wavelength, and frequency and substitute in for the frequency in term of the period of the wave ($f = 1/T$).

$$v = f\lambda = \frac{\lambda}{T}$$

Writing the expression this way we see that dimensionally we have a wavelength in meters divided by a period in seconds, giving us the units of m/s. So dimensionally this expression is correct, but does it make sense? Well, the wavelength is the repeat distance and the period is the repeat time. If you imagine yourself standing along side a wave as is passes you by, one wavelength of the wave would pass you in a time of one period. This means that the speed of the wave must be this the distance of the wave that passes you (its wavelength) divided by the time it took to pass you (its period), So, the speed of the wave is the wavelength divided by the period. Using our expression for frequency, the speed is also the wavelength times the frequency.

2. *How does the tension in a string, rope, or cable affect the speed of the wave?*

Let's begin by writing the expression for the speed of a wave on a string in terms of the tension in the string. Recall the expression,

$$v = \sqrt{\frac{F_T}{\mu}}$$

where F_T is the tension in the string and μ is the string's mass per unit length. Note that we use F_T to represent the tension so as not to confuse it with the period T. Looking at this expression we see that as we increase the tension in the string, the speed of the wave also increases. However, this is not a linear relationship since the tension appears under the square root. So, if you double the tension, the speed will increase by a factor of $\sqrt{2}$; if you triple the tension, the speed will increase by a factor of $\sqrt{3}$; if you quadruple the tension, the speed will increase by a factor of $\sqrt{4} = 2$; and so on. It is also worth noting that the speed is independent of the frequency of the wave. That means that all waves traveling along this string will have the same speed, independent of their frequencies.

3. *When determining the number of nodes present in a standing wave, do we include the points at the end of the wave?*

For a string fixed at both ends, there are indeed nodes at the ends and we should include them in the total number of nodes. So, for a string fixed at both ends the third harmonic ($n = 3$) would look as follows,

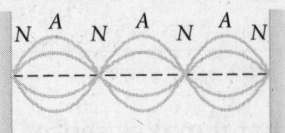

For this standing wave there are four nodes and three antinodes. Note that *n* represents the harmonic *not* the number of nodes.

4. *For a periodic wave if I plot the displacement versus time it looks the same as when I plot the displacement versus the position. Why is that?*

That's a good question, and helps to illustrate the symmetry in our expression for the displacement of a periodic wave. As was discussed in your textbook, the displacement along the *y* direction for a periodic wave traveling in the positive *x* direction can be represented by the following expression,

$$y = A \sin\left(2\pi f t - \frac{2\pi x}{\lambda}\right)$$

where *A* is the amplitude, *f* is the frequency, and λ is the wavelength of the wave. Looking carefully at this expression, we see that the displacement depends on the sine of the time *t* and the sine of the position *x*. So the displacement is symmetric in both space (position) and time. In other words, if we focus our attention at a particular point in space (i.e., a particular value of *x*), the displacement will vary as the sine of the time,

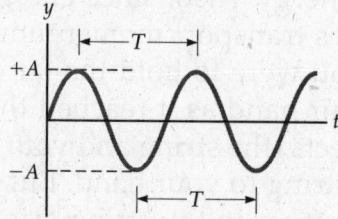

On the other hand, if we take a snapshot of the wave at a particular moment in time we will see that the displacement varies as the sine of the position,

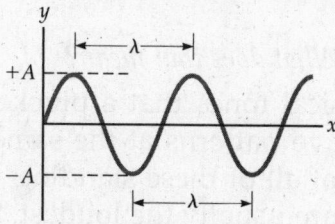

Therefore, since both time and position appear within the sine function in our expression for the displacement, both the *y-t* and *y-x* diagrams look the same.

5. *How do I know if a wave is periodic or nonperiodic.*

Sometimes it's quite difficult to tell if a wave is periodic or not. For complex waves, the displacement of the wave as a function of position and time is not sinusoidal and therefore cannot be described by a simple mathematical expression. However, these complex waves can still be periodic as long as they have a well-defined period. That is, as long as you can see a pattern in the wave which repeats then the wave is periodic. The following figure illustrates a complex wave with a well-defined period and is therefore periodic.

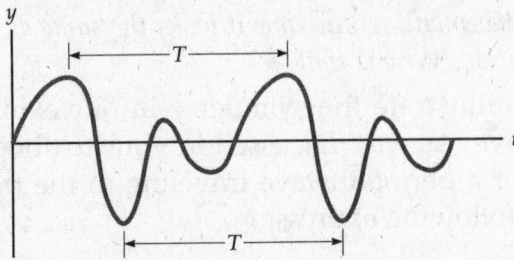

6. *What really travels with the wave?*

Though you may think that the medium through which the wave is traveling actually travels along with the wave, this notion would be wrong. The medium is displaced by the wave but returns to its original position once the wave passes by. Imagine a wave traveling along a string. If the string was carried along with the wave, the string would end up piled up at its end. Clearly this does not happen, so matter (the string) is not transported by a wave. So, waves do not transport matter, but they do transport energy. Also, since energy and momentum are related we can also say that waves transport momentum. Consider again a wave traveling along a string. If you were to hold the far end of the string you could feel the wave bump into your hand as it reached the end. In essence, this is like a collision between two objects (the string and your hand) where momentum is being transferred from the string to your hand. But where did this momentum come from? Well, the momentum in the string originated at the other end of the string, perhaps by another person shaking the string. So, some of the momentum (and therefore energy) of the other person's hand was transported along the string and eventually transferred to your hand at the other end of the string.

7. *I've heard of guitar players "playing harmonics." What does that mean?*

Recall from the section in your textbook on musical tones that a plucked guitar string vibrates at all the possible standing wave patterns at the same time. The total motion of the string is a superposition of all of these standing waves, although the vibrations at the lowest frequencies are usually the loudest. So, the sound of a musical note played on a guitar is composed of many frequencies corresponding to the various standing waves (harmonics). When a guitar player "plays harmonics" he places his finger gently at a point along the string so that some of these harmonics are eliminated. This has the effect of changing the tone of the sound produced. For example, if you were to gently place your finger at a point halfway along the length of the string, you would prevent the fundamental, 3rd, 5th, 7th, etc. harmonic from vibrating. Only the even harmonics would be produced so the tone of the sound would be quite different. The reason the even harmonics would still be produced is because they have a node at the point halfway along the string so the placement of your finger at this point does not prevent them from vibrating.

8. *Can two transverse waves travel at different speeds along the same string?*

No. Recall the expression describing the speed of a wave on a string,

$$v = \sqrt{\frac{F_T}{\mu}}$$

where F_T is the tension in the string and μ is the string's mass per unit length. This expression shows that the speed of a wave depends only on the properties of the string (the tension and mass per unit length), not on the properties of the wave (amplitude, frequency, and wavelength). So, two waves that differ in amplitude, frequency, or wavelength will travel at the same speed along the string.

9. *How can we describe a wave with a trig function? I thought trig functions were the ratios of two sides of a triangle?*

Though it is true that trig functions are simply the ratio of two sides of a right-angle triangle, we can also think of them as functions of position and time. To see how this works imagine a point on the edge of a wheel as the wheel rotates at a constant angular speed. As illustrated in the figure we can describe the y position of this point in terms of the sine of the angle above the x axis. As the wheel rotates the angle changes with time and therefore the value of y changes as well. If we were to plot how y changes with time we would see that it looks exactly like how the displacement of a transverse wave varies with time. So it makes sense that the mathematical description of a transverse wave is a trig function.

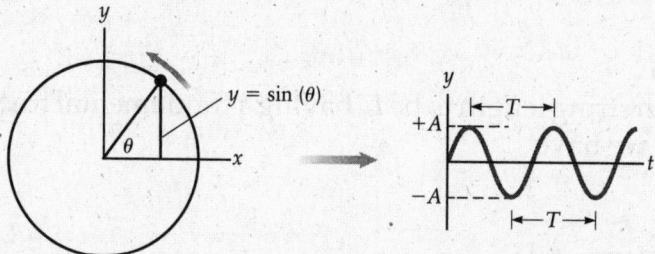

10. *What's the difference between the density of a string and its mass per unit length?*

Mass density, or simply density, of an object is how much mass there is in a given volume, $\rho = m/V$. The unit of density is kg/m^3. On the other hand, mass per unit length of a string is how much mass there is in a given length, $\mu = m/L$, and has the unit of kg/m. If we assume that the string has a uniform cross-sectional area, then its length, area, and volume are related by $V = AL$. Combining these expressions we can derive a relationship between the density of the string and its mass per unit length,

$$\rho = \frac{m}{V} = \frac{m}{AL} = \frac{1}{A}\left(\frac{m}{L}\right) = \frac{\mu}{A}$$

$$\rightarrow \mu = A\rho$$

So the density of a string and its mass per unit length are related by the cross-sectional area of the string.

Part C. Selection of End-of-Chapter Answers and Solutions

QUESTIONS

Q12.8 The neck of a guitar is designed with *frets* as shown in Figure Q12.8. A player can hold a string against one of the frets and thus shorten the vibrating length L of the string. In this way, a particular string can be used to play notes with different frequencies. The frequency of a note is determined by the fundamental standing wave mode (Fig. 12.24), and this frequency depends on L. In a musical scale, the frequencies of adjacent notes (i.e., between C and C-sharp) are in the ratio of $2^{1/12}/1 \approx 1.059$; this ratio produces the commonly used 12-tone musical scale. The artisans who make guitars position the frets according to the "rule of 18ths." According to this rule, each fret is positioned so that depressing adjacent frets shortens the vibrating length of the string by $L/18$. (See Fig. Q12.8.) Show that the rule of 18ths produces note frequencies that fit approximately on the desired musical scale.

Frets

© Teleimages / Jupiterimages

Figure Q12.8

Answer

To begin, let's define the open string length to be L_0 having a fundamental frequency f_0 and wavelength λ_0. Then, we have

$$\lambda_0 f_0 = v$$

The wavelength for the fundamental is

$$\lambda_0 = 2L_0$$

Now, let the first fretted position have length L_1, fundamental frequency f_1, and wavelength λ_1. Again this wave must satisfy

$$\lambda_1 f_1 = v$$

and we have

$$\lambda_1 = 2L_1$$

The rule of 18ths tells us that we are reducing the length of the string by $1/18$ so that

$$L_1 = \frac{17}{18}L_0$$

Both waves will have the same speed so

$$\lambda_0 f_0 = \lambda_1 f_1 = v$$

or

$$2L_0 f_0 = 2L_1 f_1$$

$$L_0 f_0 = \frac{17}{18} L_0 f_1$$

$$\boxed{f_1 = \frac{18}{17} f_0 \approx 1.059 f_0}$$

So, reducing the length by 1/18 will give the desired change in frequency.

Q12.13 A woman wakes up just after 2 AM. Wondering what woke her up, she goes to the window and looks around. Noting nothing out of the ordinary, she sits back down on her bed and all of a sudden a loud explosion rattles her window. The next day, she reads that an underground munitions dump exploded about 10 km from her home. She thinks that perhaps she woke up because she experienced a psychic precognition of the explosion. Can you think of a physical explanation for what might have woken her up the few moments before the sound from the blast hit her window? (See Problem 69.)

Answer

The explosion also shook the ground, creating sound waves in the Earth's surface. Since sound travels much faster in (most) solids than in air, the vibrations induced in her home from the waves in the Earth's surface produced shaking that occurred a few seconds before the sound waves in the air arrived. This shaking is likely to have woken her up. There might also have been a flash of light from the explosion that woke her up, where the light traveled much faster than the sound, as in the case of a lightening strike.

PROBLEMS

P12.1 Figure P12.1 shows several snapshots of a wave pulse as it travels along a string. Estimate the speed of the wave pulse.

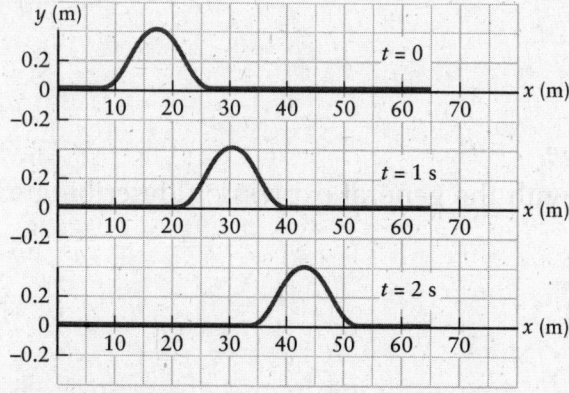

Figure P12.1

Solution

Recognize the principle

The speed of a wave pulse is the distance traveled per unit time. We can obtain information about the position and time of the wave pulse directly from the figure.

Sketch the problem

The sketch is provided in the problem.

Identify the relationships

Following Example 12.1 in your textbook let's examine the position of the peak of this pulse as a function of time. Then we can divide the distance traveled by the time interval to find the speed of the pulse. The average speed can be calculated from

$$v_{av} = \frac{\Delta x}{\Delta t}$$

Solve

From the figure, the position of the peak is at $x \approx 17.0$ m when $t = 0.0$ s and has moved to $x \approx 43.0$ m when $t = 2.0$ s. The speed of this wave is then given by

$$v_{av} = \frac{\Delta x}{\Delta t} \approx \frac{(43 - 17)\ \text{m}}{(2.0 - 0.0)\ \text{s}} = \boxed{13\ \text{m/s}}$$

What does it mean?

The wave pulse propagates to the right at a speed of about 13 m/s. This value seems reasonable from the diagram.

P12.10 A wave on a string is described by the relation $y = A \sin(35t - 0.025x)$, where t is measured in seconds and x in meters, with $A = 0.15$ m. Find the frequency, wavelength, and speed of this wave.

Solution

Recognize the principle

The parameters in the expression describing the wave are related to the frequency and wavelength of the wave. We can compare this expression with the general expression for a wave to find the frequency and wavelength, and from these we can calculate the speed.

Sketch the problem

No sketch needed.

Identify the relationships and Solve

Comparing the given expression with the general expression describing a wave, we see that

$$2\pi f = 35\ \text{rad/s}$$
$$f = \frac{35}{2\pi} = \boxed{5.6\ \text{Hz}}$$

and that

$$\frac{2\pi}{\lambda} = 0.025\ \text{rad/m}$$
$$\lambda = \frac{2\pi}{0.025} = \boxed{250\ \text{m}}$$

The speed of the wave is given by

$$v = \lambda f = (250 \text{ m})(5.6 \text{ Hz}) = \boxed{1400 \text{ m/s}}$$

What does it mean?

Note that the minus sign in the wave representation tells us that the wave is propagating in the *positive* x direction.

P12.25   Consider a violin string with a vibrating length of 30 cm. If the speed of a wave on this string is 250 m/s, what is the tension in the string? *Hint*: You will have to estimate the diameter of the string. Assume the density is the same as that of steel, 7800 kg/m³.

Solution

Recognize the principle

The speed of a wave on a string is related to the tension and the mass per unit length. This relation can be used to solve the problem.

Sketch the problem

No sketch required.

Identify the relationships

The equation for speed can be solved for tension in terms of speed and mass per unit length. The speed is given, but the mass per unit length is not. It will be necessary to use the density of steel, the given length of the string, and an estimate of the diameter of the string to find the mass per unit length. Starting with our expression for the speed of a wave on a string and solving for the tension F_T, we have,

$$v = \sqrt{\frac{F_T}{\mu}}$$

$$v^2 = \frac{F_T}{\mu}$$

$$F_T = v^2 \mu$$

The mass of a string of length L and density ρ is

$$m = \pi r^2 L \rho$$

Then, solving for μ and substituting in $r = d/2$ we have,

$$\mu = \frac{m}{L} = \pi r^2 \rho = \frac{\pi d^2 \rho}{4}$$

So, our expression for the tension becomes,

$$F_T = v^2 \frac{\pi d^2 \rho}{4}$$

We estimate the diameter of a violin string to be around 0.50 mm which is 0.50×10^{-3} m.

Solve

Substituting this and the other numerical values into our expression for the tension we have,

$$F_T = (250 \text{ m/s})^2 \frac{\pi(0.50 \times 10^{-3})^2(7800 \text{ kg/m}^3)}{4} = 96 \text{ N} \approx 100 \text{ N}$$

$$\boxed{F_T \approx 100 \text{ N}}$$

What does it mean?

This is the tension in just one string of a violin which is why violins must be quite strong.

P12.35 A lightbulb emits a spherical wave. If the intensity of the emitted light is 1.0 W/m^2 at a distance of 2.5 m from the bulb, what is the intensity at a distance of 4.0 m?

Solution

Recognize the principle

The intensity of a wave drops as the square of the distance from a point source.

Sketch the problem

No sketch required.

Identify the relationships

We can use the relationship between power, intensity, and distance, and examine what happens when the distance is changed from 2.5 m to 4.0 m.

Solve

The intensity at a distance r from a point source emitting power P is

$$I = \frac{P}{4\pi r^2}$$

For a different distance r',

$$I' = \frac{P}{4\pi(r')^2}$$

Dividing the second equation by the first we have,

$$I' = I\left[\frac{r}{r'}\right]^2 = (1.0 \text{ W/m}^2)\left[\frac{2.5}{4.0}\right]^2 = 0.39 \text{ W/m}^2$$

$$\boxed{I' = 0.39 \text{ W/m}^2}$$

What does it mean?

Since the power is spreading out over a larger spherical surface, it makes sense that the intensity, the power per unit area, should be smaller.

P12.39 Two waves of equal frequency are traveling in opposite directions on a string. Constructive interference is found at two spots on the string that are separated by a distance of 1.5 m. (a) What is the longest possible wavelength of the waves? (b) Give two other possible values of the wavelength.

Solution

Recognize the principle

Constructive interference may occur at intervals on the string separated by multiples of one-half wavelength. That is, when two crest align or when two troughs align, constructive interference occurs.

Sketch the problem

We begin by drawing a sketch of a standing wave and labeling the distance between points of constructive interference.

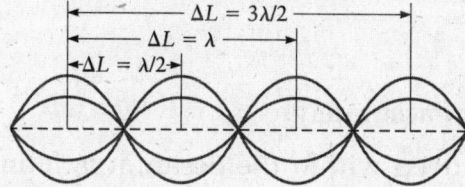

Identify the relationships

Since the points of constructive interference can occur every half wavelength, we can write the following expression,

$$\Delta L = n\frac{\lambda}{2}; n = 1, 2, 3, \ldots$$

Solve

(a) For the maximum wavelength, $n = 1$ and

$$\lambda_{max} = 2\Delta L = 2(1.5 \text{ m}) = \boxed{3.0 \text{ m}}$$

(b) Other possibilities result from other values for n. For $n = 2$ we have

$$\lambda_2 = \Delta L = 1.5 \text{ m} = \boxed{1.5 \text{ m}}$$

and $n = 3$ gives

$$\lambda_3 = \frac{2}{3}\Delta L = \frac{2}{3}(1.5 \text{ m}) = \boxed{1.0 \text{ m}}$$

What does it mean?

Given the separation of points of constructive interference, only certain wavelengths can fit the boundary conditions. This is the origin of standing waves.

P12.44 A radar system can measure the time delay between the transmitted and reflected wave pulses from a distant airplane with an accuracy of 1.0 picosecond. What is the accuracy in the calculated distance to the plane?

Recognize the principle

Radio waves are electromagnetic waves and thus travel at the speed of light. The waves must make a round trip to the plane and back.

Sketch the problem

No sketch needed.

Identify the relationships

The distance traveled by a radio wave during a time interval Δt is given by

$$\Delta d = c\Delta t$$

Solve

The distance traveled by the waves in the time interval given is

$$\Delta d = (3.0 \times 10^8 \text{ m/s})(1.0 \times 10^{-12} \text{ s}) = 3.0 \times 10^{-4} \text{ m}$$

Since an error in the roundtrip distance traveled by the waves results in an error in the distance to the airplane of half this distance, the distance can be measured to an accuracy of 1.5×10^{-4} m or $\boxed{0.15 \text{ mm}}$.

What does it mean?

A radar system can measure distances very accurately!

P12.45 When sound waves are studied deep within the ocean, it is found that a plane wave that is initially parallel to the ocean floor will gradually bend upward and have a propagation direction that is tilted slightly toward the surface (Fig. P12.45). Does the speed of sound at these depths increase or decrease as one goes deeper within the ocean?

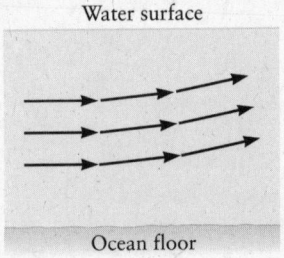

Water surface

Ocean floor

Figure P12.45

Solution

Recognize the principle

Refraction is caused by the relationship between wavelength and speed for a wave. If frequency is constant, wavelength will increase as speed increases, usually due to a change in the properties of the medium. This can cause a wave to "bend."

Sketch the problem

See Figure P12.45.

Identify the relationships

If a wave is seen to bend in a vertical direction, then the speed of the wave at the lowest part of the wave is different than the speed of the wave at the highest part. This difference in speed causes a difference in wavelength.

Solve

Since the wave appears to bend upward, the wavelength in the upper part of the wave is shorter than the wavelength in the lower part. We know that $v = f\lambda$, therefore, the speed is lower in the upper part of the wave. $\boxed{\text{This means that the speed of sound increases with depth}}$.

What does it mean?

The speed of sound is faster at greater depths in the water. This is due to changes in pressure, temperature, and salinity (if sea water is involved) as depth changes.

P12.49 Consider a guitar string that has a tension of 90 N, a length of 65 cm, and a mass per unit length of 7.9×10^{-4} kg/m. What is the fundamental frequency of the string? That is, what is the frequency of the standing wave with the lowest possible frequency? What note does this string emit when played open (i.e., without being held against a fret)? *Hint*: You may want to consult a table that gives musical notes and their corresponding frequencies for standard tuning.

Solution

Recognize the principle

The fundamental frequency (or first harmonic) is related to the length of the string and speed of propagation of the wave. The speed of a wave is also related to the tension and the mass per unit length of the string.

Sketch the problem

We begin by drawing a sketch of a guitar string vibrating at its fundamental frequency.

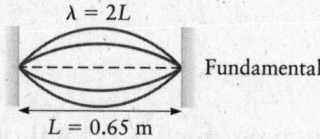

$\lambda = 2L$ Fundamental

$L = 0.65$ m

Identify the relationships

The fundamental frequency is given by the expression

$$f_1 = \frac{v}{2L}$$

and the speed of a wave on a string is given by

$$v = \sqrt{\frac{F_T}{\mu}}$$

Solve

Combining these expressions, the fundamental frequency is,

$$f_1 = \frac{\sqrt{\frac{F_T}{\mu}}}{2L} = \frac{\sqrt{\frac{90 \text{ N}}{7.9 \times 10^{-4} \text{ kg/m}}}}{2(0.65 \text{ m})} = \boxed{260 \text{ Hz}}$$

This note is essentially $\boxed{\text{middle C}}$.

What does it mean?

For the given tension, length, and mass per unit length of the string, this is the fundamental frequency of standing waves on the string.

P12.62 You are operating a seismic laboratory when an earthquake occurs. Your seismographs indicate that the P wave generated by the earthquake arrives at precisely 1:55 PM, while the S wave arrives 5 minutes and 35 seconds later. If the speed of the P wave is 6.0 km/s and the speed of the S wave is 3.5 km/s, how far away is the epicenter of the earthquake?

Solution

Recognize the principle

The P and S waves travel at different speeds through the Earth. Given the speed of the P wave, and the difference in time of arrival, the S wave speed can be determined.

Sketch the problem

No sketch needed.

Identify the relationships

Both the S and P waves travel the same distance to the lab, so

$$d_P = d_S$$

The distances traveled by the waves can be written in terms of the speeds and times

$$d = vt$$

Solve

Combining these expressions we have

$$v_P t_P = v_S t_S$$

The S wave arrives 5 min and 35 s later than the P wave which is a total of 355 s, so we can write

$$t_S = t_P + 335 \text{ s}$$

Inserting this into our expression we have

$$v_P t_P = v_S(t_P + 335 \text{ s})$$

Now, solving for the time it takes the P wave to arrive

$$v_P t_P = v_S t_P + v_S(335 \text{ s})$$
$$v_P t_P - v_S t_P = v_S(335 \text{ s})$$
$$t_P(v_P - v_S) = v_S(335 \text{ s})$$
$$t_P = \frac{v_S(335 \text{ s})}{(v_P - v_S)} = \frac{(3.5 \times 10^3 \text{ m/s})(335 \text{ s})}{(6.0 \times 10^3 \text{ m/s} - 3.5 \times 10^3 \text{ m/s})} = 469 \text{ s}$$

Knowing the speed of the P wave and the time it takes to reach the lab we can determine the distance to the epicenter of the earthquake,

$$d_P = v_P t_P = (6.0 \times 10^3 \text{ m/s})(469 \text{ s}) = \boxed{2.8 \times 10^6 \text{ m}}$$

What does it mean?

Measuring the time difference for the arrival of the P and S waves is a useful tool in estimating the location of the epicenter of an earthquake.

P12.72 Figure P12.72A shows two wave pulses at time $t = 0$ s. Both pulses travel at 5.0 cm/s in opposite directions. (a) Sketch the resulting waveform for the times $t = 1$ s, 2 s, and 3 s. (b) Figure 12.72B shows a different set of two pulses. Sketch the resulting waveform for these pulses at the times $t = 1$ s, 2 s, and 3 s.

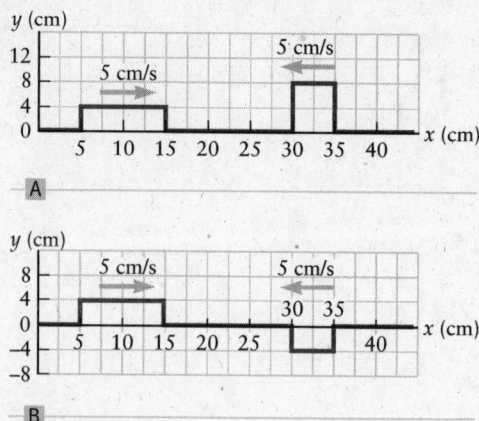

Figure P12.72

Solution

Recognize the principle

Pulses can be added by the principle of superposition. If the two pulses are drawn at the given times, they can be added algebraically.

Sketch the problem

(a) The figure shows the original waves at $t = 0$ s and at 1 s intervals. As shown, the waves add according to the principle of superposition.

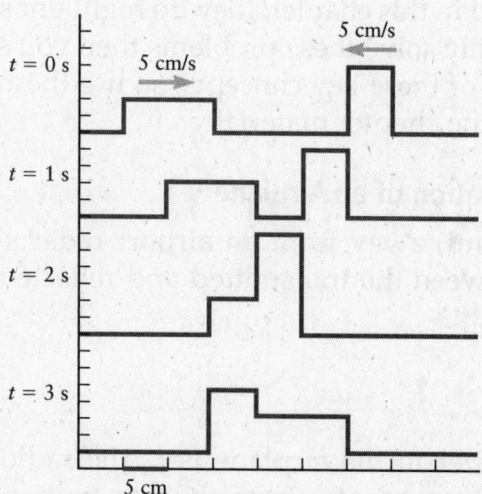

(b) Similarly, this figure shows the superposition of the waves depicted in Figure P12.72(B).

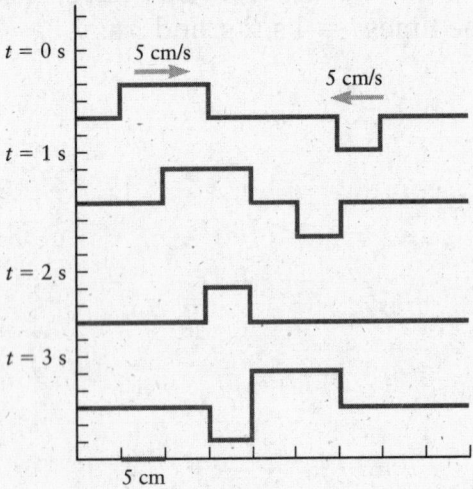

What does it mean?

The resultant wave can be found by simply applying the principle of superposition at any instant in time.

Part D. Additional Worked Examples and Capstone Problems

The following worked example and capstone problems provide you with practice relating the speed, distance, and time for a wave, dealing with standing waves, and the interference of sound waves. In addition, the capstone problems serve as a review of material from previous chapters. Although these three problems do not incorporate all the material discussed in this chapter, they do highlight several of the key concepts. If you can successfully solve these problems then you should have confidence in your understanding of these key concepts, so use these problems as a test of your understanding of the chapter material.

WE 12.1 Using Radar to Measure the Location of an Airplane

An airplane is $L = 10,000$ m (about 6 mi) away from an airport radar system. Find the value of the delay time Δt between the transmitted and reflected wave pulses.

Solution

Recognize the principle

If the distance from the radar transmitter to the airplane is L, the radio wave must travel a distance of $L_{tot} = 2L$ as it propagates from the transmitter to the airplane and back. Knowing the speed of the radar wave, we can find the time required.

Sketch the problem

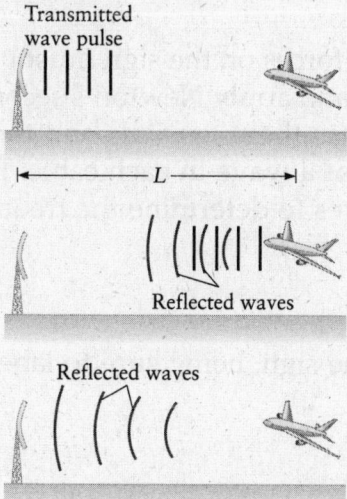

Identify the relationships

A radio wave is a type of electromagnetic wave and travels at the speed of light. The speed of a radio wave is thus $c = 3.0 \times 10^8$ m/s. We can write $2L = c\Delta t$ and solve for Δt.

Solve

Solving for the time delay, we find

$$\Delta t = \frac{2L}{v} = \frac{2L}{c} = \frac{2(10{,}000 \text{ m})}{3.0 \times 10^8 \text{ m/s}}$$

$$\Delta t = 6.7 \times 10^{-5} \text{ s} = \boxed{67 \ \mu s}$$

What does it mean?

Because one would usually like to determine L with an accuracy of a few percent or better, a radar system must be able to measure Δt with a resolution of better than $1 \ \mu s$ (10^{-6} s). In fact, resolution at the picosecond (10^{-12} s) level is possible with current systems.

CP 12.1 Standing Waves on a Cable

Consider the following configuration in which a 100 kg sign is suspended 2 m below an overhang by two solid-steel cables which are each 2 mm in diameter. (a) Determine the tension in each cable. Now, suppose that the cables are plucked resulting in a transverse wave within each cable. (b) What is the speed of the wave in each cable? (c) Determine the frequencies of the fundamental, second, and third harmonic waves produced in each cable.

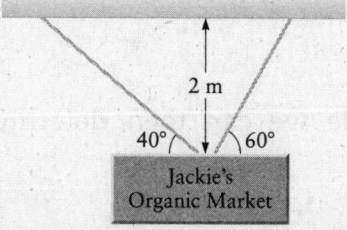

Solution

Recognize the principle

Since the sign is stationary (static), the sum of the forces on the sign must be zero. We should draw a free-body diagram for the sign and apply Newton's second law to determine the tensions in the cables. We can use these tensions and the mass and length of each cable to determine the speed of a wave in each cable. Finally, we can apply our understanding of standing waves to determine the frequencies of the various harmonics.

Sketch the problem

Let's begin by drawing a free-body diagram of the sign, being sure to label all of the forces involved.

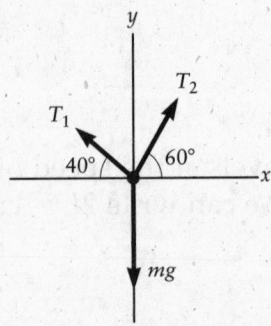

Identify the relationships

(a) Applying Newton's second law in both the x and y directions, we have the following,

$$\Sigma F_x = -T_1 \cos(40°) + T_2 \cos(60°) = 0$$
$$\Sigma F_y = T_1 \sin(40°) + T_2 \sin(60°) - mg = 0$$

(b) Once we know the tensions, we can use this information to determine the speed of a wave in each cable. We'll also need to calculate the mass per unit length of each cable. From FAQ #10 we determined that the mass per unit length of a cable is related to the density of the cable and its cross-sectional area. Since the area of the cable is $A = \pi r^2$ (where r is the radius of the cable) we have,

$$\mu = A\rho = \pi r^2 \rho$$

where ρ is the density of steel. Our expression for the speed of a wave in the cable is given by,

$$v = \sqrt{\frac{F_T}{\mu}}$$

where F_T is the tension in each cable. Combining these expressions we have,

$$v = \sqrt{\frac{F_T}{\mu}} = \sqrt{\frac{F_T}{\pi r^2 \rho}}$$

(c) Knowing the speed of the wave in a cable we can then determine the fundamental frequency,

$$f_1 = v\lambda = v(2L)$$

where we have used $\lambda = 2L$ for the fundamental vibration. The length of each cable must be determined from the geometry of the problem. Finally, the higher order harmonics can be determine from the fundamental frequency using the expression,

$$f_n = nf_1$$

where $n = 2, 3$, etc.

Solve

(a) Combining the two expressions we obtained by applying Newton's second law, we can solve for the tension in each cable.

$$-T_1\cos(40°) + T_2\cos(60°) = 0 \rightarrow T_1 = \frac{T_2\cos(60°)}{\cos(40°)}$$

and

$$T_1\sin(40°) + T_2\sin(60°) - mg = 0 \rightarrow T_1 = \frac{mg - T_2\sin(60°)}{\sin(40°)}$$

Equating the two we have,

$$\frac{T_2\cos(60°)}{\cos(40°)} = \frac{mg - T_2\sin(60°)}{\sin(40°)}$$

Solving for T_2 we have,

$$T_2 = \frac{mg}{\cos(60°)\left(\dfrac{\sin(40°)}{\cos(40°)}\right) + \sin(60°)}$$

$$= \frac{mg}{\cos(60°)\tan(40°) + \sin(60°)}$$

Substituting in for the numerical values we have,

$$T_2 = \frac{mg}{\cos(60°)\tan(40°) + \sin(60°)} = \frac{(100\text{ kg})(9.8\text{ m/s}^2)}{\cos(60°)\tan(40°) + \sin(60°)}$$

$$= \boxed{762\text{ N}}$$

Therefore,

$$T_1 = \frac{T_2\cos(60°)}{\cos(40°)} = \frac{(762\text{ N})\cos(60°)}{\cos(40°)}$$

$$= \boxed{497\text{ N}}$$

(b) Using these values for the tension in each cable, $r = 0.01$ m, and $\rho = 7800$ kg/m^3, we can solve for the speed of the wave in each cable,

$$v_1 = \sqrt{\frac{T_1}{\pi r^2 \rho}} = \sqrt{\frac{497\text{ N}}{\pi(0.001\text{ m})^2(7800\text{ kg/m}^3)}} = \boxed{142\text{ m/s}}$$

$$v_2 = \sqrt{\frac{T_2}{\pi r^2 \rho}} = \sqrt{\frac{762\text{ N}}{\pi(0.001\text{ m})^2(7800\text{ kg/m}^3)}} = \boxed{176\text{ m/s}}$$

(c) From the geometry of the problem we can determine the length of each cable,

$$L_1 = \frac{2}{\sin(40°)} = 3.11 \text{ m}$$

$$L_2 = \frac{2}{\sin(60°)} = 2.31 \text{ m}$$

So, the frequencies of the fundamental, second, and third harmonic waves produced in each cable are,

Cable 1: $f_1 = \dfrac{v_1}{2L_1} = \dfrac{142 \text{ m/s}}{2(3.11 \text{ m})} = \boxed{23 \text{ Hz}}$; $f_2 = 2f_1 = \boxed{46 \text{ Hz}}$; $f_3 = 3f_1 = \boxed{69 \text{ Hz}}$

Cable 2: $f_1 = \dfrac{v_2}{2L_2} = \dfrac{176 \text{ m/s}}{2(2.31 \text{ m})} = \boxed{38 \text{ Hz}}$; $f_2 = 2f_1 = \boxed{76 \text{ Hz}}$; $f_3 = 3f_1 = \boxed{114 \text{ Hz}}$

What does it mean?

These frequencies are all within the range which we can hear. On particularly windy days you can often hear cables vibrating in this way.

CP 12.2 Interference of Sound Waves

During a physics demonstration two speakers are driven by the same frequency generator operating at 500 Hz. A student is told to walk from one speaker directly toward the other speaker (see figure) and comment on when the sound is loud, indicating that the sound from the two speakers is constructively interfering at his location, and when the sound is quiet, indicating that it is destructively interfering. (a) Derive a general expression for the positions (values of x) between the two speakers for which the sound should be loud and when it should be quiet. (b) Solve this expression for the first three loud and quiet positions.

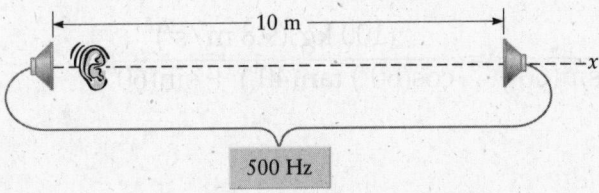

Solution

Recognize the principle

The key principle involved in this problem is the interference of waves. When two sound waves constructively interfere, the sound gets louder and when they destructively interfere, the sound gets quieter. Constructive interference is when the crest of one wave superimposes on the crest of the other wave. Destructive interference is when the crest of one wave superimposes on the trough of the other wave. We can set up this problem by recognizing that the difference in distance from the speakers to the student will provide us with information as to how the crests of the sound waves align.

Sketch the problem

We can begin by drawing a sketch of the situation and labeling the distance to the student from each of the speakers.

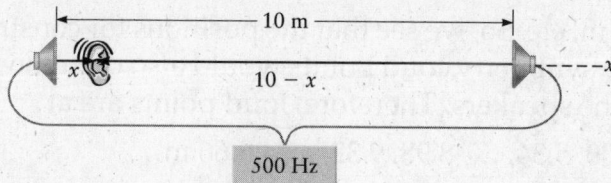

Identify the relationships

For the two sound waves to constructively interfere the difference in the two distances must be a multiple of the wavelength of the sound. This ensures that the crest from one wave always superimposes on the crest of the other sound wave. So, we can write the following expression,

For Constructive Interference:

Difference $= (10 - x) - x = n\lambda$; $n = 0, 1, 2, \ldots$

Similarly, for the two sound waves to destructively interfere the difference in the two distances must be ½ a wavelength, 1½ wavelengths, etc. This ensures that the crest from one wave always superimposes on the trough of the other sound wave. So, we can write the following expression,

For Destructive Interference:

Difference $= (10 - x) - x = \left(n + \dfrac{1}{2}\right)\lambda$; $n = 0, 1, 2, \ldots$

Since we are given the frequency and speed of the sound waves, we can calculate the wavelength of sound using the expression,

$$v = \lambda f \rightarrow \lambda = \frac{v}{f}$$

Solve

(a) Solving our expressions for x and substituting in for the wavelength in terms of the speed and frequency, we have,

For Constructive Interference:

$$x = \frac{10 - \left(\dfrac{nv}{f}\right)}{2}; n = 0, 1, 2, \ldots$$

For Destructive Interference:

$$x = \frac{10 - \left(\dfrac{(n + \frac{1}{2})v}{f}\right)}{2}; n = 0, 1, 2, \ldots$$

(b) We can now solve these expressions for the positions of the loud and quiet locations.

For Constructive Interference:

$$x_0 = \frac{10 - \left(0\dfrac{v}{f}\right)}{2} = 5.00 \text{ m}$$

$$x_1 = \frac{10 - \left(1\dfrac{v}{f}\right)}{2} = \frac{10 - \left(\dfrac{340 \text{ m/s}}{500 \text{ Hz}}\right)}{2} = 4.66 \text{ m}$$

Similarly, $x_2 = 4.32$ m, $x_3 = 3.98$ m, etc. So, we see that the positions for constructive interference differ by 0.34 m. By symmetry, loud points would also occur on the far side of the midpoint between the speakers. Therefore, loud points are at

$x = 0.24, 0.58, 0.92, \ldots, 4.66, 5.00, 5.34, \ldots, 8.98, 9.32,$ and 9.66 m.

For Destructive Interference:

$$x_0 = \frac{10 - \left(\dfrac{(0 + \frac{1}{2})v}{f}\right)}{2} = \frac{10 - \left(\dfrac{(0 + \frac{1}{2})340 \text{ m/s}}{500 \text{ Hz}}\right)}{2} = 4.83 \text{ m}$$

$$x_0 = \frac{10 - \left(\dfrac{(1 + \frac{1}{2})v}{f}\right)}{2} = \frac{10 - \left(\dfrac{(1 + \frac{1}{2})340 \text{ m/s}}{500 \text{ Hz}}\right)}{2} = 4.49 \text{ m}$$

So, we see that the positions for destructive interference occur midway between the points for constructive interference, as expected. Therefore, quiet points are at

$x = 0.07, 0.41, 0.75, \ldots, 4.49, 4.83, 5.17, 5.51, \ldots, 9.25, 9.59,$ and 9.93 m.

What does it mean?

When constructing recording studios and concert halls, architects consider the effects of interference to avoid placing seats on positions which may be points where constructive or destructive interference of the sound occurs.

Part E. MCAT Review Problems and Solutions

PROBLEMS

1. What is the wavelength of a transverse wave traveling at 15 m/s with a frequency of 5.0 Hz?
   (a) 3.0 m
   (b) 10.0 m
   (c) 20.0 m
   (d) 45.0 m

2. Standing waves on a string can be formed if two waves traveling along a string
   (a) are traveling in the same direction.
   (b) have the same frequency.
   (c) have different amplitudes.
   (d) have different wavelengths.

3. Consider a wave traveling on a long string. If you increase the tension in the string by a factor of 4, by what factor will the wave speed change?
   (a) 2.0
   (b) 4.0
   (c) 16.0
   (d) $\sqrt{2.0}$

4. Consider a string with a fixed tension. A wave is sent along the string with a frequency of 200 Hz and has a speed of 40.0 m/s. A second wave with a frequency of 400 Hz is sent along the same. What is the speed of the second wave?
   (a) 20.0 m/s
   (b) 40.0 m/s
   (c) 50.0 m/s
   (d) 80.0 m/s

5. The principle of superposition has to do with which of the following?
   (a) How waves reflect off surfaces.
   (b) How the energy of a wave changes over time.
   (c) How the displacements of interacting waves add together.
   (d) How the direction of a wave changes as its speed changes.

6. A wave pulse traveling along a string reflects when it reaches the end of the string which is tied to a wall. Which statement best describes the reflected wave pulse?
   (a) Its amplitude remains unchanged.
   (b) Its amplitude becomes inverted.
   (c) Its amplitude doubles.
   (d) Its amplitude is halved.

7. Two water waves meet at a particular point in the middle of a pond. One of the waves has an amplitude of 50 cm while the other has an amplitude of 60 cm. At the moment they meet, what is the resulting displacement of the combined wave?
   (a) 10 cm
   (b) 110 cm
   (c) 55 cm
   (d) 0 cm

8. An electromagnetic wave has a speed of 3.0×10^8 m/s and a frequency of 100 MHz. What is its wavelength?
   (a) 0.33 m
   (b) 3.0 m
   (c) 3.0×10^{16} m
   (d) 0.33×10^{-16} m

9. If a guitar string has a fundamental frequency of 110.0 Hz, what is the frequency of its second harmonic?
   (a) 55.0 Hz
   (b) 110.0 Hz
   (c) 220.0 Hz
   (d) 330.0 Hz

10. Which of the following statements is true?

(a) Waves transport matter from one place to another.

(b) Waves transport matter and transmit energy from one place to another.

(c) Waves transmit energy from one place to another.

(d) Waves do not transport nor transmit anything from one place to another.

SOLUTIONS

1. *MCAT strategies*

To answer this problem we simply need to remember the relationship among the frequency, wavelength, and speed of a wave,

$$v = \lambda f$$

Rearranging this expression to solve for the wavelength and substituting in the numerical values, we have,

$$\lambda = \frac{v}{f} = \frac{15 \text{ m/s}}{5.0 \text{ Hz}} = 3.0 \text{ m}$$

So the correct answer is (a). Note: If you can't quite remember the form of this equation, think about the units!

2. *MCAT strategies*

This problem requires no calculations but simply an understanding of how standing waves are formed. Recall that a standing wave on a string is actually the interference of a wave with itself as it reflects off a fixed end of the string. Both the original and reflected waves have the same frequency and this is necessary to create the stationary points called nodes on the standing wave. So, the correct answer is (b).

3. *MCAT strategies*

This problem requires us to understand the expression for the speed of a wave on a string, so let's apply our general problem-solving strategy.

Recognize the principle

As the tension in a string increases, so does the speed of the wave.

Sketch the problem

No sketch is need.

Identify the relationships

The speed is related to the tension in the string by the following expression,

$$v = \sqrt{\frac{F_T}{\mu}}$$

where F_T is the tension in the string and μ is the mass per unit length of the string.

Solve

We see that if we increase the tension in the string by a factor of 4 then the speed will double. So, the correct answer is (a).

What does it mean?

Note that since all of the answers are greater than 1, none could be easily eliminated without thinking more carefully about the expression for speed.

4. MCAT strategies

Careful with this question! We know that the speed of a wave depends on the tension in the string and the mass per length of the string. The speed does not depend on the frequency of the wave. Therefore, the second wave will travel at the same speed as the first wave, 40.0 m/s. So, the correct answer is (b).

5. MCAT strategies

No analysis is required for this problem. It's simply testing our understanding of the term *superposition*. Superposition has to do with how interacting waves superimpose (or add) together. So, the correct answer is (c).

6. MCAT strategies

Once again, no analysis is required for this problem, we only need to consider how waves reflect. When a wave pulse traveling along a string reflects when it reaches the end of the string, the nature of the reflection depends on whether the end of the string is fixed or free to move. For a string which is fixed at the end, the reflected wave is inverted. So, the correct answer is (b).

7. MCAT strategies

The principle of superposition tells us that when two waves interact, the waves will add at the point of interaction. If the amplitudes of both waves are above the equilibrium point (i.e., both are crests) when they interact, then the amplitudes will add and make a wave amplitude which is the sum of the two amplitudes. In this case, the wave amplitudes will sum to 110 cm (50 cm + 60 cm). So, the correct answer is (b).

8. MCAT strategies

To answer this problem we simply need to remember the relationship among the frequency, wavelength, and speed of a wave,

$$v = \lambda f$$

Rearranging this expression to solve for the wavelength and substituting in the numerical values, we have,

$$\lambda = \frac{v}{f} = \frac{3.0 \times 10^8 \text{ m/s}}{1.0 \times 10^8 \text{ Hz}} = 3.0 \text{ m}$$

So the correct answer is (b). Remember: Think about the units if you don't remember the form of the expression.

9. MCAT strategies

If you don't remember that the frequency of higher order harmonics are multiples of the fundamental frequency, then following our problem-solving strategy and sketching the problem may help.

Recognize the principle

The harmonics are the possible standing waves produced on the string. Sketching the different standing waves will help us solve this problem.

Sketch the problem

The fundamental and second harmonics produce standing waves that look like:

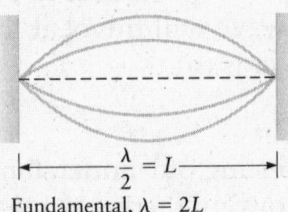

$$\frac{\lambda}{2} = L$$
Fundamental, $\lambda = 2L$

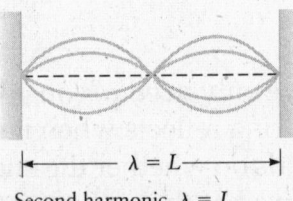

$$\lambda = L$$
Second harmonic, $\lambda = L$

Identify the relationships

We see from the sketch that the wavelength of the second harmonic is half the wavelength of the fundamental. Also, we know that the speed, frequency, and wavelength are related by the expression,

$$v = \lambda f$$

Finally, we note that the speed of the wave is the same in these two situations since wave speed depends on the string, which has not changed.

Solve

Therefore, since the speed of the wave is the same for these two situations and the wavelength of the second harmonic is half the wavelength of the fundamental, the frequency of the second harmonic must be twice that of the fundamental. So, the correct answer is (c).

What does it mean?

Just by drawing some pictures and thinking about some basic relationships we were able to solve this problem.

10. MCAT strategies

This problem requires no calculations but simply a basic understanding of the properties of waves. Waves do not transport matter; however they do transport energy from one place to another. So, the correct answer is (c).

13 Sound

Part A. Summary of Key Concepts and Problem-Solving Strategies

KEY CONCEPTS AND PRINCIPLES

Sound is a longitudinal wave

The molecular motion associated with a sound wave is parallel to the wave velocity, so sound is a longitudinal wave. Sound can also be viewed as a *pressure wave*. Regions of high pressure are regions of condensation, whereas low-pressure regions are regions of rarefaction. The frequency and wavelength of a periodic sound wave are related by

$$v_{sound} = f\lambda$$

The speed of sound depends on the properties of the medium and is generally high in a solid and low in gases. The speed of sound in air at atmospheric pressure and room temperature is approximately 343 m/s.

Human hearing

The normal range of human hearing is approximately 20–20,000 Hz. Sounds with lower frequencies are called *infrasonic*, and sounds with higher frequency are *ultrasonic*.

Sound intensity

The *intensity* of a sound wave is proportional to the square of the amplitude of the wave. Typically, the human ear can hear sounds with intensities as small as 1×10^{-12} W/m^2, which corresponds to a pressure amplitude of about 2×10^{-5} Pa. Sound intensity is often measured in *decibels* (dB). The intensity level of a sound in decibels is given by

$$\beta = 10 \log \left(\frac{I}{I_0} \right)$$

where $I_0 = 1.00 \times 10^{-12}$ W/m^2. The human ear can function (without damage) at intensity levels up to about 120 dB.

APPLICATIONS

Sound in pipes

Sound propagating back and forth within a pipe forms *standing waves* when the wavelength of the sound (or a multiple of the wavelength) matches the length of the pipe. Standing waves have pressure *nodes* where the pressure amplitude is zero and pressure *antinodes* where the pressure amplitude is largest. The open end of a pipe is always a pressure node, and the closed end is always a pressure antinode.

BEHAVIOR OF PRESSURE FOR STANDING WAVES

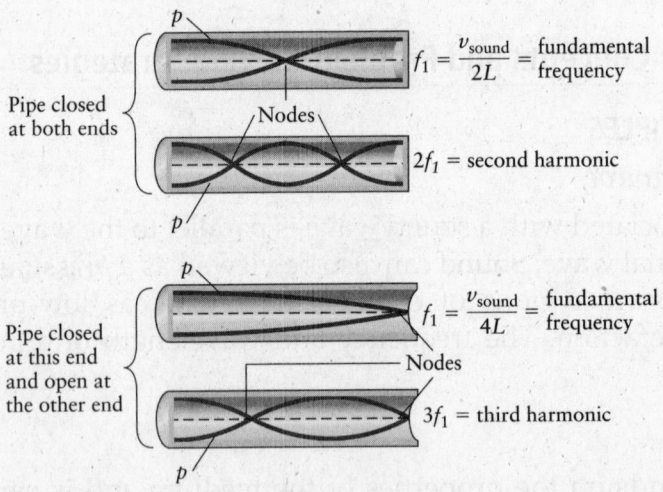

For a pipe closed at both ends, the possible standing waves have frequencies

$$f_n = n \frac{v_{\text{sound}}}{2L} = n f_1, \quad n = 1, 2, 3, \ldots$$

where $f_1 = v_{\text{sound}}/2L$ is called the *fundamental frequency*, the lowest possible standing wave frequency. For a pipe that is closed at one end and open at the other (as in many musical instruments), the standing wave frequencies are

$$f_n = n \frac{v_{\text{sound}}}{4L} = n f_1, \quad n = 1, 3, 5, \ldots$$

Beats

When two sound waves that have nearly the same frequency are combined, a listener will hear *beats*. Beats are an oscillation of the overall sound intensity at a beat frequency equal to $|f_1 - f_2|$, where f_1 and f_2 are the frequencies of the two sound waves.

Doppler shift

When the source of a sound is moving, the frequency observed by a listener is *Doppler-shifted*. If the frequency of the source when stationary is f_{source}, then when the source moves at speed v_{source} and the listener moves at speed v_{obs}, the frequency heard by the listener is

$$f_{\text{obs}} = f_{\text{source}} \left(\frac{1 \pm (v_{\text{obs}}/v_{\text{sound}})}{1 \pm (v_{\text{source}}/v_{\text{sound}})} \right)$$

The choices of plus and minus signs in this expression depend on whether the source and observer are moving toward or away from each other and can always be remembered by noting that the frequency measured by the observer f_{obs} is increased when the observer and source are moving toward each other, and f_{obs} is decreased when they move away from each other.

PROBLEM-SOLVING STRATEGIES

In this chapter we further developed our understanding of waves and wave phenomena as they relate to sound. In addition, two particularly challenging concepts were introduced. The first was standing waves of sound in columns of air. This concept has applications in understanding musical instruments. The second concept is the Doppler shift which describes how the frequency of a sound wave changes if there is relative motion between the source of the sound and the listener. The following two schematics outline the thought process we should follow when tackling problems involving standing waves of sound and the Doppler shift.

Problem Solving: Standing Waves of Sound

Recognize the principle

A standing wave can occur when two waves travel in opposite directions in the same medium such as a string or column of air in a pipe. Displacement nodes (pressure antinodes) occur at closed ends of a pipe and displacement antinodes (pressure nodes) occur at the open ends of a pipe. The frequency and wavelength of a standing wave are related by the speed according to the expression $v_{\text{sound}} = f\lambda$.

Sketch the problem

We usually begin by sketching the standing wave. We can either sketch the displacement wave or the pressure wave, being sure to draw the nodes at the appropriate points. Label the length of the air column (pipe) so that it is easy to determine the wavelength of the standing wave.

Identify the relationships

From the diagram it's easy to determine the wavelength of the standing wave in terms of the length of the pipe. For example, for a pipe closed at both ends the sketch of the pressure wave for the first three harmonics would look like the figure shown.

Since the length of the pipe is L, the wavelengths for each of these standing waves are $\lambda_1 = 2L$, $\lambda_2 = L$, and $\lambda_3 = 2L/3$. The corresponding frequencies are

$$f_1 = \frac{v_{\text{sound}}}{\lambda_1} = \frac{v_{\text{sound}}}{2L},$$

$$f_2 = \frac{v_{\text{sound}}}{\lambda_2} = \frac{v_{\text{sound}}}{L}, \text{ and } f_3 = \frac{v_{\text{sound}}}{\lambda_3} = \frac{3\,v_{\text{sound}}}{2L}.$$

For a pipe open at both ends or open at just one end the expressions are different but can be derived in a similar way.

Solve

Solve for the unknown quantities. These may be the wavelengths, frequencies, speed of sound, length of the pipe, or other quantities which are not given in the problem.

What does it mean?

Always *consider what your answer means*, and check that it makes sense.

Problem Solving: Applying the Doppler Shift

Recognize the principle

When the source of sound is moving, the frequency of the sound observed by a listener is Doppler shifted to a higher or lower frequency depending on the relative motion of the source and observer.

Sketch the problem

You should begin with a sketch of the problem, being sure to label the source, observer, their velocities, and the frequency of the sound produced by the source and observed by the listener.

Identify the relationships

The key relationship relating the frequency of sound produced by the source and observed by the listener is given by,

$$f_{obs} = f_{source}\left(\frac{1 \pm (v_{obs}/v_{sound})}{1 \pm (v_{source}/v_{sound})}\right)$$

The choice of the plus or minus sign depends on the motion of the source and observer. Imagine for a moment that the source were stationary. If the observer is moving away from the source then we use a negative sign in the numerator, and if it's moving toward the source then we use a positive sign in the numerator. Similarly, if we imagine the observer is stationary, then if the source is moving away we use a positive sign and if toward we use a negative sign in the denominator of the expression. It's easy to remember if we recall our own experience with the sound of a siren as an ambulance passes us.

Solve

Solve for the unknown quantities. These may be the wavelengths, frequencies, speed of sound, velocity of the source or the observer, or other quantities which are not given in the problem.

What does it mean?

Always *consider what your answer means*, and check that it makes sense.

Part B. Frequently Asked Questions

1. *Does the speed of a sound wave depend on its frequency?*

No. The speed of sound depends on the medium through which the sound is traveling. Recall from the last chapter that the speed of a wave on a string depends on the tension in the string and the mass per unit length of the string. In the same way, the speed of sound in solids and liquids depends on the "stiffness" of the medium which is related to the modulus (Young's or bulk) of the medium, and also depends on the density of the medium. The speed of sound in a gas also depends on the temperature of the gas. It's interesting to think about what would happen if the speed of sound in air did depend on frequency. Imagine yourself at a concert where the orchestra is composed of many instruments simultaneously playing many sounds with different frequencies. If those sounds all traveled at different speeds they would reach you at different times, creating a noisy combination of sounds rather than music. Fortunately this does not happen since the sound waves all travel at the same speed determined by the properties of the air not the wave!

2. *Given the intensity level of a sound, how do I solve for the intensity?*

This is a good question which requires us to review a little mathematics. The intensity level of sound measured in decibels is related to the intensity by the following expression,

$$\beta = 10 \log \left(\frac{I}{I_0} \right)$$

where $I_0 = 1.00 \times 10^{-12} \, \text{W/m}^2$. To invert this expression and solve for the intensity I, we first divide both sides of the expression by 10.

$$\frac{\beta}{10} = \log \left(\frac{I}{I_0} \right)$$

Now, to eliminate the log function we take the inverse log of both sides which is the same as taking 10^x of both sides. We can then solve for I.

$$10^{\frac{\beta}{10}} = 10^{\log\left(\frac{I}{I_0}\right)}$$

$$10^{\frac{\beta}{10}} = \frac{I}{I_0}$$

$$I = I_0 10^{\frac{\beta}{10}}$$

This expression can now be used to determine the intensity of sound given its intensity level.

3. *Why is it that as you age, you tend to lose the ability to hear high-frequency sounds?*

There can be several factors which contribute to hearing loss as you age, but the most significant upper frequency hearing loss is likely due to a stiffening of the ear drum and other components within the ear. We can understand this effect by considering a drum. The head of a drum is a thin membrane which vibrates when struck with a drumstick. If you were to use a thicker and therefore stiffer membrane, it would vibrate at a lower frequency. Much in the same way, as you age your ear drum stiffens so that it can no longer vibrate at higher frequencies. Therefore, you lose the ability to detect high frequency sounds.

4. *Does the Doppler shift depend on the distance between the source and the observer?*

No. Recall our expression for the Doppler shift,

$$f_{obs} = f_{source} \left(\frac{1 \pm (v_{obs}/v_{sound})}{1 \pm (v_{source}/v_{sound})} \right)$$

This expression relates the frequency of sound produced by a source to that observed by a listener when the source and observer are moving relative to each other. The key factors affecting the shift in frequency are the velocities of the source and observer. Note that the expression does not include the positions of the source and observer so these are not important. Of course, knowing the change in position of an object may allow us to determine its velocity; however, the distance between the source and the observer is not important.

5. *Does the speed of the sound emitted by a source change if the source is moving?*

That's a good question. Since the Doppler shift tells us that the frequency of sound produced by a source will shift to a higher or lower frequency if there is relative motion between the source and observer, it seems logical to conclude that the speed of sound produced by the source will change as well. However, this conclusion would be wrong. Remember, the speed of sound depends on the medium through which the sound is traveling, not on the frequency, wavelength, or source of the sound. Once the sound has been produced by a source it travels through the medium (often air) at a speed determined by the properties of that medium.

6. *Suppose two identical sound waves interact so that their amplitudes add. Does the intensity of the sound double? Does the sound intensity level double?*

The intensity of a sound wave is proportional to the square of the amplitude of the wave. So, if two identical sound waves added together so that the resultant amplitude was doubled, then the intensity of the resulting wave would be 4 times the intensity of an individual wave. To see how the sound intensity level would change let's look carefully at our expression,

$$\beta = 10 \log \left(\frac{I}{I_0} \right)$$

By doubling the amplitude, the intensity I would increase by a factor of 4. So let's substitute $4I$ into our expression and determine how β will change.

$$\beta_{added} = 10 \log \left(\frac{4I}{I_0} \right) = 10 \left(\log \left(\frac{I}{I_0} \right) + \log (4) \right)$$

$$= 10 \log \left(\frac{I}{I_0} \right) + 10 \log (4)$$

$$= \beta + 10 \log (4)$$

So, the sound intensity level would increase by $10 \log (4)$ which is approximately 6 dB.

7. *If sound is a longitudinal wave, why do we draw it as a transverse wave?*

Actually, we don't. It is true that sound waves are longitudinal waves, however, longitudinal waves are difficult to draw. So, to help simplify the analysis and visualize sound waves we draw either the displacement of the air molecules versus their position along the wave or the pressure of the air versus the position along the wave. When we sketch these graphs they appear to be transverse, however we need to remember that these graphs describe how these variables change with position not how the wave travels.

8.　*As sound waves go from air into water, what changes?*

Since the medium changes as we go from air into water, the speed of the wave will change as well. As it turns out, the frequency of the sound will not change. Since the speed, frequency, and wavelength of a wave are related by the expression $v = \lambda f$, the wavelength must change as well. There's a good analogy to help us understand this process. Imagine you are walking along a road while listening to music on your iPod. Suppose you are stepping to the beat (frequency) of the music, and try to maintain that beat as the road turns into a muddy field. As you enter the muddy field you try to maintain the same frequency with your steps, however, to do so you would find that you'd need to take shorter steps. As a result, your speed would decrease. Much in the same way, as a sound wave enters a medium where its speed is less, the wavelength of the sound decreases but its frequency remains unchanged.

9.　*Does a louder sound necessarily mean it is more intense?*

No. Remember that loudness is a subjective quality that depends on how the ear functions at different frequency sounds. At low intensities, low-frequency sounds are often perceived as louder than higher frequency sounds so it's possible to have a low-frequency, low intensity sound which is perceived as louder than a higher-frequency, high intensity sound. Figure 13.4 in your textbook illustrates the relationship between perceived loudness and frequency for normal human hearing.

10.　*What is reverberation?*

You may have noticed that sounds seem to bounce around (echo) in an empty room where there are no objects such as furniture to absorb the sound energy. This persistence of sound is known as reverberation or reverb. When listening to music, sometimes reverberation is a benefit providing a fuller, richer sound. Other times, this can produce a muddy, noisy sound. People often say that they sound better when they sing in the shower. This is because the sound waves reverberate in the shower stall, producing the effect of a richer voice. When designing large concert halls, architects need to carefully consider the amount of reverberation that occurs so that the audience will hear the best quality sound.

Part C. Selection of End-of-Chapter Answers and Solutions

QUESTIONS

Q13.12　A buzzer generates sound by vibrating with a frequency of 440 Hz in air. If this buzzer is placed underwater, what frequency would a fish hear? How does placing the buzzer underwater affect the wavelength of the sound?

Answer

The frequency of the sound is unchanged by the medium. Since the mechanism that vibrates is still making 440 cycles/s, the medium (now water), is still compressed and stretched at the same rate. However, from Table 13.1 in your textbook, the speed of sound waves in water is much higher than air. Since we know that $v = \lambda f$ for a sound wave, the wavelength much increase by the same factor as the wave speed since the frequency remains constant. The values in Table 13.1 ($v_{water} = 1480$ m/s, $v_{air} = 343$ m/s) imply that the wavelength of sound in water is over 4 times longer than the same sound in air.

Q13.17 When you blow across the top of a cola bottle, you can generate sound (Figure Q13.17). As you drink more and more of the cola and lower the liquid level, how does the frequency of the sound change?

Figure Q13.17

Answer

The bottle is essentially a tube closed at one end. The lowest frequency corresponds to the longest wavelength, and therefore the longest tube. The bottle is "longest" when it is empty. A cola bottle has a height of about 25 cm, and for an open-closed tube, this corresponds to one-fourth of the wavelength of the fundamental. The wavelength of these longest waves is therefore about 1 m. Given the speed of sound in air is about 343 m/s, we can find the corresponding frequency, since $v = \lambda f$. We can solve this expression for the frequency (f) and insert values,

$$f = \frac{v}{\lambda} = \frac{343 \text{ m/s}}{1.0 \text{ m}} = \boxed{343 \text{ Hz}}$$

PROBLEMS

P13.8 Consider two sounds of frequency 300 Hz, one traveling in carbon dioxide gas and the other in hydrogen gas. What is the ratio of the wavelengths in the two cases?

Solution

Recognize the principle

The speed of sound differs in different media. Since the frequency is the same in both, we can find the ratio of the wavelengths from the ratio of speeds.

Sketch the problem

No sketch needed.

Identify the relationships

The speed, frequency, and wavelength of a sound wave are related by the expression,

$$v_{\text{sound}} = \lambda f$$

The frequency in both media is 300 Hz, so we have

$$v_{\text{sound}_{CO_2}} = \lambda_{CO_2} f$$
$$v_{\text{sound}_{H_2}} = \lambda_{H_2} f$$

The speed of sound in each gas can be found in Table 13.1 of your textbook.

Solve

Dividing these two equations gives the ratio of wavelengths,

$$\frac{v_{\text{sound}_{CO_2}}}{v_{\text{sound}_{H_2}}} = \frac{\lambda_{CO_2} f}{\lambda_{H_2} f} = \frac{\lambda_{CO_2}}{\lambda_{H_2}} = \frac{269 \text{ m/s}}{1330 \text{ m/s}} = \boxed{0.20}$$

What does it mean?

The wavelength of these waves in hydrogen gas is about 5 times longer than the wavelength in carbon dioxide gas.

P13.17 You are at a rock concert, and the sound intensity reaches levels as high as 130 dB. The sound pressure produces an oscillating force on your eardrum. Estimate the amplitude of this oscillating force.

Solution

Recognize the principle

The intensity of sound is proportional to the square of the sound pressure.

Sketch the problem

No sketch needed.

Identify the relationships

This problem is similar to Example 13.2 in your textbook. First solve for the intensity of this sound wave. Calling the intensity of the rock concert I_{concert}, we can find the intensity from the given sound level intensity using the expression,

$$\beta = 10 \log \left(\frac{I_{\text{concert}}}{I_0} \right)$$

Also, we know that the intensity and pressure are related by the expression,

$$I \propto p^2$$

At the intensity I_0 the pressure amplitude p_0 is 2×10^{-5} Pa. Also, your eardrum has an area of about 100 mm^2.

Solve

Inserting our values into our expression for the sound level intensity, we can find the intensity at the concert in term of I_0:

$$130 = 10 \log \left(\frac{I_{concert}}{I_0} \right)$$

$$\log \left(\frac{I_{concert}}{I_0} \right) = 13$$

$$\frac{I_{concert}}{I_0} = 10^{13}$$

$$I_{concert} = 10^{13} I_0$$

Then, after replacing the intensities with the square of their pressure amplitudes, we get:

$$p_{concert}^2 = 10^{13} p_0^2$$

We can then solve for the pressure amplitude at the concert:

$$p_{concert} = p_0 \sqrt{10^{13}}$$

And inserting the lowest pressure amplitude gives:

$$p_{concert} = (2 \times 10^{-5} \, \text{Pa}) \sqrt{10^{13}} = 63 \, \text{Pa}$$

Pressure is defined as force per unit area, so we can solve for force on our eardrums:

$$F = p_{concert} (A) = (63 \, \text{Pa})(1 \times 10^{-4} \, \text{m}^2) = \boxed{0.006 \, \text{N}}$$

What does it mean?

The force on your eardrum, even from very loud sounds, is less than one-hundredth of a Newton.

P13.30 A person has severe hearing loss which reduces the intensity in his inner ear by 40 dB. To compensate, a hearing aid can amplify the sound pressure amplitude by a certain factor to bring the intensity level back to the value for a healthy ear. What amplification factor is needed for this person?

Solution

Recognize the principle

A reduction in intensity level can be translated to a reduction in intensity. Since intensity is proportional to pressure amplitude, we can use this relationship to find the needed amplification factor.

Sketch the problem

No sketch needed.

Identify the relationships

The intensity level is related to the intensity through the expression:

$$\beta_{normal} = 10 \log \left(\frac{I_{normal}}{I_0} \right) \text{ and } \beta_{reduced} = 10 \log \left(\frac{I_{reduced}}{I_0} \right)$$

We know that:

$$\beta_{\text{normal}} - \beta_{\text{reduced}} = 40 \text{ dB}$$

We must also remember the logarithmic identity:

$$\log x - \log y = \log\left(\frac{x}{y}\right)$$

We can create an equality by using our expression relating the intensity to the pressure ($I \propto p^2$) to find the corresponding pressure amplitude, that is:

$$\frac{I_{\text{normal}}}{I_{\text{reduced}}} = \left(\frac{p_{\text{normal}}}{p_{\text{reduced}}}\right)^2$$

The amplification factor we're seeking is $\frac{p_{\text{normal}}}{p_{\text{reduced}}}$, since this is the number of times the reduced pressure would need to be amplified.

Solve

We can insert our expressions for each sound level,

$$\beta_{\text{normal}} - \beta_{\text{reduced}} = 10\log\left(\frac{I_{\text{normal}}}{I_0}\right) - 10\log\left(\frac{I_{\text{reduced}}}{I_0}\right)$$

Then we combine the log expressions using our logarithmic identity:

$$\beta_{\text{normal}} - \beta_{\text{reduced}} = 10\log\left(\frac{\frac{I_{\text{normal}}}{I_0}}{\frac{I_{\text{reduced}}}{I_0}}\right) = 10\log\left(\frac{I_{\text{normal}}}{I_{\text{reduced}}}\right)$$

Inserting the equivalent expression in terms of the pressure amplitude:

$$\beta_{\text{normal}} - \beta_{\text{reduced}} = 10\log\left(\frac{p_{\text{normal}}}{p_{\text{reduced}}}\right)^2$$

And finally, inserting the sound level change and solving for the amplification factor:

$$40 \text{ dB} = 10\log\left(\frac{p_{\text{normal}}}{p_{\text{reduced}}}\right)^2$$

$$4 \text{ dB} = \log\left(\frac{p_{\text{normal}}}{p_{\text{reduced}}}\right)^2$$

$$10^4 = \left(\frac{p_{\text{normal}}}{p_{\text{reduced}}}\right)^2 \rightarrow \frac{p_{\text{normal}}}{p_{\text{reduced}}} = 10^2 = \boxed{100}$$

What does it mean?

The amplifier must increase the pressure amplitude by a factor of 100 in order to provide the equivalent of normal hearing.

P13.36 An organ pipe is open at one end and closed at the other. The pipe is designed to produce the note middle C (262 Hz). (a) How long is the pipe? (b) What is the frequency of the third harmonic? (c) How many pressure nodes of the third harmonic are found inside the pipe? Do not count nodes that may be at the ends of the pipe.

Solution

Recognize the principle

The length of the pipe determines the wavelength. The wavelength can be found from the speed of sound and the given frequency. Harmonic frequencies are multiples of the fundamental frequency.

Sketch the problem

We begin by drawing a sketch of the pipe with the fundamental pressure wave and the third harmonic.

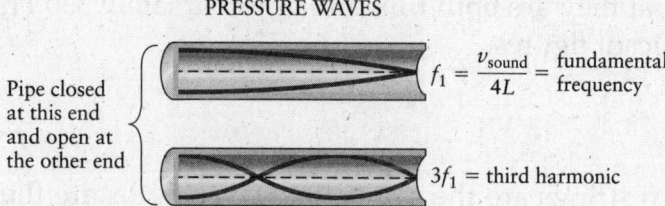

PRESSURE WAVES

Pipe closed at this end and open at the other end

$f_1 = \dfrac{v_{sound}}{4L} =$ fundamental frequency

$3f_1 =$ third harmonic

Identify the relationships

(a) A closed-open pipe must have a pressure antinode node at the closed end and a pressure node at the open end of the pipe, so the pipe is one-fourth of the fundamental wavelength. That is:

$$L = \frac{\lambda}{4}$$

Our expression relating the wavelength to the wave speed and frequency is,

$$v = \lambda f$$

Since the wave is a sound wave in air, its speed is 343 m/s.

(b) The third harmonic is 3 times the fundamental frequency.

(c) The third harmonic occurs when ¾ of a wavelength "fits" in the pipe. A pressure node occurs at the open end, and a pressure antinode at the closed end, with one pressure node and one pressure antinode in between.

Solve:

(a) We can solve both expressions for the wavelength, and set them equal:

$$\frac{v}{f} = \lambda \text{ and } 4L = \lambda \rightarrow 4L = \frac{v}{f}$$

Then solving for the length and inserting the numerical values,

$$L = \frac{v}{4f} = \frac{343 \text{ m/s}}{4(262 \text{ Hz})} = \boxed{0.33 \text{ m}}$$

(b) The frequency of the third harmonic is then:

$$f_2 = 3(262 \text{ Hz}) = \boxed{786 \text{ Hz}}$$

(c) The third harmonic therefore has $\boxed{1 \text{ pressure node within the tube,}}$ not including the one at the open end.

What does it mean?

We can see that the fundamental has no pressure nodes (or antinodes) within the pipe, and the third harmonic has one of each. The pattern continues, as the nth harmonic will always have $n-2$ pressure nodes (and antinodes) within the pipe.

P13.47 Two similar guitar strings have the same length and mass, but slightly different tensions. If the beat frequency is 1.5 Hz, what is the ratio of the tensions of the two strings? Assume that they are both tuned to approximately 330 Hz. Give your answer to three significant figures.

Solution

Recognize the principle

Since the lengths of the two strings are the same, the wavelengths are the same. The slightly different frequencies that cause the beats are from slightly different wave velocities due to the different tensions.

Sketch the problem

No sketch needed.

Identify the relationships

Let's begin by labeling the strings A and B. The fundamental frequency on each string is a standing wave given by the expression,

$$f_A = \frac{v_A}{2L}, f_B = \frac{v_B}{2L}$$

Each of these can be solved for the speed of a wave. Then using our expression for the speed of a wave on a string from Chapter 12, each can be expressed in terms of the tension and the mass per unit length (μ):

$$v_A = 2Lf_A = \sqrt{\frac{F_A}{\mu}}, v_B = 2Lf_B = \sqrt{\frac{F_B}{\mu}}$$

Note that since the mass and length of the strings are the same, μ and L are the same for both strings.

Solve

We can then divide these two equations:

$$\frac{2Lf_A}{2Lf_B} = \frac{\sqrt{\frac{F_A}{\mu}}}{\sqrt{\frac{F_B}{\mu}}}$$

And cancel to reduce to:

$$\frac{f_A}{f_B} = \sqrt{\frac{F_A}{F_B}} \text{ or } \left(\frac{f_A}{f_B}\right)^2 = \frac{F_A}{F_B}$$

If we then assume that one string is at 330 Hz, the other must be either 1.5 Hz higher (331.5 Hz) or 1.5 Hz lower to get a 1.5 Hz beat frequency:

$$\left(\frac{328.5 \text{ Hz}}{330 \text{ Hz}}\right)^2 = \frac{F_A}{F_B} = \boxed{0.991} \text{ or } \left(\frac{331.5 \text{ Hz}}{330 \text{ Hz}}\right)^2 = \frac{F_A}{F_B} = \boxed{1.009}$$

What does it mean?

A beat frequency of 1.5 Hz means that the string tension differs by only 1%!

P13.52 An owl is chasing a squirrel and is using echolocation (the reflection of sound) to aid in the hunt. If the squirrel is 25 m from the owl, how long does it take sound to travel from the owl to the squirrel and then back to the owl?

Solution

Recognize the principle

The speed of sound in air is 343 m/s. Since the distance the sound travels is also known, we can find the time for the trip.

Sketch the problem

No sketch needed.

Identify the relationships

The sound travels at a constant speed, so we can relate the distance and time using the equation:

$$d = v_{sound}\, t$$

We need to remember to consider both the trip from the owl to the squirrel and back.

Solve

We can solve this equation for the time it takes for the sound to reach the squirrel:

$$t = \frac{x}{v_{sound}}$$

Inserting values,

$$t = \frac{25 \text{ m}}{343 \text{ m/s}} = 0.07 \text{ s}$$

The return time is equal to the arrival time, so

$$t_{total} = 2t$$

$$t_{total} = 2(0.07) = \boxed{0.14 \text{ s}}$$

What does it mean?

The sound from the owl makes the round trip in just over one-tenth of a second.

P13.54 A siren has a frequency of 950 Hz when it and an observer are both at rest. The observer then starts to move and finds that the frequency he hears is 1000 Hz. (a) Is the observer moving toward or away from the siren? (b) What is the speed of the observer?

Solution

Recognize the principle

The frequency is increased by the Doppler effect when an observer is moving toward the source.

Sketch the problem

No sketch needed.

Identify the relationships

The speed of the observer can be found using the general expression for the Doppler shift,

$$f_{obs} = f_{source} \left| \frac{1 \pm \frac{v_{obs}}{v_{sound}}}{1 \pm \frac{v_{source}}{v_{sound}}} \right|$$

In our situation, the source is not moving, so $v_{source} = 0$. Since the observer is moving toward the source, we need to choose the top $(+)$ signs.

Solve

Our Doppler equation therefore reduces to:

$$f_{obs} = f_{source} \left(1 + \frac{v_{obs}}{v_{sound}} \right)$$

Solving this equation for the speed of the observer, we have:

$$v_{obs} = \left(\frac{f_{obs}}{f_{source}} - 1 \right) v_{sound}$$

Inserting values:

$$v_{obs} = \left(\frac{1000}{950} - 1 \right) (343 \text{ m/s}) = \boxed{18 \text{ m/s}}$$

What does it mean?

This observer is moving toward the siren at about 40 mi/h.

P13.63 A child drops a rock into a vertical mineshaft that is precisely 406 m deep. The sound of the rock hitting the bottom of the shaft is heard 10.3 s after the child drops the rock. What is the temperature of the air in the shaft? Assume the temperature is the same throughout the mine.

Recognize the principle

The speed of sound varies with the temperature of the air in the shaft.

Sketch the problem

No sketch needed.

Identify the relationships

From kinematics, the time it takes for the rock to fall is:

$$t = \sqrt{\frac{2\Delta y}{g}}$$

Whatever time remains must be used by the sound traveling back up the mine shaft, so we can find the speed of sound.

We can then use our relationship between the speed of sound and the temperature of the gas in which it travels to find the temperature of the air in the mine:

$$v_{\text{sound}} \approx 343 + 0.6\,(T - 20°C)\ \text{m/s}$$

Solve

Inserting values for our kinematic equation gives the time for the rock to fall:

$$t = \sqrt{\frac{2(406\ \text{m})}{9.8\ \text{m/s}^2}} = 9.103\ \text{s}$$

This means the time it took the sound to travel up the mine is

$$t_{\text{sound}} = 10.3\ \text{s} - 9.103\ \text{s} = 1.197\ \text{s}$$

Then the velocity of sound is

$$v_{\text{sound}} = \frac{x}{t} = \frac{406\ \text{m}}{1.197\ \text{s}} = 339\ \text{m/s}$$

Then inverting our expression for speed as a function of temperature to solve for the temperature we have,

$$\frac{v_{\text{sound}} - 343}{0.6} + 20 \approx T\ (\text{in °C})$$

Inserting our value of 339 m/s gives $\boxed{T = 13.4°C}$

What does it mean?

The mine temperature is about 56° Fahrenheit.

P13.66 The lowest frequency an average human ear can perceive is about 15 Hz, and the highest about 18 kHz. How tall would an organ pipe have to be to produce the lowest note? How short a tube would be needed to play the highest note? What harmonic (i.e., what value of *n* in Eq. 13.15) would the smaller tube excite in the larger one?

Solution

Recognize the principle

An organ pipe is modeled as closed at one end and open on the other. The length of an open-closed pipe determines its fundamental frequency.

Sketch the problem
No sketch needed.

Identify the relationships
The fundamental frequency of an open-closed pipe is given by:

$$f_1 = \frac{v_{\text{sound}}}{4L}$$

If the whistle has enough energy to excite a standing wave in the large organ pipe it would correspond to a high resonance of that long pipe. In order to be a resonance, the high frequency must be a multiple of the low frequency. That is:

$$f_{\text{Hi}} = n f_{\text{Lo}}$$

Solve
We can solve this expression for the length of the pipe:

$$L = \frac{v_{\text{sound}}}{4f_1}$$

Then, inserting the values for each case:

$$L_{\text{Lo}} = \frac{343 \text{ m/s}}{4(15 \text{ Hz})} = \boxed{5.7 \text{ m}}$$

$$L_{\text{Hi}} = \frac{343 \text{ m/s}}{4(18{,}000 \text{ Hz})} = \boxed{4.8 \times 10^{-3} \text{ m} = 4.8 \text{ mm}}$$

Solving for the harmonic number, we have:

$$n = \frac{f_{1 \text{ Hi}}}{f_{1 \text{ Lo}}} = \frac{18{,}000 \text{ Hz}}{15 \text{ Hz}} = 1200$$

What does it mean?
It would be difficult to manufacture this shorter pipe, which is the size of a tiny whistle, very similar to a dog whistle, designed for high frequencies beyond the range of human hearing. The whistle's frequency corresponds to the tall organ pipe's 1200th harmonic.

Part D. Additional Worked Examples and Capstone Problems

The following worked example provides you with addition practice in applying the expression for the Doppler shift. In this problem both the source of the sound and the listener are moving, so it's important to think carefully about your choice of the plus/minus signs in the expression. Two capstone problems are also included in this section and draw together concepts from these chapter and previous chapters. If you successfully solve these capstone problems you should feel confident in your understanding of some of the material in this chapter. They will also serve as a review of earlier material.

WE 13.1 Doppler Shift When Both Source and Observer are Moving

A siren in an ambulance has a frequency of 1000 Hz as heard when the ambulance is stopped. (a) Suppose you are driving down the road in your car at 10 m/s when this ambulance approaches from behind with its siren turned on. If the speed of the ambulance is 20 m/s, what frequency for the siren do you observe? (b) Determine the observed frequency of the siren once the ambulance has passed you by. (Use 343 m/s for the speed of sound.)

Solution

Recognize the principle

As the ambulance approaches us from behind we will hear a frequency which is higher than 1000 Hz and once it passes us we will hear a frequency less than 1000 Hz. Since both the source and observer are moving, we'll need to be careful applying our expression for the Doppler shift.

Sketch the problem

It is helpful to sketch the two situations, being sure to label all the variables.

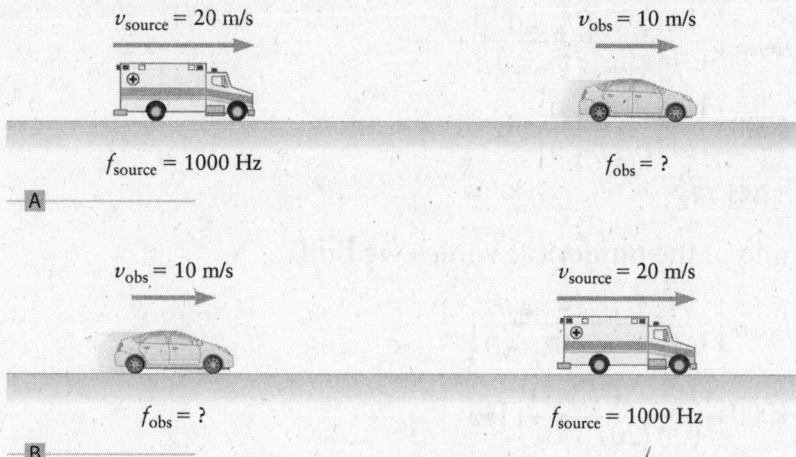

Identify the relationships

The general expression for the Doppler shift of sound is,

$$f_{obs} = f_{source} \left(\frac{1 \pm (v_{obs}/v_{sound})}{1 \pm (v_{source}/v_{sound})} \right)$$

The challenge when applying this expression is in determining whether to use the plus or minus signs. This choice is made easier if we think about our own experience with the sound of a siren when we are standing still as the ambulance passes by. As the ambulance approaches we hear a higher frequency sound than when it recedes. With this in mind we can tackle this problem.

(a) To choose the correct signs in the Doppler shift expression let's first imagine that the ambulance is not moving. We would then be moving away from the ambulance and our experience tells us that we should hear a lower frequency. Therefore, we choose the minus sign for v_{obs}. To choose the correct sign for the source velocity, imagine that we are not moving but the ambulance is moving. The ambulance would then be moving toward us and we would hear a higher frequency. Therefore, we choose a negative sign for v_{source}. So, combining these two scenarios we can write the complete expression for the ambulance approaching us from behind while we are moving in the same direction.

$$f_{\text{obs}} = f_{\text{source}} \left(\frac{1 - (v_{\text{obs}}/v_{\text{sound}})}{1 - (v_{\text{source}}/v_{\text{sound}})} \right)$$

(b) By going through the same thought process as part (a) we come to the conclusion that we must choose the positive sign for v_{obs} and the positive sign for v_{source}. So the correct expression is

$$f_{\text{obs}} = f_{\text{source}} \left(\frac{1 + (v_{\text{obs}}/v_{\text{sound}})}{1 + (v_{\text{source}}/v_{\text{sound}})} \right)$$

Solve

(a) Substituting in the numerical values we find,

$$f_{\text{obs}} = f_{\text{source}} \left(\frac{1 - (v_{\text{obs}}/v_{\text{sound}})}{1 - (v_{\text{source}}/v_{\text{sound}})} \right)$$

$$= 1000 \left(\frac{1 - (10/343)}{1 - (20/343)} \right) \text{Hz}$$

$$= \boxed{1031 \text{ Hz}}$$

(b) Substituting in the numerical values we find,

$$f_{\text{obs}} = f_{\text{source}} \left(\frac{1 + (v_{\text{obs}}/v_{\text{sound}})}{1 + (v_{\text{source}}/v_{\text{sound}})} \right)$$

$$= 1000 \left(\frac{(1 + (10/343))}{1 + (20/343)} \right) \text{Hz}$$

$$= \boxed{972 \text{ Hz}}$$

What does it mean?

As expected we hear a drop in frequency as the ambulance passes us, regardless of whether or not we are moving. If we had been stationary, the numerical results would have been different but the frequency would still be less after the ambulance has passed by compared to when it approaches.

CP 13.1 U-tube Flute: Combining Sound Waves with Fluids

A unique wind instrument utilizes a U-shaped tube open at both ends and partially filled with water. As you blow across the top of one end of the tube, a standing wave is produced in the column of air above the surface of the water. This is much the same as how sound is produced when you blow across the top of a soda bottle. As you blow harder the increase in speed reduces the air pressure across the top of the tube so that the water level shifts, thus changing the length of the column of air in which the standing wave is produced. Different notes are played by blowing with different speeds. The instrument is tuned by filling the correct amount of water so that the length of the air column L corresponds to the note C at 262 Hz, when you blow softly ($v \approx 0$). (a) Determine the length of the air column for which the note C is played. (b) Determine the length of the air column required to produce the note A at 440 Hz. (c) How fast must you blow to produce the note A?

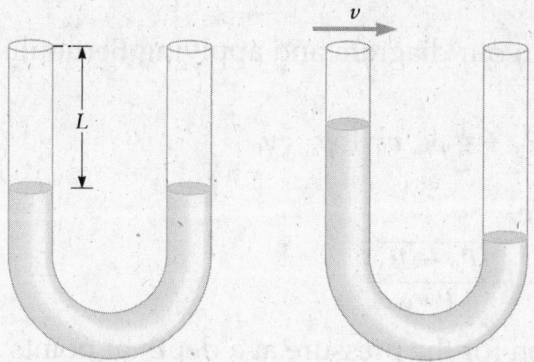

Solution

Recognize the principle

Knowing the frequency and the speed of sound we can determine the wavelength of the sound. The standing wave of sound produced in the column of air is the fundamental frequency, so we can use the wavelength to determine the length of the column of air. As we blow across the top of the tube the pressure is related to the speed of the air through Bernoulli's principle. In all of these calculations we can assume the speed of sound in air to be 343 m/s.

Sketch the problem

For part (c) of the problem it is useful to label the different points in the tube as follows:

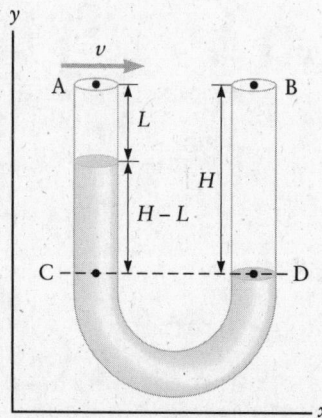

Identify the relationships

(a) and (b) The speed, frequency, and wavelength of sound are related by the expression,

$$v = \lambda f$$

The fundamental standing wave for a column of air open at one end and closed at the other has a wavelength of $\lambda = 4L$. Substituting this into our expression for the speed of a wave and solving for the length of the column, we have,

$$v = \lambda f = 4Lf \rightarrow L = \frac{v}{4f}$$

(c) Comparing points A and B in our diagram and applying Bernoulli's equation we have,

$$P_A + \frac{1}{2}\rho_{air} v_A^2 + \rho_{air} g y_A = P_B + \frac{1}{2}\rho_{air} v_B^2 + \rho_{air} g y_B$$

But $v_B = 0$ and $y_A = y_B$. Therefore,

$$P_A + \frac{1}{2}\rho_{air} v_A^2 = P_B \rightarrow v_A = \sqrt{\frac{2(P_B - P_A)}{\rho_{air}}}$$

Now, by applying our expression for the pressure at a depth at points (c) and (d), we have,

$$P_C = P_A + \rho_{air} gL + \rho_{water} g(H - L) \text{ and } P_D = P_B + \rho_{air} gH$$

Pascal's principle tell us that $P_C = P_D$, therefore,

$$P_A + \rho_{air} gL + \rho_{water} g(H - L) = P_B + \rho_{air} gH$$
$$P_B - P_A = -\rho_{air} g(H - L) + \rho_{water} g(H - L)$$
$$P_B - P_A = g(\rho_{water} - \rho_{air})(H - L)$$

Finally, we know that since the water on each side of the tube started at the same level, then

$$\frac{L + H}{2} = L_{original} \rightarrow H = 2L_{original} - L$$

Therefore,

$$H - L = 2(L_{original} - L)$$

Solve
(a) Plugging in the numerical values, we have,

$$L = \frac{v}{4f} = \frac{343 \text{ m/s}}{4(262 \text{ Hz})} = \boxed{0.33 \text{ m}}$$

(b) Plugging in the numerical values, we have,

$$L = \frac{v}{4f} = \frac{343 \text{ m/s}}{4(440 \text{ Hz})} = \boxed{0.19 \text{ m}}$$

(c) Combining our expressions, we have,

$$v_A = \sqrt{\frac{2(P_B - P_A)}{\rho_{air}}}$$
$$= \sqrt{\frac{2g(\rho_{water} - \rho_{air})(H - L)}{\rho_{air}}}$$
$$= \sqrt{\frac{4g(\rho_{water} - \rho_{air})(L_{original} - L)}{\rho_{air}}}$$

Substituting in the numerical values of $\rho_{air} = 1.0$ kg/m^3, $\rho_{water} = 1000$ kg/m^3, $L_{original} = 0.33$ m, and $L = 0.19$ m, we have,

$$v_A = \sqrt{\frac{4g\,(\rho_{water} - \rho_{air})(L_{original} - L)}{\rho_{air}}}$$

$$= \sqrt{\frac{4(9.8 \text{ m/s}^2)(1000 \text{ kg/m}^3 - 1.0 \text{ kg/m}^3)(0.33 \text{ m} - 0.19 \text{ m})}{1.0 \text{ kg/m}^3}}$$

$$= \boxed{74 \text{ m/s}}$$

What does it mean?

This is a very fast speed so it is not likely you would be able to play the A-note on this instrument. However, you should be able to create enough wind speed to play several different lower-frequency notes.

CP 13.2 Speed Detector: Combining the Doppler Effect with Freefall

In a clever physics experiment, two students drop an electronic buzzer off the top of a 30-m tall building in order to determine the acceleration of gravity. Placed on the ground below they have a microphone connected to a computer which records the frequency of the sound of the buzzer as a function of time. The following data is recorded:

Frequency Measured (Hz)	Time Measured (s)
1000.00	0.0
1011.56	0.4
1023.39	0.8
1035.50	1.2
1047.90	1.6
1060.61	2.0
1073.62	2.4

(a) Using the data provided and the fact that the temperature of the air was 20°C, determine the speed of the buzzer for the given points in time. (b) Using the speeds you've calculated for the various points in time, calculate the acceleration of gravity and show that it is a constant throughout the motion of the buzzer. *Note*: Neglect the time it takes for the sound to travel from the buzzer to the microphone, as this will introduce only a small error.

Solution

Recognize the principle

As the buzzer falls, it is accelerating so its speed is increasing. This speed relative to the microphone Doppler shifts the frequency of sound. We'll need to correctly apply the expression for the Doppler shift.

Sketch the problem

No sketch is required; however, it is useful to define a coordinate system for the situation. Let's choose downward as the y axis with the origin at the top of the building.

Identify the relationships

Since the buzzer is the source of sound and is moving <u>toward</u> the microphone, which is stationary, we can apply the Doppler shift expression as follows,

$$f_{obs} = f_{source}\left(\frac{1 - (0/v_{sound})}{1 - (v_{source}/v_{sound})}\right)$$

$$= f_{buzzer}\left(\frac{1}{1 - (v_{buzzer}/v_{sound})}\right)$$

Since the initial frequency recorded was 1000 Hz, this is the frequency of the buzzer. Solving this expression for the speed of the buzzer we have,

$$v_{buzzer} = v_{sound}\left(1 - \frac{f_{buzzer}}{f_{obs}}\right)$$

Also, since the temperature on the day of the experiment was 20°C, the speed of sound was 343 m/s. So, we can solve this expression for the speed of the buzzer at the various points in time. Once we know the speeds, we can use that information to calculate the acceleration of the buzzer at different points in time using the expression

$$a = \frac{\Delta v}{\Delta t}$$

Solve

Putting this all together we can create the following table:

Time (s)	Calculated Speed (m/s)	Calculated Acceleration (m/s²)
0	0	
0.4	3.92	$(3.92 - 0)/0.4 = 9.8$
0.8	7.84	$(7.84 - 3.92)/0.4 = 9.8$
1.2	11.76	$(11.76 - 7.84)/0.4 = 9.8$
1.6	15.68	$(15.68 - 11.76)/0.4 = 9.8$
2.0	19.60	$(19.60 - 15.68)/0.4 = 9.8$
2.4	23.52	$(23.52 - 19.60)/0.4 = 9.8$

What does it mean?

The acceleration for each of the time intervals is exactly what we would expect for an object in free fall. Furthermore, the results confirm that the acceleration of gravity remains constant throughout the motion.

Part E. MCAT Review Problems and Solutions

PROBLEMS

1. The foghorn of a ship echoes off an iceberg in the distance. If the echo is heard 5.0 s after the horn is sounded, and the air temperature is −50°C, then how far away is the iceberg?
   (a) 200 m
   (b) 750 m
   (c) 1000 m
   (d) 1500 m

2. What is the intensity, in W/m^2, of sound with a sound intensity level of 20 dB?
   (a) $10^{-12}\ W/m^2$
   (b) $10^{-10}\ W/m^2$
   (c) $1\ W/m^2$
   (d) $10\ W/m^2$

3. What is the sound level of a wave with an intensity of $10^{-3}\ W/m^2$?
   (a) 30 dB
   (b) 60 dB
   (c) 90 dB
   (d) 120 dB

4. A small explosive device is dropped down a deep mining shaft that is 500 m deep. The device explodes upon impact with the bottom of the shaft. If the air temperature is 15°C, which is the closest estimate of how long it takes to hear the explosion after the device is dropped?
   (a) 5.0 s
   (b) 10 s
   (c) 12 s
   (d) 15 s

5. If the speed of a transverse wave on a piano string is 330 m/s and the frequency of the note played is 440 Hz, what is the wavelength of the sound?
   (a) 75 m
   (b) 15 m
   (c) 7.5 m
   (d) 0.75 m

6. Two wave pulses travel toward each other along a string as shown. The phenomenon that occurs when the pulses meet is called

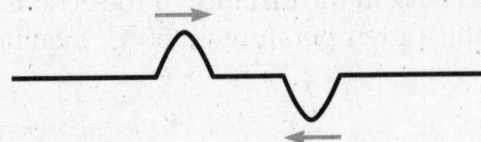

   (a) refraction.
   (b) reflection.
   (c) diffraction.
   (d) interference.

7. In which medium will the speed of sound of frequency 10^3 Hz be greatest?
   (a) air at 20°C
   (b) water at 20°C
   (c) gelatin at 20°C
   (d) iron metal at 20°C

8. If two identical sound waves interact so that they constructively interfere, the resulting wave will have
   (a) a shorter period.
   (b) a larger amplitude.
   (c) a higher frequency.
   (d) a greater velocity.

9. If two sounds emitted from two sources have frequencies 54 Hz and 48 Hz, what beat frequency is heard?
   (a) 3
   (b) 6
   (c) 9
   (d) 12

10. The frequency of the sound heard by a detector is higher than the frequency of sound emitted by the source. Which statement must be true?
   (a) The source must be moving away from the detector.
   (b) Both the source and detector must be stationary.
   (c) The source and the detector must be getting closer together.
   (d) The source and the detector must be getting further apart.

SOLUTIONS

1. MCAT strategies

At first glance this problem appears to be a sound problem, however, at closer look it is simply an application of the definition of speed. Speed is defined as distance divided by time, $v = d/t$. Recall that the speed of sound in air is around 340 m/s at 20°C. At colder temperatures the speed is lower, so we can estimate the speed of the sound to be around 300 m/s. The echo is heard 5.0 s after the horn is sounded, so the time for the sound to get to the iceberg is 2.5 s. Using this information we can determine the distance to the iceberg,

$$d = vt = (300 \text{ m/s})(2.5 \text{ s}) = 750 \text{ m}$$

So the correct answer is (b).

2. MCAT strategies

No answers can be easily eliminated so we proceed with our usual problem-solving strategy.

Recognize the principle

We'll need to correctly apply our expression for the sound level in terms of sound-level intensity.

Sketch the problem

No sketch is required.

Identify the relationships

The intensity level of a sound in decibels is given by

$$\beta = 10 \log \left(\frac{I}{I_0} \right)$$

where $I_0 = 1.00 \times 10^{-12}$ W/m². Inverting this expression we have,

$$I = I_0 \, 10^{\frac{\beta}{10}}$$

Solve

Substituting in the numerical values we have,

$$I = I_0 \, 10^{\frac{\beta}{10}} = 10^{-12} \, 10^2 = 10^{-10} \text{ W/m}^2$$

So the correct answer is (b).

What does it mean?

Note, if we remember that 0 dB = 10^{-12} W/m², and for every 10 dB the intensity increases by one order of magnitude, then it's easy to see that for every 20 dB corresponds to an increase of two orders of magnitude in the intensity, therefore $I = 10^{-10}$ W/m². With this in mind we can estimate the intensity for any sound level without having to do the calculation!

3. MCAT strategies

Let's use our logic from the previous question to answer this question without doing the calculation. Since the intensity 10^{-3} W/m^2 is 9 orders of magnitude larger than 10^{-12} W/m^2, the sound intensity level must be 9×10 dB = 90 dB. So the correct answer is (c).

4. MCAT strategies

No answers can be quickly eliminated so we proceed with our usual problem-solving strategy.

Recognize the principle

As the device falls freefall, we can apply our kinematic expressions for free fall to find the time it takes for the device to reach the bottom of the shaft. Once the device reaches the bottom, the sound of the explosion must travel back up the shaft to the top. By estimating the speed of sound in the air we can determine the time for the sound to travel the 500 m.

Sketch the problem

No sketch is required.

Identify the relationships

Defining down as the y direction, we can apply the following kinematic expression,

$$y = v_0 \, t_{\text{down}} + \frac{1}{2} g t^2_{\text{down}}$$

Since $v_0 = 0$, we can solve this expression for t,

$$t_{\text{down}} = \sqrt{\frac{2y}{g}}$$

The time for the sound to travel back up the shaft can be determined using the following expression:

$$t_{\text{up}} = \frac{y}{v}$$

Since the speed of sound in air is around 340 m/s at 20°C, we estimate $v \approx 340$ m/s at 15°C.

Solve

Plugging in the numerical values of $g \approx 10$ m/s^2, $y = 500$ m, and $v \approx 340$ m/s, we have

$$t_{\text{down}} = \sqrt{\frac{2y}{g}} = \sqrt{\frac{2(500 \text{ m})}{10 \text{ m/s}^2}} = \sqrt{100 \text{ s}^2} = 10 \text{ s}$$

$$t_{\text{up}} = \frac{y}{v} = \frac{500}{340} = 1.5 \text{ s}$$

Therefore, the total time is around $11.5 \approx 12$ s. So the correct answer is (c).

What does it mean?

It was important to consider the time it takes the sound to travel back up the shaft. If we had neglected this time, we would have incorrectly chosen (b) as the answer!

5. *MCAT strategies*

To solve this problem we must simply know the relationship among speed, frequency, and wavelength, and recognize that the wavelength of the sound created by the string is the same as the wavelength of the wave on the string. The speed, wavelength, and frequency of a wave are related by

$$v = \lambda f$$

Solving this expression for the wavelength, we have

$$\lambda = \frac{v}{f}$$

Substituting in the numerical values, we have

$$\lambda = \frac{v}{f} = \frac{330 \text{ m/s}}{440 \text{ Hz}} = 0.75 \text{ m}$$

So the correct answer is (d).

6. *MCAT strategies*

This problem is simply testing our understanding of various wave phenomena. When two wave pulses meet while traveling along the same string, they superimpose. This interaction is called interference. So the correct answer is (d).

7. *MCAT strategies*

We know that sound travels fastest in solids, so the correct answer must be (d).

8. *MCAT strategies*

When two sound waves constructively interfere, their amplitudes add so that the combined wave is larger in amplitude than the individual waves. So the correct answer is (b).

9. *MCAT strategies*

The key principle being tested with this problem is the concept of beat frequency. When two sound waves are close in frequency, they superimpose to create a combine sound wave which oscillates in amplitude. This beat frequency is equal to the difference in frequency of the two waves. Therefore,

$$f_{\text{beat}} = |f_1 - f_2| = |54 \text{ Hz} - 48 \text{ Hz}| = 6 \text{ Hz}.$$

So the correct answer is (b).

10. *MCAT strategies*

This problem is testing our understanding of the Doppler shift. We can answer this problem by reflecting on our own experience of hearing a siren on an ambulance or fire truck as it approaches and passes by. Upon its approach the siren has a higher pitch (frequency) than when it recedes. In general, if the source of the sound and the listener are getting closer together, the frequency of the sound will be Doppler shifted to a higher frequency. If the two are getting further apart, then the sound will be shifted to a lower frequency. Since the sound heard by the detector is higher in frequency than the sound emitted by the source, the source and detector must be getting closer together. So the correct answer is (c).

14 Temperature and Heat

> **CONTENTS**

Part A. Summary of Key Concepts and Problem-Solving Strategies

KEY CONCEPTS

Zeroth law of thermodynamics

If two systems A and B are in thermal equilibrium ($T_A = T_B$) and systems B and C are in thermal equilibrium ($T_B = T_C$), systems A and C are also in thermal equilibrium ($T_A = T_C$).

In words, the zeroth law of thermodynamics asserts that temperature is a unique property of a system that determines when two systems are in thermal equilibrium. Three different *temperature scales* are in common use: *Fahrenheit*, *Celsius*, and *Kelvin*.

Heat

Heat flow is the transfer of energy from one object to another by virtue of a temperature difference. The energy transferred is also called heat energy or simply heat. Heat flow can occur in three ways: *conduction*, *convection*, and *radiation*.

- *Heat conduction* is determined by the *thermal conductivity* κ, with

$$\frac{Q}{t} = \kappa A \frac{\Delta T}{L}$$

where Q is the heat energy transferred in a time t, L is the length of the system, and A is the cross-sectional area of the system.

- *Convective heat flow* involves the movement of mass and is caused by thermal expansion and contraction.

- *Radiative heat transfer* involves electromagnetic radiation. The total amount of energy an object radiates depends on its temperature. An object with a surface area A and temperature T radiates energy at a rate Q/t given by the *Stefan–Boltzmann law*,

$$\frac{Q}{t} = \sigma e A T^4$$

In this expression $\sigma = 5.67 \times 10^{-8}\,\text{W}/(\text{m}^2 \cdot \text{K}^4)$ and is called Stefan's constant, and e is the emissivity of the object.

The wavelength at which this radiation is strongest depends on temperature through *Wien's law*,

$$\lambda_{\max} = (2.90 \times 10^{-3}\,\text{m} \cdot \text{K})/T$$

That is why the color of a glowing object depends on its temperature.

APPLICATIONS

Specific heat and latent heat

If an amount of heat energy Q is added to a substance of mass m, the temperature of the system increases by an amount ΔT, where

$$Q = cm\Delta T$$

and c is the *specific heat* of the substance. This equation applies provided that there is not a phase change. If there is a phase change such as melting or evaporation, a certain amount of energy called the *latent heat* must be added to or taken from the system. The amount of heat energy required to change the phase of a substance of mass m is,

$$Q_{\text{melt}} = +mL_{\text{fusion}}$$

$$Q_{\text{freeze}} = -mL_{\text{fusion}}$$

$$Q_{\text{vaporization}} = +mL_{\text{vaporization}}$$

$$Q_{\text{condensation}} = -mL_{\text{vaporization}}$$

Thermal expansion

When the temperature of a substance changes, its physical dimensions usually change. Changes in length are proportional to the *coefficient of thermal expansion* α, with

$$\frac{\Delta L}{L_0} = \alpha \Delta T$$

and changes in volume are proportional to the *coefficient of volume expansion* β,

$$\frac{\Delta V}{V_0} = \beta \Delta T$$

PROBLEM-SOLVING STRATEGIES

In this chapter we once again applied the principle of conservation of energy. This is an extremely powerful tool which proves useful in the study of systems which interact by exchanging heat energy. Calorimetry is the term we use for these processes; however, calorimetry is just the principle of conservation of energy as applied to these specific situations. The following two schematics will help guide you through the thought process involved in applying this principle.

Problem Solving: Dealing with Specific Heat in Calorimetry

Recognize the principle

Calorimetry problems are all based on the principle of conservation of energy.

Sketch the problem

Your sketch should show the system or systems of interest and indicate how heat flows between them. Also, be sure to label all the relevant information such as temperature, mass, and specific heat.

Identify the relationships

Determine (if possible) the initial and final temperatures of the system(s), and the heat energy Q added to each system. Q is positive if energy flows into a system, while Q is negative if energy flows out of a system.

Solve

Apply $Q = cm\Delta T$ to relate Q and ΔT for each system, and solve for the quantities of interest. This process assumes that there are no phase changes; for example, the systems do not melt or evaporate. If there is a phase change, apply the problem-solving strategy outlined in the next schematic.

What does it mean?

Always *consider what your answer means*, and check that it makes sense.

Problem Solving: Dealing with Both Specific Heat and Latent Heat in Calorimetry

Recognize the principle
Calorimetry problems are all based on the principle of conservation of energy.

Sketch the problem
Your sketch should show the system or systems of interest and indicate how heat flows between them. Also, be sure to label all relevant information such as temperature, mass, specific heat, and latent heat.

Identify the relationships
Determine (if possible) the initial and final temperatures of the system(s), and the heat energy Q added to each system. Q is positive if energy flows into a system, while Q is negative if energy flows out of a system.

Solve
- Use $Q = cm\Delta T$ to relate Q and ΔT for temperature changes that do not involve a phase change.
- *If there is a phase change,* the latent heat ($Q = \pm mL$) must be added to or subtracted from the system depending on the phase change.

What does it mean?
Always *consider what your answer means,* and check that it makes sense.

Part B. Frequently Asked Questions

1. *How do we know when a system has reached thermal equilibrium?*

Though there are more complicated answers to this question, it is sufficient to equate thermal equilibrium with the notion of equal temperature. That is, if two objects are brought in thermal contact, heat energy will flow from the object with the higher temperature to the object with the lower temperature until their temperatures become equal. Once the temperatures of the two objects are the same we say that they are in thermal equilibrium.

2. *If two objects are in thermal contact and have reached thermal equilibrium, does that mean that all heat flow stops?*

To answer this question, let's review the definition of heat or heat energy. Heat is the flow of energy from one system to another system by virtue of a temperature difference. Using this definition, once thermal equilibrium has been reached, the temperatures of the two systems are the same and thus heat flow stops. This is not to say that individual molecules within one system cannot lose energy to molecules in the other system. However, the net flow of energy between the systems is zero once thermal equilibrium has been reached. This net flow of energy between the systems is what we call heat.

3. *I'm still confused about the difference between heat, temperature, and internal energy. Are they all related?*

Indeed, heat, temperature, and internal energy are related and we will discuss this further in the next chapter. For now it's sufficient to think of these quantities in terms of the following definitions/descriptions:

- Temperature is a measure of the hotness or coldness of a system. It is what we measure with a thermometer. Another definition could be, temperature is the thing that's the same for two systems when they are in thermal equilibrium. Again, we'll develop a more specific definition in the next chapter.

- Heat is the flow of energy from one system to another.

- Internal energy is the sum of the potential energies associated with the intermolecular bonds and the kinetic energy of the molecules within a system.

Clearly as we add heat energy to a system, there will likely be a change in the system's internal energy and the temperature of the system. This relationship will be discussed in the next chapter.

4. *How are specific heat and heat capacity related?*

First, let's clearly define heat capacity. The heat capacity of a system is given by the expression,

$$C = \frac{Q}{\Delta T}$$

where Q is the heat energy added to a system and ΔT is the resulting temperature change of the system. Note that the heat capacity will depend on the composition and mass of the system. To eliminate the mass dependence of the heat capacity we can define a new term called the mass specific heat capacity, abbreviated to just specific heat. The specific heat of a system is given by the expression,

$$c = \frac{Q}{m\Delta T}$$

where Q is the heat energy added to a system, m is the mass of the system, and ΔT is the resulting temperature change of the system. Now, the specific heat is only dependent on the composition of the system and not the mass. We can rearrange this expression as follows,

$$Q = mc\Delta T$$

So, if we want to know how much heat energy Q must be added to a system to change its temperature by ΔT, we must know the specific heat of the system c and the mass of the system.

5.   *Can you change the temperature of an object without adding heat energy?*

Yes. In this chapter we discussed only one way to change the temperature of a system, i.e., by adding heat energy. However, this is not the only way the temperature of a system can be changed. In the next two chapters we will discuss other ways the temperature of system can be changed, including changing the system's pressure or volume. As we'll see, the first law of thermodynamics incorporates these other ways into a general expression of energy conservation.

6.   *Can you add heat energy to an object without changing its temperature?*

Definitely! We saw in this chapter that during a phase change, the temperature of the system does not change even though heat energy is being added or removed from the system. We call this energy the latent heat. When a solid melts, the energy we provide goes into breaking the bonds between the molecules so they can move more freely as a liquid. This does not change the temperature of the material, only its phase. The same is true when we change a liquid to a gas. The energy we provide during this phase change goes into further weakening the bonds between the atoms/molecules and does not change the temperature of the material. So, during a phase change heat energy is added or removed from the system without a change in the system's temperature.

7.   *When solving calorimetry problems, do I have to convert all the temperatures to Kelvin or can I use Celsius?*

That's a good question. Let's look carefully at the important expressions when tackling calorimetry problems:

$$Q = mc\Delta T$$
$$Q_{melt} = +mL_{fusion}, Q_{freeze} = -mL_{fusion}$$
$$Q_{vaporization} = +mL_{vaporization}, \text{ and } Q_{condensation} = -mL_{vaporization}$$

We can see from these expressions that the only time the temperature of the system is needed is when there is a change in temperature as illustrated in the first expression. Now, recall the relationship between the Celsius and Kelvin temperature scales,

$$T_K = T_C + 273.15$$

We see from this expression that a change in temperature in Celsius is equivalent to a change in temperature in Kelvin,

$$\Delta T_K = \Delta T_C$$

Therefore, in calorimetry problems we can use either Kelvin or Celsius and do not need to convert all our temperatures to Kelvin. We should be careful to note that this is only true since we're dealing with changes in temperature rather than specific temperatures.

8. *When solving calorimetry problems, I'm never sure whether my Q's should be positive or negative. Is there an easy way to remember?*

Actually, you don't need to remember if you start with the general expression of energy conservation. That is, the sum of all the Q's must equal zero. For phase changes the Q's are given with the correct sign depending on the direction of the phase change. When using the expression $Q = mc\Delta T$, the sign is given by the change in temperature. Remember $\Delta T = T_f - T_i$, so this could be positive or negative depending on whether the object warms or cools. Simply put the correct temperature in the correct place and the sign will take care of itself.

9. *Looking at Table 14.5 in the textbook, it appears that $\beta \approx 3\alpha$. Is that always true?*

Though it is only an approximation, for most materials it is true that the coefficient of volume expansion is approximately three times the coefficient of linear expansion. To see why this is the case, consider a rectangular object of length L_0, width W_0, and height H_0. Assuming the coefficient of linear expansion is the same for all three directions, the change in the volume will be given by the expression,

$$\Delta V = V - V_0$$
$$= LWH - L_0W_0H_0$$
$$= (L_0 + \Delta L)(W_0 + \Delta W)(H_0 + \Delta H) - L_0W_0H_0$$
$$= (L_0 + \alpha L_0\Delta T)(W_0 + \alpha W_0\Delta T)(H_0 + \alpha H_0\Delta T) - L_0W_0H_0$$

Expanding out and simplifying this expression, we find,

$$\Delta V = (L_0 + \alpha L_0\Delta T)(W_0 + \alpha W_0\Delta T)(H_0 + \alpha H_0\Delta T) - L_0W_0H_0$$
$$= \cancel{L_0W_0H_0} + 3\alpha L_0W_0H_0\Delta T + (L_0W_0 + L_0H_0 + W_0H_0)(\alpha\Delta T)^2$$
$$+ L_0W_0H_0(\alpha\Delta T)^3 - \cancel{L_0W_0H_0}$$
$$= 3\alpha L_0W_0H_0\Delta T + (L_0W_0 + L_0H_0 + W_0H_0)(\alpha\Delta T)^2 + L_0W_0H_0(\alpha\Delta T)^3$$

In the first term on the right-hand side we recognize that $L_0W_0H_0 = V_0$. Also, since α is small, α^2 and α^3 are very small so the second and third terms can be neglected.

Therefore,

$$\Delta V \approx 3\alpha V_0 \Delta T \rightarrow \frac{\Delta V}{V_0} \approx 3\alpha\Delta T$$

Comparing this with our expression in the textbook we see that $\beta \approx 3\alpha$. Note that this is only an approximation (albeit a good approximation!) since we neglected terms in our derivation. Also, there are some materials for which the coefficient of linear expansion is different for different directions but we will not discuss them here.

10. *The thermal conductivity is very different for different materials. What does it really depend on?*

Thermal conductivity depends on many properties of a material. Certainly the temperature and the structure of the material play an important role. For example, in general, the thermal conductivity for gases is much less than for solids. Solids have a higher density of atoms than gases, so solids can conduct a greater amount of heat by transferring the energy between vibrating atoms. Also, pure crystalline materials can exhibit very different thermal conductivities along different directions within the crystal. So, the density of a solid, how strongly its atoms interact, and the orientation of the atoms all play a role in determining the thermal conductivity of a solid.

Part C. Selection of End-of-Chapter Answers and Solutions

QUESTIONS

Q14.14 Fruit trees bloom in the spring, and their blossoms eventually turn into oranges, apples, and so forth. Blossoms are susceptible to cold weather, so when there is the possibility of a frost, orchard owners often spray their trees with water. Explain how this practice can help prevent the blossoms from freezing.

Answer

Once the water is in contact with the blossom, the water and blossom must have the same temperature. Water has a high specific heat. As the outside temperature cools, water resting on the blossom helps resist downward temperature changes as extra heat must be removed from the water in order to cool the blossom. Also, before the water will freeze, it must give up heat equivalent to its latent heat of fusion. The temperature must therefore drop below freezing and remain there for some time in order to do damage to a wet blossom.

Q14.20 A good-quality thermos bottle is double-walled and evacuated between these walls, and the internal surfaces are like mirrors with a silver coating. This configuration combats heat loss from all three transfer methods and keeps the bottle's contents—your coffee—hot. Which design feature of the thermos bottle minimizes heat transfer by radiation, by conduction, and by convection?

Answer

A partial vacuum is a very poor heat conductor, reducing loss from conduction. The evacuated space has very little substance to carry heat away via convection since there is not much gas to move around. And finally, the mirrored walls reflect some of the radiated heat back toward the contents, reducing loss by radiation.

PROBLEMS

P14.8 Consider Joule's experiment described in Insight 14.1. Assume the mass $m = 2.0$ kg and the fluid is water with a volume of 1.0 L. What is the increase in the temperature of the water when the mass moves vertically a distance $z = 1.0$ m?

Solution

Recognize the principle

The loss of potential energy of the mass is equal to the heat added to the water. Using water's specific heat, we can find the temperature rise due to this heat.

Sketch the problem

No sketch needed.

Identify the relationships

If we call the lowest (final) level of the block $PE_{grav} = 0$, then the change in potential energy of the block is $\Delta PE_{grav} = -mgz$. This energy becomes the heat ($Q = mgz$), which raises the temperature of the water according to the expression,

$$Q = m_{water}c\Delta T$$

The specific heat of water is given in Table 14.3 in your textbook as $c = 4186$ J/ (kg · K), and the mass of water can be expressed in terms of its density from Table 10.1 and the given volume,

$$m_{water} = \rho_{water}V$$

Solve

Setting these two energies equal and substituting for the mass, we have,

$$mgz = m_{water}c\Delta T$$
$$mgz = \rho_{water}Vc\Delta T$$

Solving this expression for the change in temperature and inserting the given values (noting that 1 L = 0.001 m³),

$$\Delta T = \frac{mgz}{\rho_{water}Vc} = \frac{(2.0 \text{ kg})(9.8 \text{m/s}^2)(1.0 \text{ m})}{(1000 \text{ kg/m}^3)(0.001 \text{ m}^3)(4186 \text{ J/(kg · K)})} = \boxed{0.0047 \text{ K}}$$

What does it mean?

It's likely that Joule repeatedly lowered masses or used a much larger height in order to obtain larger, more easily measurable temperature changes. This falling mass only changes the temperature of the water by less than five-thousandths of a degree!

P14.15   If you combine 35 kg of ice at 0°C with 75 kg of steam at 100°C, what is the final temperature of the system?

Solution

Recognize the principle

The ice absorbs heat (first melting and then, if necessary, warming) while the steam loses heat (first condensing and then, if necessary, cooling).

Sketch the problem

No sketch needed.

Identify the relationships

We assume that this is a closed system, so we can approach this as a calorimetry problem. The heat lost by the steam is equal to the heat gained by the ice. Applying the principle of conservation of energy, we have,

$$Q_{ice} + Q_{steam} = 0 \rightarrow Q_{ice} = -Q_{steam}$$

The heat required to melt the ice is:

$$Q_{ice} = m_{ice} L_{fusion} = (35 \text{ kg})(3.34 \times 10^5 \text{ J/kg}) = 1.17 \times 10^7 \text{ J}$$

while the heat lost in condensing *all* the steam is:

$$Q_{steam} = m_{steam} L_{vaporization} = (75 \text{ kg})(2.2 \times 10^6 \text{ J/kg}) = 1.65 \times 10^8 \text{ J}$$

Comparing these numbers, we can see that the ice is all melted long before the steam is all condensed. At the point where the ice is all melted, the water starts to warm. Raising this water all the way to 100 degrees Celsius would require additional heat equal to:

$$Q_{100} = m_{ice} c_{water} \Delta T_{water} = (35 \text{ kg})(4186 \text{ J/kg} \cdot \text{K})(100°C) = 1.47 \times 10^7 \text{ J}$$

Solve

The total heat lost by the steam in melting the ice and raising it to 100 degrees Celsius is then:

$$Q_{100} + Q_{ice} = 1.47 \times 10^7 \text{ J} + 1.17 \times 10^7 \text{ J} = 2.63 \times 10^7 \text{ J}$$

This value is still much smaller (by nearly a factor of 10) than the heat required to condense all of the steam, which is also still at 100 degrees Celsius.

Therefore, the equilibrium temperature of this system is 100 degrees Celsius.

What does it mean?

The ice melts and is raised to the temperature of the steam. Its phase change and temperature rise can both be accomplished using only the heat given off as some of the steam is condensed into water.

P14.26   You are a high-precision carpenter and use a high-precision steel ruler. If you want the dimensions of your work to be accurate to 1.0 mm over distances of 15 m, you will have to keep the temperature of your ruler constant to within an uncertainly of ΔT. Find ΔT.

Solution

Recognize the principle

We need to find the temperature range at which a steel ruler 15 m in length expands by less than 1.0 mm.

Sketch the problem

No sketch needed.

Identify the relationships

We can find this thermal expansion using our expression,

$$\frac{\Delta L}{L_0} = \alpha \Delta T$$

Solve

Rearranging this expression we find,

$$\frac{\Delta L}{L_0 \alpha} = \Delta T$$

Then, inserting the numerical values, we have,

$$\Delta T = \frac{1 \times 10^{-3}\ \text{m}}{(15\ \text{m})(12 \times 10^{-6}\ \text{K}^{-1})} = \boxed{5.6\ \text{K}}$$

What does it mean?

This is a relatively small temperature range, equal to only 10°F! It is very difficult to do work at the 1-mm level of precision over such a long scale as 15 m.

P14.39   A small dog has thick fur with thermal conductivity 0.040 W/(m · K). The dog's metabolism produces heat at a rate of 40 W, and its internal (body) temperature is 38°C. (a) If all this heat flows out through the dog's fur, what is the outside temperature? Assume the dog has a surface area of 0.50 m² and the length of the dog's hair is 1.0 cm. (b) Now assume it is a hot summer day with an outdoor temperature of 32°C (about 90°F). What is the body temperature of the dog now? Your answer will explain why a dog can quickly overheat on a warm day.

Solution

Recognize the principle

The temperature difference determines the rate of heat flow through a conductive surface, such as the dog's fur.

Sketch the problem

No sketch needed.

Identify the relationships

The heat flow for this transition can be represented by the expression,

$$\frac{Q}{t} = \kappa A \frac{\Delta T}{L} = \kappa A \frac{T_{\text{dog}} - T_{\text{out}}}{L}$$

Solve

(a) Solving this expression for the outside temperature we have,

$$T_{\text{out}} = T_{\text{dog}} - \frac{(Q/t)L}{\kappa A}$$

Then, inserting the numerical values, and assuming the dog's fur is about 1 cm thick, we have,

$$T_{\text{out}} = 38°C - \frac{(40 \text{ W})(0.01 \text{ m})}{(0.040 \text{ W/m} \cdot \text{K})(0.5 \text{ m}^2)} = \boxed{18°C}$$

(b) We can solve the equation for the dog's temperature:

$$T_{\text{dog}} = T_{\text{out}} + \frac{(Q/T)L}{\kappa A} = 32°C + \frac{(40 \text{ W})(0.01 \text{ m})}{(0.040 \text{ W/m} \cdot \text{K})(0.5 \text{ m}^2)} = \boxed{52°C}$$

What does it mean?

The temperature of the dog is more than 125 degrees Fahrenheit! To survive, the dog will have to resort to panting and other methods to increase the rate of heat loss.

P14.42   When a dog "pants," it exhausts water vapor through its mouth (Fig. P14.42). This process converts liquid water (inside the dog) into water vapor, removing heat from the dog via the evaporation of water. If 10 g of water evaporates from the dog each minute, how much heat energy does it lose during this time?

© Pure Stock
Jupiterimages

Figure P14.42

Solution

Recognize the principle

The latent heat of vaporization of water gives the energy per kilogram needed to vaporize water.

Sketch the problem

No sketch needed.

Identify the relationships

The energy needed to vaporize the water is given by the expression,

$$Q_{\text{vaporization}} = m L_{\text{vaporization}}$$

The latent heat of vaporization of water is given in Table 14.4 in your textbook.

Solve

Inserting values:

$$Q = (0.010 \text{ kg})(2200 \text{ kJ/kg}) = \boxed{22 \text{ kJ}}$$

What does it mean?

The dog's panting is a very effective way of getting rid of heat, as long as he has enough water.

P14.50 An incandescent lightbulb emits 100 W of radiation. If the surface filament is at a temperature of 3000 K, what is the area of the filament?

Solution

Recognize the principle

This problem can be solved using the Stefan–Boltzmann law, which relates the power emitted to the temperature and surface area of an object.

Sketch the problem

No sketch needed.

Identify the relationships

The Stefan–Boltzmann law is given by the expression,

$$\frac{Q}{t} = \sigma e A T^4$$

We will assume the filament is a black body radiator, so the emissivity $e = 1$.

Solve

Inserting this emissivity, we can solve our expression for the filament surface area

$$A = \frac{Q/t}{\sigma e T^4}$$

Then inserting the numerical values, we have,

$$A = \frac{100 \text{ W}}{(5.67 \times 10^{-8} \text{ W/m}^2 \cdot \text{K}^4)(1)(3000 \text{ K})^4} = \boxed{2.2 \times 10^{-5} \text{ m}^2}$$

What does it mean?

This very small value is consistent with the small filaments we see in lightbulbs which have lengths around 1 cm (10^{-2} m) and diameters of around 1 mm (10^{-3} m).

P14.54 An ice-cube tray containing 600 g of water at room temperature is placed into a freezer. It is found that the water becomes solid after 60 minutes. What is the average rate at which heat energy flows out of the water into the freezer?

Solution

Recognize the principle

We can use the specific heat to find the amount of heat energy that must be removed from the water to lower the temperature to the freezing point, and then use the latent heat of fusion to find the amount of heat that must be removed to freeze the water. The sum of these two heat energies divided by the total time will give us the needed rate.

Sketch the problem

No sketch needed.

Identify the relationships

We can use our expression for specific heat to determine how much energy must be released to bring 600 g of water from room temperature to the freezing point:

$$Q_{cool} = mc\Delta T$$

We estimate room temperature to be around 25°C.

The energy that must be removed to freeze 600 g of water is given by the expression,

$$Q_{freeze} = -mL_{fusion}$$

Solve

Inserting values for each of these cases,

$$Q_{cool} \approx (0.6 \text{ kg})(4186 \text{ J/kg} \cdot \text{K})(-25 \text{ K}) = -6.3 \times 10^4 \text{ J}$$

$$Q_{freeze} = -(0.6 \text{ kg})(334000 \text{ J/kg}) = -2.0 \times 10^5 \text{ J}$$

Adding these two together gives the total energy lost by the ice,

$$Q_{total} = -(6.3 \times 10^4 + 2.0 \times 10^5)\text{J} = -2.6 \times 10^5 \text{ J}$$

This energy must flow from the water to the freezer in one hour, so the rate is

$$\frac{Q}{t} = \frac{-2.6 \times 10^5 \text{ J}}{3600 \text{ s}} = -72.2 \text{ J/s} \approx \boxed{-72 \text{ W}}$$

What does it mean?

If you could capture the heat energy leaving the ice, you could power a 60-W lightbulb with some power left over!

P14.59 *The Rankine scale.* Named after William John Macquorn Rankine (a Scottish engineer and physicist who proposed it in 1859), the Rankine scale is similar to the Kelvin scale in that the zero point is placed at absolute zero, but the size of temperature differences is the same as that of the Fahrenheit scale (e.g., a change in temperature of 1°F corresponds to a change of 1°R). (a) Determine the conversion formula to go from the Fahrenheit to the Rankine scale. (b) Find the formula to convert from Kelvin to Rankine. (c) What is the temperature of the freezing point of water on the Rankine scale? What is room temperature on this scale?

Solution

Recognize the principle

Since we have expressions relating the Fahrenheit, Celsius, and Kelvin temperatures scales, we can combine them with the information provided in the problem to derive an expression to convert from the Fahrenheit to the Rankine scale and from the Kelvin to Rankine scale.

Sketch the problem

No sketch needed.

Identify the relationships

We begin by writing down the expressions we already know relating the various temperature scales,

$$T_F = \frac{9}{5} T_C + 32 \text{ and } T_K = T_C + 273.15$$

Working with these expressions and looking at temperature differences, we see that,

$$\Delta T_F = \frac{9}{5} \Delta T_C \text{ and } \Delta T_K = \Delta T_C$$

Now, we are told in the problem that a temperature difference in Fahrenheit is the same as in Rankine. Therefore,

$$\Delta T_R = \Delta T_F$$

Combining these expressions we have,

$$\Delta T_R = \Delta T_F = \frac{9}{5} \Delta T_C = \frac{9}{5} \Delta T_K$$
$$\rightarrow \Delta T_R = \frac{9}{5} \Delta T_K$$

Now, we also know that the zero point of the Rankine scale is at absolute zero, therefore,

$$0 \text{ K} = 0°\text{R} = -273.15°\text{C} = -459.67°\text{F}$$

Solve

(a) Since the Rankine and Fahrenheit scales are shifted by 459.67°F and $\Delta T_R = \Delta T_F$, we have,

$$\boxed{T_R = T_F + 459.67°\text{R}}$$

(b) Since $0 \text{ K} = 0°\text{R}$ and $\Delta T_R = \frac{9}{5} \Delta T_K$, we have,

$$\boxed{T_R = \left(\frac{9°\text{R}}{5 \text{ K}} \right) T_K}$$

(c) The freezing point of water is at 0°C = 32°F and room temperature is around 20°C = 68°F. Using our conversion from Fahrenheit to Rankine, we have,

$$T_{\text{Freezing}, R} = T_F + 459.67°R$$

$$= (32 + 459.67)°R$$

$$= \boxed{491.67°R}$$

$$T_{\text{Room}, R} = T_F + 459.67°R$$

$$= (68 + 459.67)°R$$

$$= \boxed{527.67°R}$$

What does it mean?

The Rankine temperature scale is not frequently used, but shows that temperature can be measured in many different ways.

Part D. Additional Worked Examples and Capstone Problems

The following worked examples and capstone problems provide you with additional practice with thermal expansion and calorimetry. Working these problems will give you greater insight into these concepts. Also, the capstone problems incorporate material from previous chapters, thereby serving as a review. Although these problems do not incorporate all the material discussed in this chapter, they do highlight several of the key concepts. If you can successfully solve these problems then you should have confidence in your understanding of these key concepts, so use these problems as a test of your understanding of the chapter material and a review of previous chapter material.

WE 14.1 Volume Expansion of a Container and Its Contents

Suppose you are working in a factory which manufactures chemicals for various applications. In the morning, you completely fill a 55-gal steel drum with ethyl-alcohol when the temperature in the factory is 20°C. By the afternoon, the factory has warmed to 25°C and you notice that some of the ethyl-alcohol has spilled out of the steel drum. Neglecting evaporation, how much ethyl-alcohol has spilled?

Solution

Recognize the principle

As the temperature increases, the steel drum and the ethyl-alcohol will thermally expand at different rates depending on their coefficients of volume expansion. Since the ethyl-alcohol has a larger coefficient it will expand more and therefore spill out of the steel drum.

Sketch the problem

No sketch is required for this problem.

Identify the relationships

The amount the volume of a material changes due to a change in temperature is given by the following expression,

$$\frac{\Delta V}{V_0} = \beta \Delta T$$

Rearranging this expression and writing it explicitly for the drum and the alcohol, we have,

$$\Delta V_{drum} = V_{0, drum} \beta_{steel} \Delta T$$

$$\Delta V_{alcohol} = V_{0, alcohol} \beta_{alcohol} \Delta T$$

So, the amount of alcohol that will spill out of the drum is the difference,

$$V_{spilled} = \Delta V_{alcohol} - \Delta V_{drum} = V_{0, alcohol} \beta_{alcohol} \Delta T - V_{0, drum} \beta_{steel} \Delta T$$

$$= (V_{0, alcohol} \beta_{alcohol} - V_{0, drum} \beta_{steel}) \Delta T$$

However, since the drum is initially completely filled with the alcohol, we have $V_{0, alcohol} = V_{0, drum} = V_0$. Therefore, we can simplify the expression to be,

$$V_{spilled} = (\beta_{alcohol} - \beta_{steel}) V_0 \Delta T$$

Solve

Substituting in the numerical values of $\beta_{alcohol} = 1120 \times 10^{-6} \, K^{-1}$, $\beta_{steel} = 36 \times 10^{-6} \, K^{-1}$, $V_0 = 55$ gal, and $\Delta T = 5°C$, we find,

$$V_{spilled} = (\beta_{alcohol} - \beta_{steel}) V_0 \Delta T$$

$$= (1120 \times 10^{-6} \, K^{-1} - 36 \times 10^{-6} \, K^{-1})(55 \text{ gal})(5°C)$$

$$= 0.30 \text{ gal}$$

What does it mean?

Notice that we treated the volume of the inside of the steel drum as if it were filled with steel when we determined the change in volume of the steel drum. We were able to do this since all dimensions of an object expand, including the inside diameter of a container.

WE 14.2 Calorimetry with Multiple Objects

A 150-g aluminum bowl containing 220 g of water is at 20°C. A very hot 300-g aluminum cylinder is dropped into the bowl, causing the water to boil, with 5.0 g being converted to steam. The final temperature of the system is 100°C. Neglect energy transfers with the environment. What is the original temperature of the cylinder?

Solution

Recognize the principle

Since all calorimetry problems are based on the principle of conservation of energy, all the heat energies Q must sum to zero. The aluminum cylinder loses heat energy while the water and aluminum bowl gain heat energy. We must also consider the heat energy required to turn 5.0 g of water into steam.

Sketch the problem

Though a sketch is not really necessary it does help us organize the given information. Let's consider the situation before and after the aluminum cylinder is dropped into the water.

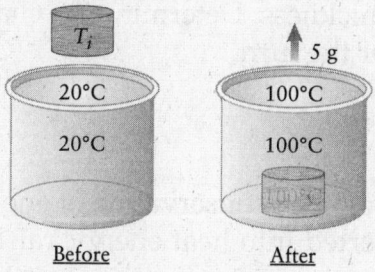

Identify the relationships

We begin by writing down our expression for the conservation of energy,

$$Q_{bowl} + Q_{water} + Q_{steam} + Q_{cylinder} = 0$$

Next, we can write the heat energies for each of these processes in terms of the specific heats, temperature change, and latent heat,

$$Q_{bowl} = m_{bowl} c_{aluminum} \Delta T_{bowl}$$
$$Q_{water} = m_{water} c_{water} \Delta T_{water}$$
$$Q_{cylinder} = m_{cylinder} c_{aluminum} \Delta T_{cylinder}$$
$$Q_{steam} = m_{steam} L_{vaporization, water}$$

Substituting these into our expression for energy conservation, we find,

$$Q_{bowl} + Q_{water} + Q_{steam} + Q_{cylinder} = 0$$

$$m_{bowl} c_{aluminum} \Delta T_{bowl} + m_{water} c_{water} \Delta T_{water} + m_{steam} L_{vaporization, water} + m_{cylinder} c_{aluminum} \Delta T_{cylinder} = 0$$

Solve

Substituting in our numerical values and solving for the initial temperature of the cylinder, we have,

$$m_{bowl} c_{aluminum} \Delta T_{bowl} + m_{water} c_{water} \Delta T_{water} + m_{steam} L_{vaporization, water} + m_{cylinder} c_{aluminum} \Delta T_{cylinder} = 0$$

$$(0.150 \text{ kg})(900 \text{ J/kg} \cdot \text{K})(80°C) + (0.220 \text{ kg})(4186 \text{ J/kg} \cdot \text{K})(80°C)$$

$$+ (0.005 \text{ kg})(2200 \times 10^3 \text{ J/kg}) + (0.300 \text{ kg})(900 \text{ J/kg} \cdot \text{K})(100°C - T_i) = 0$$

$$\rightarrow T_i = 454°C = \boxed{450°C}$$

What does it mean?

Notice that we were able to work in Celsius rather than convert to Kelvin since we were dealing with changes in temperature. Also, this answer seems reasonable since the melting temperature of aluminum is around 660°C.

CP 14.1 Heating During Braking

Brakes on an automobile consist of brake pads which are hydraulically pressed against a rotating steel disk (one at each wheel). The majority of the energy lost in slowing the vehicle is transferred to the steel disks during the braking process. Consider a 1000-kg race car traveling at 200 mi/h. As the driver approaches a turn, he applies the brakes and quickly slows to 100 mi/h. The braking disks on this race car are 30 cm in diameter and 3 cm in thickness. Determine the change in temperature of the disks as the race car slows for the turn.

Solution

Recognize the principle

We can tackle this problem by applying the principle of conservation of energy. As the car loses kinetic energy, this energy is converted into heat energy, raising the temperature of the steel disks.

Sketch the problem

$v_i = 200$ mi/h $\qquad\qquad\qquad v_f = 100$ mi/h

Identify the relationships

We begin by writing down our expression for the conservation of energy,

$$E_{\text{lost by car}} + Q_{\text{gained by disks}} = 0$$

Since the car loses kinetic energy during the braking, the energy lost by the car is the change in kinetic energy of the car,

$$E_{\text{lost by car}} = \Delta KE = KE_f - KE_i$$
$$= \frac{1}{2} m_{\text{car}} v_f^2 - \frac{1}{2} m_{\text{car}} v_i^2$$

The expression for the heat energy gained by the disks is given by,

$$Q_{\text{gained by disks}} = m_{\text{disks}} c_{\text{steel}} \Delta T$$

Substituting these expressions into our expression for the conservation of energy and solving for the temperature change, we have,

$$E_{\text{lost by car}} + Q_{\text{gained by disks}} = 0$$

$$\frac{1}{2}m_{\text{car}}\,v_f^2 - \frac{1}{2}m_{\text{car}}\,v_i^2 + m_{\text{disks}}\,c_{\text{steel}}\,\Delta T = 0$$

$$\Delta T = \frac{\frac{1}{2}m_{\text{car}}\,v_i^2 - \frac{1}{2}m_{\text{car}}\,v_f^2}{m_{\text{disks}}\,c_{\text{steel}}}$$

Solve

To solve this expression we need to convert the speeds of the car to m/s,

$$v_i = 200 \text{ mi/h}\left(0.447\,\frac{\text{m/s}}{\text{mi/h}}\right) = 89 \text{ m/s}$$

$$v_f = 100 \text{ mi/h}\left(0.447\,\frac{\text{m/s}}{\text{mi/h}}\right) = 45 \text{ m/s}$$

Also, we need to determine the total mass of the disks. We can determine the mass of each disk using the density of steel and the volume of a disk. Therefore,

$$m_{\text{disks}} = 4\,m_{\text{disk}} = 4\rho_{\text{steel}}V_{\text{disk}} = 4\rho_{\text{steel}}\,t(\pi d^2/4) = \rho_{\text{steel}}\,t\pi d^2$$

where t is the thickness of a disk and d is its diameter. Substituting in the numerical values, we have,

$$m_{\text{disks}} = \rho_{\text{steel}}\,t\pi d^2 = \left(7800 \text{ kg/m}^3\right)(0.03 \text{ m})\,\pi\,(0.3 \text{ m})^2 = 66 \text{ kg}$$

Now, substituting this mass and the other numerical values into our expression for the temperature change, we have,

$$\Delta T = \frac{\frac{1}{2}m_{\text{car}}\,v_i^2 - \frac{1}{2}m_{\text{car}}\,v_f^2}{m_{\text{disks}}\,c_{\text{steel}}} = \frac{(0.5)\,(1000 \text{ kg})\left[(89^2 - 45^2)\text{ m}^2/\text{s}^2\right]}{(66 \text{ kg})\,(450 \text{ J/kg}\cdot\text{K})} = \boxed{99 \text{ K}}$$

What does it mean?

Since a temperature change in Kelvin is the same as in Celsius, the temperature of the disks will increase by 99°C during the braking process. This is a rather large change in temperature. Repeated braking can raise the temperature of the disks to the point where they glow red hot.

CP 14.2 Out of Tune: Thermal Expansion of a Guitar String

Changes in temperature can cause a musical instrument to become out of tune. Consider a guitar which is perfectly tuned in a room whose temperature is 20°C. When tuned, the A-string on the guitar vibrates at a fundamental frequency of 110 Hz and has a length of 80 cm. Later, when the guitar is played at an outdoor concert, the A-string vibrates at a frequency of 112 Hz. What is the temperature of the outdoors? (*Note*: The A-string on this guitar is 9.0×10^{-4} m in diameter and made of steel. Neglect the thermal contraction of the body of the guitar.)

Solution

Recognize the principle

As the temperature changes, the string thermally contracts, thus increasing the tension in the string. With an increase in tension, the string vibrates at a higher frequency. We can combine our understanding of thermal contraction from this chapter with our analysis of a vibrating string from a previous chapter to determine the change in temperature of the string

Sketch the problem

No sketch is required.

Identify the relationships

Since this problem involves combining our understanding of vibrating strings, stresses and strains, and thermal contraction, let's begin by writing down or calculating all the parameters which may be important.

$$\rho_{steel} = 7800 \text{ kg/m}^3$$

$$\alpha_{steel} = 12 \times 10^{-6} \text{ k}^{-1}$$

$$Y_{steel} = 2.0 \times 10^{11} \text{ Pa}$$

$$A_{string} = \frac{\pi D^2}{4} = \frac{\pi (9 \times 10^{-4})^2}{4} = 6.4 \times 10^{-7} \text{ m}^2$$

$$\mu_{string} = \frac{m_{string}}{L_{string}} = \frac{\rho_{steel} V_{string}}{L_{string}} = \frac{\rho_{steel} A_{string} L_{string}}{L_{string}} = \rho_{steel} A_{string}$$

$$= (7800 \text{ kg/m}^3)(6.4 \times 10^{-7} \text{ m}^2) = 5.0 \times 10^{-3} \text{ kg/m}$$

Now that we have all the relevant parameters let's consider the vibration of the string. The speed of a wave on a string is given by the expression $v = \sqrt{\frac{F_T}{\mu}}$, and the speed, wavelength, and frequency are related by $v = \lambda f$. Combining these expressions and considering the initial and final frequencies, we can write the following,

$$\lambda f_i = \sqrt{\frac{F_{T,i}}{\mu}} \text{ and } \lambda f_f = \sqrt{\frac{F_{T,f}}{\mu}}$$

Since the length of the string does not change, the wavelength is the same for the two situations. Combining these expressions, we have,

$$\frac{F_{T,i}}{f_i^2} = \frac{F_{T,f}}{f_f^2}$$

Now let's consider the thermal contraction of the string. When the guitar is taken outside the string tries to contract, however, the body of the guitar prevents this contraction from occurring. This results in the tension in the string increasing. We can calculate the change in tension of the string by equating the change in length that would have occurred if the string were allowed to contract to the change in length that would result by a tensile stress. Proceeding with this approach, we have,

$$\Delta L_{\text{thermal contraction}} = \Delta L_{\text{tensile stress}}$$

$$L_0 \alpha_{\text{steel}} \Delta T = \frac{L_0 F}{Y_{\text{steel}} A_{\text{string}}}$$

Rearranging this expression to solve for the temperature change, we have,

$$\Delta T = \frac{F}{\alpha_{\text{steel}} Y_{\text{steel}} A_{\text{string}}}$$

In this expression, F is the change in tension the string experiences, so $F = F_{T,f} - F_{T,i}$. Combining these expressions, we have,

$$\Delta T = \frac{F}{\alpha_{\text{steel}} Y_{\text{steel}} A_{\text{string}}} = \frac{F_{T,f} - F_{T,i}}{\alpha_{\text{steel}} Y_{\text{steel}} A_{\text{string}}} = \frac{F_{T,i} \left(\frac{f_f^2}{f_i^2} - 1 \right)}{\alpha_{\text{steel}} Y_{\text{steel}} A_{\text{string}}}$$

The initial tension of the string we can get from the following,

$$\lambda f_i = \sqrt{\frac{F_{T,i}}{\mu_{\text{string}}}} \rightarrow F_{T,i} = \mu_{\text{string}} \lambda^2 f_i^2 = \mu_{\text{string}} (2L)^2 f_i^2$$

Solve
Substituting in the numerical values, we have,

$$F_{T,i} = \mu_{\text{string}} (2L)^2 f_i^2 = (5.0 \times 10^{-3} \text{ kg/m}) (2 \times 0.80 \text{ m})^2 (110 \text{ Hz})^2 = 155 \text{ N}$$

And substituting this and the other numerical values into our expression for the temperature change, we have,

$$\Delta T = \frac{F_{T,i} \left(\frac{f_f^2}{f_i^2} - 1 \right)}{\alpha_{\text{steel}} Y_{\text{steel}} A_{\text{string}}} = \frac{155 \text{ N} \left(\frac{112^2}{110^2} - 1 \right)}{(12 \times 10^{-6} \text{ K}^{-1})(2.0 \times 10^{11} \text{ Pa})(6.4 \times 10^{-7} \text{ m}^2)}$$

Therefore, the outside temperature is $20.0 - 3.7 = 16.3°C = \boxed{16°C}$.

What does it mean?
This problem illustrates the importance of tuning instruments in the environment they'll be played. Also, the temperature of a concert hall often changes during a concert due to the heat generated by the audience. As a result, musicians retune their instruments part way through a concert.

Part E. MCAT Review Problems and Solutions

PROBLEMS

1. Equal amounts of heat are added to two different objects of the same mass and initial temperature. The object which experiences the greatest temperature change will

   (a) have the greatest specific heat.

   (b) have the smallest specific heat.

   (c) have the largest latent heat of fusion.

   (d) Both objects will experience the same temperature change.

2. The coefficient of thermal expansion of steel is 1.2×10^{-5} K^{-1}. By how much will a steel rod 1 m in length change if it is heated from 300 K to 400 K?

   (a) 1.2×10^{-2} m

   (b) 1.2×10^{-3} m

   (c) 1.2×10^{-5} m

   (d) 1.2×10^{-6} m

3. By what factor will the radiant energy output of an object change if its temperature is doubled?

   (a) 0.5

   (b) 2

   (c) 4

   (d) 16

4. On a warm summer day the temperature is 90°F. What is the temperature in Celsius?

   (a) 18°C

   (b) 32°C

   (c) 50°C

   (d) 194°C

5. If 1 cal of heat is added to 1 g of water, by what amount will its temperature change?

   (a) 1°C

   (b) 10°C

   (c) 100°C

   (d) 1000°C

6. Which of the following diagrams best describes the temperature versus heat energy added graph for a typical solid object as it warms and melts?

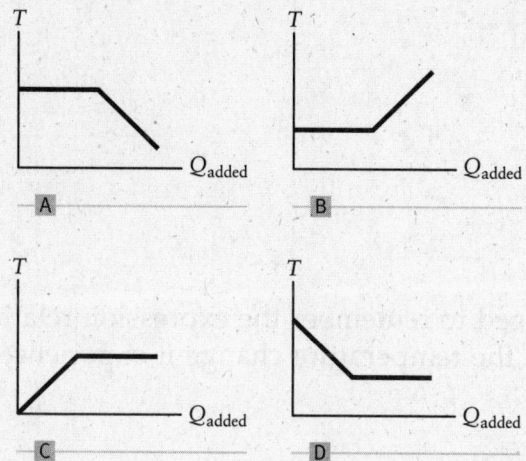

(a) A

(b) B

(c) C

(d) D

7. Two objects A and B have equal mass but are made of different materials. Object A has a specific heat which is twice that for object B. Object A is initially at a temperature of 100 K and object B is at 400 K. The two objects are brought into thermal contact and insulated from their surroundings. Determine the final equilibrium temperature of the two objects.

(a) 100 K

(b) 200 K

(c) 250 K

(d) 300 K

8. In order to minimize the heat conducted through the walls and ceiling of your home you should do the following:

(a) Place material in the walls and ceiling which has a high thermal conductivity.

(b) Decrease the difference in temperature between the inside and outside of your home.

(c) Both (a) and (b).

(d) Neither (a) nor (b).

9. A temperature change of 10 K corresponds to a corresponding temperature change of

(a) −273.15°C

(b) 10°C

(c) 273.15°C

(d) 1/273.15°C

10. If you double the surface area of a window pane and double its thickness, by what factor will the heat flow through the window change?

    (a) It will not change.

    (b) It will decrease by half.

    (c) It will double.

    (d) It will quadruple.

SOLUTIONS

1. *MCAT Strategies*

To answer this question we simply need to remember the expression relating the heat energy added to a material and the temperature change it experiences. The expression is,

$$Q = cm\Delta T$$

If two objects of the same mass receive the same heat energy Q, then the object with the smaller specific heat c will experience the larger temperature change ΔT. So, the correct answer is (b).

2. *MCAT Strategies*

Since all these answers are similar, we should apply our problem-solving strategy to obtain the correct answer.

Recognize the principle

The rod thermally expands as it is heated. We can calculate the change in its length by applying the appropriate expression for thermal expansion.

Sketch the problem

No sketch is required.

Identify the relationships

The change in length of a material due to thermal expansion is given by the expression,

$$\Delta L = L_0 \alpha \Delta T$$

Solve

Substituting in the numerical values, we have,

$$\Delta L = L_0 \alpha \Delta T = (1 \text{ m})(1.2 \times 10^{-5} \text{ K}^{-1})(400 \text{ K} - 300 \text{ K}) = 1.2 \times 10^{-3} \text{ m}$$

So, the correct answer is (b).

What does it mean?

Notice that the numbers provided in the question allowed us to calculate the answer without the need for a calculator. Remember, calculators are not allowed on the MCAT exams, so either the calculations must be straightforward or we'll need to make some approximations.

3. MCAT Strategies

To answer this question, we simply need to remember our expression relating the radiant energy output to the temperature of an object. The expression is,

$$\frac{Q}{t} = \sigma e A T^4$$

The important part of this expression to remember is that the radiant energy output (or absorbed) by an object is proportional to T^4. So, if we double the temperature of an object, then its radiant energy output increases by a factor of $2^4 = 16$. So, the correct answer is (d).

4. MCAT Strategies

The easiest way to answer questions relating the various temperature scales is to remember a few examples. One example is normal body temperature which is $98.6°F = 37°C$. Another good example is the freezing point of water which is $32°F = 0°C$. With these in mind we estimate that $90°F$ must be around $30°C$, so the correct answer must be (b). Just to be sure let's do the formal calculation.

Recognize the principle

We can use the exact expression relating the two temperature scales.

Sketch the problem

No sketch is required.

Identify the relationships

The Fahrenheit and Celsius temperature scales are related by the following expression,

$$T_F = \frac{9}{5}T_C + 32$$

Inverting this expression to solve for T_C, we have,

$$T_C = \frac{5}{9}(T_F - 32)$$

Solve

Plugging in the numerical value, we have,

$$T_C = \frac{5}{9}(90 - 32) = \frac{5}{9}(58) \approx 30°C$$

What does it mean?

Notice that in this last step we made the approximation that $\frac{5}{9} \approx \frac{1}{2}$ and $58 \approx 60$.

5. MCAT Strategies

This question is easy to answer if we remember how the calorie is defined. Recall that 1 calorie is the heat energy required to raise 1 g of water by $1°C$. Therefore, by adding 1 cal of heat energy to 1 g of water, we will raise its temperature by $1°C$. So, the correct answer is (a).

6. *MCAT Strategies*

To correctly answer this question we need to understand how adding heat energy affects the temperature of a solid object while warming and melting the object. While warming an object there is a linear relationship between the heat energy added and the temperature of the object. In other words, as you add heat energy to an object, its temperature increases proportionally. During a phase transition such as when a solid melts, the heat energy goes into breaking the molecular bonds in the solid and not into changing the temperature of the object. So the temperature remains constant during this process. With this in mind we see that the correct answer must be (c).

7. *MCAT Strategies*

Since we know that the equilibrium temperature will be somewhere between the initial temperatures of the two objects we can eliminate (a) as a correct answer. Also, we know that the masses of the two objects are the same so the difference in their specific heats will determine the final equilibrium temperature. Since object A has a larger specific heat, we know that the final equilibrium temperature will be closer to 100 K than 400 K. This brings us to the conclusion that the correct answer must be (b). Just to be sure, let's do the formal calculation.

Recognize the principle

We can apply the principle of conservation of energy.

Sketch the problem

No sketch is required.

Identify the relationships

Conservation of energy tells us that,

$$Q_A + Q_B = 0$$

Our expression relating heat energy, mass, specific heat, and temperature change is given by,

$$Q = cm\Delta T$$

Solve

Combining these expressions we have,

$$mc_A(T_f - 100) + mc_B (T_f - 400) = 0$$
$$(2c_B)(T_f - 100) + c_B (T_f - 400) = 0$$
$$2T_f - 200 + T_f - 400 = 0$$
$$3T_f = 600$$
$$T_f = 200 \text{ K}$$

What does it mean?

With this formal calculation we obtain the same answer as before, confirming that our reasoning was correct.

8. *MCAT Strategies*

To correctly answer this question we need to remember the expression for energy transfer through conduction,

$$\frac{Q}{t} = \kappa A \frac{\Delta T}{L}$$

Examining this expression we see that if we place material in the walls which has a higher thermal conductivity we will increase the amount of heat conduction through the walls, so answer (a) cannot be correct. On the other hand, if we decrease the temperature difference between the inside and outside of our home we will decrease the heat conducted through the walls. So, answer (b) is correct.

9. *MCAT Strategies*

To correctly answer this question we simply need to remember that a temperature difference is the same for the Celsius scale as for the Kelvin scale. So a temperature change of 10 K is the same as a temperature change of 10°C. So, the correct answer is (b).

10. *MCAT Strategies*

Once again, to correctly answer this question we need to understand the expression for heat transfer through conduction,

$$\frac{Q}{t} = \kappa A \frac{\Delta T}{L}$$

Looking at this expression we see that by doubling the area of a window pane A while also doubling its thickness L, the heat flow through the window will not be affected. So, the correct answer is (a).

15 Gas and Kinetic Theory

Part A. Summary of Key Concepts and Problem-Solving Strategies

KEY CONCEPTS

Kinetic theory

Kinetic theory is the application of Newton's laws to the mechanics of a gas of particles. This theory leads to the *ideal gas law*: for a dilute gas composed of any substance,

$$PV = nRT$$

where P is the pressure, V is the volume, n is the number of particles as measured in units of moles, and R is the *universal gas constant* with the value

$$R = 8.31 \frac{J}{mole \cdot K}$$

The ideal gas law can also be written in the form

$$PV = Nk_BT$$

where N equals the number of particles (i.e., molecules) and k_B is Boltzmann's constant,

$$k_B = 1.38 \times 10^{-23} J/K$$

Internal energy

The temperature of a dilute gas is related to the average kinetic energy of a molecule by

$$KE = \frac{3}{2} k_B T$$

The *internal energy* of a monatomic ideal gas is equal to the total kinetic energy of the molecules and is given by

$$U = \frac{3}{2} N k_B T$$

For an ideal monatomic gas, U is independent of all other factors such as the molecular mass and pressure.

APPLICATIONS

Gas laws

The ideal gas law is a general relation that contains several other gas laws as special cases.

Avogadro's law: For a sample of gas at constant pressure and temperature, the volume is proportional to the number of molecules in the sample.

Boyle's law: For a sample of gas at constant temperature, the product of the pressure and the volume is constant:

$$PV = \text{constant}$$

Charles's law: For a sample of gas at constant pressure, if the temperature is changed by a small amount ΔT, the volume changes by an amount ΔV, with

$$\Delta V \propto \Delta T$$

Gay–Lussac's law: For a sample of gas held in a container with constant volume, changes in pressure are proportional to changes in temperature:

$$\Delta P \propto \Delta T$$

Specific heat of a gas

The specific heat per mole of a monatomic ideal gas at constant volume is given by

$$C_V = \frac{3}{2} R$$

Diffusion

The motion of individual gas molecules is described by diffusion. The average net distance that a molecule diffuses in a time t is given by

$$\Delta r = \sqrt{Dt}$$

where D is the diffusion constant.

PROBLEM-SOLVING STRATEGIES

One of the key concepts discussed in this chapter is the application of the ideal gas law. The ideal gas law applies to a dilute gas; however, it is also a good approximation for many other systems including air at room temperature. As such, application of the ideal gas law provides us with a very powerful tool for studying and understanding gases and their interactions. The following schematic will help guide you through the thought process involved in applying this principle.

Problem Solving: Applying the Ideal Gas Law

Recognize the principle

The ideal gas law allows us to relate the temperature, pressure, volume, and number of particles or moles of an ideal gas. It is also a good approximation for other systems of gases.

Sketch the problem

We are often interested in studying the change in the parameters of a gas as it undergoes a particular process. Therefore, it is helpful to sketch the system before and after the process, being sure to label all the relevant parameters including the volume, pressure, temperature, and number of particles or moles.

Identify the relationships

The ideal gas law can be written in the following ways,

$$PV = nRT \text{ and } PV = Nk_BT$$

where P is pressure in Pa, V is volume in m^3, T is temperature in K, n is the number of moles, N is the number of particles (i.e., molecules), $R = 8.31$ J/mole $\cdot$ K, and $k_B = 1.38 \times 10^{-23}$ J/K.

Solve

Since the universal gas constant R and Boltzmann's constant k_B are both constants, we can write the following expression for any process through which the ideal gas undergoes,

$$\frac{P_iV_i}{n_iT_i} = \frac{P_fV_f}{n_fT_f} \text{ and } \frac{P_iV_i}{N_iT_i} = \frac{P_fV_f}{N_fT_f}$$

Using these expressions we can solve for an unknown using the given information.

What does it mean?

Always *consider what your answer means*, and check that it makes sense.

Part B. Frequently Asked Questions

1. *If I compress an ideal gas to half its initial volume, what changes?*

It depends on what you hold constant. Let's look at the ideal gas law,

$$PV = Nk_BT$$

We can rearrange this expression so that the volume is by itself,

$$V = \frac{Nk_BT}{P}$$

If we cut the volume in half, then this will result in the right-hand side of this expression being cut in half as well. This can be accomplished in a number of ways. If we hold the number of particles constant and the pressure constant, then the temperature must reduce by half as well. If we hold the number of particles and the temperature constant, then the pressure will double. If we hold the temperature and pressure constant, then we'll need to reduce the number of particles by half. Finally, we could adjust the temperature, number of particles, and the pressure to accommodate the reduction in volume. So there are many possible consequences to compressing an ideal gas, each depending on what parameters we allow to change. It's all governed by the ideal gas law.

2. *If I double the pressure of an ideal gas while holding its volume constant, what changes?*

Once again, it depends on what else you hold constant. Suppose you maintain the number of particles N in the gas. The ideal gas law can be rearranged as follows,

$$\frac{P}{T} = \frac{Nk_B}{V} = \text{constant}$$

So in order to double the pressure you must double the temperature. Or, if we allow the number of particles to change while holding the temperature constant, then we have,

$$\frac{P}{N} = \frac{k_BT}{V} = \text{constant}$$

Now we see that in order to double the pressure we must double the number of particles in the gas. Of course, we could also adjust both the temperature and the number of particles in order to double the pressure of the gas.

3. *Does heating up an ideal gas always cause it to expand?*

That's a good question. Often, heating up an ideal gas causes it to expand to a larger volume; however, this is not always the case. Consider a sealed container filled with an ideal gas. If the container is rigid and not able to expand, then as we heat the gas inside the container the pressure will increase. Again, it is governed by the ideal gas law and depends on which variables are allowed to change.

4. *What's the relationship between the translational kinetic energy of a molecule and the internal energy of an ideal gas?*

If we determine the average translational kinetic energy of all the molecules in an ideal gas, we would find that it depends on the temperature of the gas according to the expression,

$$KE_{trans} = \frac{3}{2} k_B T$$

If there are N particles in this gas, then the total translational kinetic energy is simply given by,

$$KE_{total} = N(KE_{trans}) = \frac{3}{2} Nk_B T$$

If this gas is a monatomic ideal gas, then this total translational kinetic energy of the atoms in the gas is the internal energy of the gas,

$$U = KE_{total} = \frac{3}{2} Nk_B T$$

It's important to note that this expression for internal energy is only true for a monatomic ideal gas. For molecules comprised of more than one atom the internal energy is given by a different expression.

5. *According to the Maxwell–Boltzmann speed distribution, as you heat up a gas, the curve shifts to the right meaning that the average speed of the molecules within the gas increases. Does that mean that hotter gases always have faster moving molecules than cooler gases?*

No. It is true that as you heat up a gas, the curve shifts to the right, therefore the average speed of the molecules within a gas increases. However, when comparing different gases we must also consider the different masses of the particles. Remember, the speed of a typical molecule in a gas is related to temperature by the following expression,

$$v = \sqrt{\frac{3k_B T}{m}}$$

We see from this expression that mass is important as well. For a given temperature, more massive particles move slower. So, it's possible to have two gases at different temperatures where the cooler gas has less massive particles, but whose average speeds are greater than the more massive particles in the hotter gas.

6. *Can I use temperature in Celsius in the ideal gas law?*

No, definitely not. When using the ideal gas law we need to express the temperatures in Kelvin. Using Celsius can result in nonsensical answers, particularly if we use negative temperatures. It's easy to remember to use Kelvin since both the universal gas constant R and Boltzmann's constant k_B are expressed in Kelvin. Remember, the only time you can use either Celsius or Kelvin is when dealing with differences in temperature, since $\Delta T_C = \Delta T_K$.

7.   *How does diffusion depend on temperature?*

That's a good question since temperature does not explicitly appear in any of our expressions for diffusion. Recall our expression for the time t required for an average molecule to diffuse a distance Δr,

$$t = \frac{(\Delta r)^2}{D}$$

The quantity D is called the diffusion constant and depends on many factors including temperature. It also depends on the mass of the particles as well as properties of the medium through which the particles are diffusing. So, temperature certainly affects the diffusion constant, and therefore, the process of diffusion.

8.   *According to the textbook, nitrogen molecules in the air are colliding into me at 510 m/s. Why don't I feel them?*

The collisions of the nitrogen molecules (along with the oxygen and other molecules in the air) collide with our bodies producing a force over the surface of our bodies which we call a pressure—atmospheric pressure! Pressure comes from the collisions of gas molecules. We aren't aware of this pressure because of the outward pressure of the cell membranes in our bodies. The two pressures cancel and we feel nothing.

9.   *How do I determine the mass of an atom from its atomic mass?*

The atomic mass (or molecular mass) is measured in grams per mole, g/mol. So, to determine the mass of an atom or molecule from its atomic or molecular mass, we simply need to divide by the number of atoms/molecules per mole which is given by Avogadro's number, $N_A = 6.02 \times 10^{23}$. Therefore, we have the following conversion,

$$m = \frac{M}{N_A}$$

For example, the atomic mass of oxygen is 15.999 g/mole. So, the mass of an oxygen atom is,

$$m = \frac{M}{N_A} = \frac{15.999 \text{ g/mole}}{6.02 \times 10^{23} \text{ atoms/mole}} = 2.66 \times 10^{-23} \text{ g} = 2.66 \times 10^{-26} \text{ kg}$$

10.   *What is the relationship between the universal gas constant R and Boltzmann's constant k_B?*

This is easily answered by comparing our two equivalent expressions for the ideal gas law,

$$PV = nRT \text{ and } PV = Nk_BT$$

From these we see that $nR = Nk_B$. We can rearrange this expression to get the following,

$$k_B = \frac{n}{N}R$$

Since $N = nN_A$, we can also write the relationship between k_B and R as,

$$k_B = \frac{R}{N_A}$$

Part C. Selection of End-of-Chapter Answers and Solutions

QUESTIONS

Q15.6 In Chapter 11, we learned about the bulk modulus B of a substance. The bulk modulus is related to changes in the volume by

$$\frac{\Delta V}{V} = -\frac{\Delta P}{B}$$

Use the ideal gas law to calculate the bulk modulus for an ideal gas. Assume that the temperature is held constant.

Answer

To answer this question we can first solve the given equation for the bulk modulus:

$$B = -\frac{\Delta P V_1}{\Delta V}$$

where we've used V_1 for the initial volume of the substance, and

$$\Delta P = P_2 - P_1$$
$$\Delta V = V_2 - V_1$$

Therefore,

$$B = -\frac{(P_2 - P_1)V_1}{(V_2 - V_1)}$$

We can write the ideal gas law for each state,

$$P_1 = \frac{nRT}{V_1} \text{ and } P_2 = \frac{nRT}{V_2}$$

This allows us to write ΔP as,

$$P_2 - P_1 = nRT\left(\frac{1}{V_2} - \frac{1}{V_1}\right)$$

Multiplying both sides by V_1 gives,

$$(P_2 - P_1)V_1 = nRT\left(\frac{V_1}{V_2} - 1\right)$$

$$(P_2 - P_1)V_1 = nRT\left(\frac{V_1 - V_2}{V_2}\right)$$

If we then divide both sides by the change in volume and reduce the expression, we can get an expression for the bulk modulus,

$$B = -\frac{(P_2 - P_1)V_1}{(V_2 - V_1)} = -\frac{nRT\left(\dfrac{V_1 - V_2}{V_2}\right)}{(V_2 - V_1)}$$

$$B = -nRT\left(\frac{V_1 - V_2}{V_2(V_2 - V_1)}\right)$$

Writing this in terms of volume changes,

$$B = -nRT\left(\frac{-\Delta V}{V_2 \Delta V}\right) = \frac{nRT}{V_2}$$

And this expression, according to the ideal gas law, is,

$$P_2 = \frac{nRT}{V_2} \rightarrow \boxed{B = P_2}$$

The bulk modulus of an ideal gas is therefore equal to the final pressure of the gas as it is compressed. The accepted bulk modulus of room-temperature air, for example, is very close to standard pressure! (1.01×10^5 Pa)

Q15.9   A container of gas under pressure has a pinhole leak. If the content of the container is oxygen, the pressure decreases at a certain rate as the gas escapes. If the content is hydrogen gas, however, we find that the rate of pressure drop is greater under the same conditions. Why?

Answer

The pinhole leak allows the molecules of gas to escape. Gas molecules are moving in many random directions at any given time. When the velocity vector of a molecule lines up with the pin hole, the molecule moves out of the container. The average speed of the escaping molecules is just their average speed—at least until they run into something outside of the container (like a room air molecule). Remember that temperature is proportional to kinetic energy, which depends on mass and speed. H_2 molecules are much less massive than O_2 molecules. This implies that in order to have the same temperature, H_2 gas molecules must have a higher overall average molecular speed than O_2 molecules. Faster moving H_2 molecules are more likely to find the opening and escape much more quickly.

PROBLEMS

P15.5   You have 6.0 moles of particles of an unknown (pure) substance with a mass of 240 g. What might the substance be?

Solution

Recognize the principle

A pure substance contains all of the same types of particles, be they atoms or molecules. The mass of one mole of any substance is equal to its molecular mass, and the mass of one mole of atoms is equal to its atomic mass.

Sketch the problem

No sketch required.

Identify the relationships

We know that 6.0 moles has a mass of 240 g, which implies that a single mole has a mass of 40 g. This mass must be equal to the atomic mass (which can be found in the periodic table) or molecular mass.

Solve

We can look at the periodic table and find that the atomic mass of Ca is 40.01 u, so this substance could be pure calcium. It could also be any number of compounds with a molecular mass of 40 grams, such as magnesium oxide (MgO).

What does it mean?

The substance could be pure $\boxed{\text{calcium or magnesium oxide}}$, among other possibilities.

P15.14 The gas in a cylindrical container has a pressure of 1.0×10^5 Pa and is at room temperature (293 K). A piston (a moveable wall) at one end of the container is then adjusted so that the volume is reduced by a factor of three, and it is found that the pressure increases by a factor of five. What is the final temperature of the gas?

Solution

Recognize the principle

The ideal gas law can be used to relate pressure, temperature, and volume for a constant amount of gas.

Sketch the problem

No sketch required.

Identify the relationships

We begin by writing our expression for the ideal gas law,

$$PV = Nk_BT$$

With a constant amount of gas (N), we can rearrange this equation so that the right side is constant:

$$\frac{PV}{T} = Nk_B = \text{constant}$$

This implies that the left side will also be constant, which allows us to write,

$$\frac{P_1V_1}{T_1} = \frac{P_2V_2}{T_2}$$

Solve

Solving this expression for T_2:

$$T_2 = \frac{P_2V_2T_1}{P_1V_1}$$

The problem statement gives us a ratio for the volume and the pressure,

$$V_2 = \frac{V_1}{3} \rightarrow \frac{V_2}{V_1} = \frac{1}{3}$$

$$P_2 = 5P_1 \rightarrow \frac{P_2}{P_1} = 5$$

Inserting these values for each ratio in our expression for T_2 yields,

$$T_2 = \left(\frac{1}{3}\right)(5)T_1$$

And finally, inserting the value for the initial temperature,

$$T_2 = \left(\frac{5}{3}\right)(293 \text{ K}) = \boxed{490 \text{ K}}$$

What does it mean?

This compression causes a temperature increase of almost 200 K, assuming no heat escapes!

P15.23 The pressure in a gas thermometer is 5000 Pa at the freezing point of water. (a) What is the pressure in this thermometer at the boiling point of water? (b) If the pressure is 6000 Pa, what is the temperature?

Solution

Recognize the principle

The pressure and temperature of an ideal gas are directly proportional when the volume and amount of substance are both constant.

Sketch the problem

No sketch required.

Identify the relationships

The freezing point of water is 0°C or 273 K. Since the pressure must also be zero when the temperature is zero, we can define a calibration constant for this thermometer from the ratio of temperature to pressure:

$$\frac{5000 \text{ Pa}}{273 \text{ K}} = 18.3 \text{ Pa/K}$$

The boiling temperature of water is 100°C, or 373 K.

Solve

(a) The pressure at the boiling point of water is therefore,

$$373 \text{ K}(18.3 \text{ Pa/K}) = \boxed{6830 \text{ Pa}}$$

(b) If the pressure is 6000 Pa, then the corresponding temperature is,

$$\frac{6000 \text{ Pa}}{18.3 \text{ Pa/K}} = \boxed{328 \text{ K} = 55°C}$$

What does it mean?

Once the calibration constant for the thermometer is found, the pressure at any temperature follows easily.

P15.26 You place 80 moles of hydrogen gas in a balloon of volume 2.5 m³, and find the pressure to be 1.5 times atmospheric pressure. What is the typical speed of a hydrogen molecule?

Solution

Recognize the principle

The average speed of the molecules depends on the temperature, which can be found using the ideal gas law from the given pressure, volume, and amount of gas.

Sketch the problem

No sketch required.

Identify the relationships

Since we are given a number of moles of gas, we find the temperature using the ideal gas law in the form,

$$PV = nRT$$

The average speed of a molecule in the atmosphere in terms of this temperature is given by,

$$v = \sqrt{\frac{3k_B T}{m}}$$

Solve

We first solve our ideal gas law equation for the temperature and insert the given values,

$$T = \frac{PV}{nR}$$

$$T = \frac{(1.5)(1.01 \times 10^5 \text{ Pa})(2.5 \text{ m}^3)}{(80 \text{ moles})(8.31 \text{ J/K} \cdot \text{mole})} = 570 \text{ K}$$

The mass of each hydrogen gas molecule is

$$m = 2m_H$$

$$m = 2\left(\frac{1.00 \text{ g}}{6.02 \times 10^{23}}\right) = 3.32 \times 10^{-24} \text{ g}$$

We can then solve our expression for the speed of a molecule in a gas and insert the numerical values,

$$v = \sqrt{\frac{3k_B T}{m}}$$

$$v = \sqrt{\frac{3(1.38 \times 10^{-23}\,\text{J/K})(570\,\text{K})}{3.32 \times 10^{-27}\,\text{kg}}} = 2666\,\text{m/s} \approx \boxed{2700\,\text{m/s}}$$

What does it mean?

The higher temperature and greater pressure results in a much greater particle speed than molecules at standard pressure and room temperature.

P15.32 A balloon of volume 1.5 m³ contains argon gas at a pressure of 1.5×10^5 Pa and room temperature (20°C). What is the total internal energy of the gas?

Solution

Recognize the principle

The total internal energy of a monatomic ideal gas (like argon) depends on the number of particles and the temperature. We can find the number of particles using the ideal gas law from the given pressure, temperature, and volume.

Sketch the problem

No sketch required.

Identify the relationships

The internal energy of an ideal monatomic gas is given by the expression,

$$U = \frac{3}{2} N k_B T$$

The number of gas atoms, N, can be found using the ideal gas law,

$$PV = N k_B T$$

Solve

Combining these two expressions, we find,

$$U = \frac{3}{2} PV$$

Inserting the given pressure and volume numerical values, we can then find the internal energy,

$$U = \frac{3}{2}(1.5 \times 10^5\,\text{Pa})(1.5\,\text{m}^3) = \boxed{3.4 \times 10^5\,\text{J}}$$

What does it mean?

This argon gas has an internal energy of about 340,000 J. This value is proportional to the pressure, and so the pressurized gas has higher energy than it would at standard pressure.

P15.35   How long would it take a nitrogen molecule in the atmosphere to travel a distance of 1 cm if there were no collisions with other molecules?

Solution

Recognize the principle

We can calculate the time it takes from the average speed of a nitrogen molecule.

Sketch the problem

No sketch required.

Identify the relationships

If there are no collisions then the nitrogen molecules will have a constant velocity. The speed of a nitrogen molecule in the atmosphere was found in the textbook to be

$$v_{N_2} = 510 \text{ m/s}$$

The time required to move 1 cm can be found using the relationship between speed, distance, and time,

$$v = \frac{x}{t}$$

Solve

Solving this expression for time and inserting our numerical values yields,

$$t = \frac{0.01 \text{ m}}{510 \text{ m/s}} = \boxed{2.0 \times 10^{-5} \text{ s}}$$

What does it mean?

The nitrogen molecule covers the centimeter in about 20 microseconds.

P15. 39   What is the approximate time required for a hemoglobin molecule to diffuse through a cell membrane? *Hint*: The diffusion constant D is approximately proportional to $1/r$, where r is the diameter of the diffusing particle or atom. Estimate the diffusion constant of a hemoglobin molecule from the value of D for an oxygen molecule in Table 15.1.

Solution

Recognize the principle

The time required for diffusion through this membrane can be estimated using the diffusion coefficient and thickness of the membrane.

Sketch the problem

No sketch needed.

Identify the relationships

We know the following expression for the magnitude of the average displacement of a molecule,

$$\Delta r = \sqrt{Dt}$$

The diffusion coefficient for oxygen through tissue is given as 2×10^{-11} m^2/s in Table 15.1. A hemoglobin molecule has a diameter on the order of 10 times the size of the oxygen molecule, so we can expect its diffusion constant to be smaller by a factor of 10 ($D \approx 2 \times 10^{-12}$ m^2/s).

Solve

Solving the diffusion equation for the time gives,

$$t = \frac{(\Delta r)^2}{D}$$

Then inserting our estimated values, we have,

$$t \approx \frac{(10 \times 10^{-9} \text{ m})^2}{2 \times 10^{-12} \text{ m}^2/\text{s}} = \boxed{5 \times 10^{-5} \text{ s}}$$

What does it mean?

The diffusion through this membrane happens on the order of tenths of microseconds.

P15.57   In our discussions of kinetic theory, we have focused on the motion of the molecules in a gas. The result for the typical speed in Equation 15.18, however, also applies to molecules in a liquid. Use this result to calculate the speed of the amino acid molecule glutamine ($C_5H_{10}N_2O_3$) in solution at room temperature.

Solution

Recognize the principle

The average speed of all molecules is related to their temperature.

Sketch the problem

No sketch required.

Identify the relationships

We can begin by identifying the expression for the kinetic energy of a molecule in terms of the temperature,

$$KE_{\text{trans}} = \frac{3}{2} k_B T$$

This translational kinetic energy can also be described in terms of the molecule's mass and speed,

$$KE_{\text{trans}} = \frac{1}{2} mv^2$$

We can find the mass of a glutamine molecule by summing the atomic masses and dividing by the number of moles,

$$\frac{5(.012 \text{ kg/mol}) + 10(.001 \text{ kg/mol}) + 2(.014 \text{ kg/mol}) + 3(.016 \text{ kg/mol})}{6.02 \times 10^{23}/\text{mol}}$$

$$= 2.43 \times 10^{-25} \text{ kg}$$

We can assume that room temperature is about 293 K.

Solve

We can set equal our two expressions for the translational kinetic energy of a molecule,

$$\frac{1}{2} mv^2 = \frac{3}{2} k_B T$$

Then solving for the speed and inserting the numerical values yields,

$$v = \sqrt{\frac{3k_B T}{m}} = \sqrt{\frac{3(1.38 \times 10^{-23} \text{ J/K})(293 \text{ K})}{2.43 \times 10^{-25} \text{ kg}}} = 223 \text{ m/s} \approx \boxed{220 \text{ m/s}}$$

What does it mean?

This speed is smaller (by more than half) of a typical gas at room temperature. The larger glutamine molecules result in the same kinetic energy with less speed.

Part D. Additional Worked Examples and Capstone Problems

The following worked example and capstone problems provide you with additional practice with diffusion and the application of the ideal gas law. Working these problems will give you greater insight into these concepts. Also, the capstone problems incorporate material from previous chapters, thereby serving as a review. Although these problems do not incorporate all the material discussed in this chapter, they do highlight several of the key concepts. If you can successfully solve these problems then you should feel confident in your understanding of these key concepts, so use these problems as a test of your understanding of the chapter material and a review of previous chapter material.

WE 15.1 Diffusive Motion of a Gas Molecule

Consider the diffusion of a nitrogen molecule through the atmosphere at room temperature. (a) Find the time required for an average molecule to diffuse through a net displacement $\Delta r = 1$ cm. (b) Approximately how many random walk steps does the molecule take during this time?

Solution

Recognize the principle

In an example in the textbook we found that the speed of a typical nitrogen molecule in the atmosphere is $v = 510$ m/s, and we saw that the mean free path in air is approximately $l \approx 1 \times 10^{-7}$ m. We can use these values together with the result for diffusive motion to analyze the diffusive motion of a nitrogen molecule through the air.

Sketch the problem

Figure 15.12 in your textbook describes the problem. The net displacement of the molecule is the (straight line) distance from A to B.

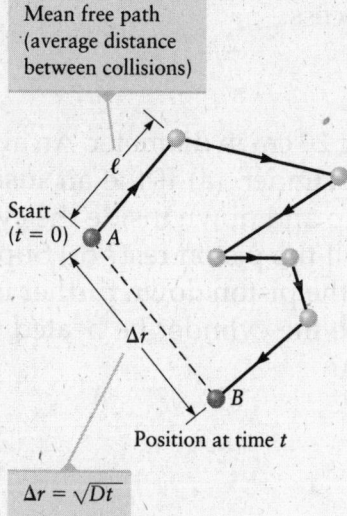

Figure 15.12

Identify the relationships

According to our expression for diffusive motion, the time required to diffuse a distance Δr is

$$t = \frac{(\Delta r)^2}{D}$$

Solve

(a) Inserting the given value of Δr along with the value of the diffusion constant from Table 15.1 in your textbook, we get

$$t = \frac{(\Delta r)^2}{D} = \frac{(0.01 \text{ m})^2}{2 \times 10^{-5} \text{ m}^2/\text{s}} = 5 \text{ s}$$

(b) Each random walk step has an average length ℓ, and because the molecule moves with a speed v, the time for one step is (on average) $t_{step} = \ell/v$. Inserting the given values of ℓ and v gives

$$t_{step} = \frac{\ell}{v} = \frac{1 \times 10^{-7} \text{ m}}{510 \text{ m/s}} = 2 \times 10^{-10} \text{ s}$$

In part (a), we found that the total time is $t = 5$ s. Because each step takes a time t_{step}, the number of steps is

$$N_{steps} = \frac{t}{t_{step}} = \frac{5 \text{ s}}{2 \times 10^{-10} \text{ s}} \approx \boxed{3 \times 10^{10}}$$

What does it mean?

The molecule must take an enormous number of steps (about 30 billion) to travel only 1 cm! If the molecule had traveled this same distance at the same speed but without any collisions, the time required would have been only $t = \Delta r / v = 2 \times 10^{-5}$ s. By comparison, diffusion is a very slow process.

CP 15.1 Ideal Gas and the Forces on a Piston

A 50-cm deep cylindrical container has an opening 20 cm in diameter. An air-tight, 10-kg piston is inserted into the opening of the cylinder. (a) If the air inside the cylinder is at 20°C and the piston is free to slide up and down inside the cylinder, at what distance below the top of the cylinder will the piston rest? (b) Suppose a 10-kg mass is placed on top of the piston, forcing the piston down further into the cylinder. To what temperature must the air inside the cylinder be heated so that the piston returns to the position found in part (a)?

Solution

Recognize the principle

Since the container is filled with air, we can treat the air as an ideal gas and apply the ideal gas law. We know that the air inside the container is originally at atmospheric pressure, P_0, before the piston is inserted and it is at a temperature of $T_0 = 20$°C. Also, we know the initial volume of the container V_0. After the piston is inserted we can calculate the pressure of the air and use this along with the ideal gas law to determine the volume of the air, and therefore the position of the piston.

Sketch the problem

Let's begin by drawing a sketch of the three situations described: (1) once the piston is released and has settled at a final position, and (2) after the mass has been added and the piston has settled at a final position, and (3) after we've heated up the gas to return the piston to its previous position.

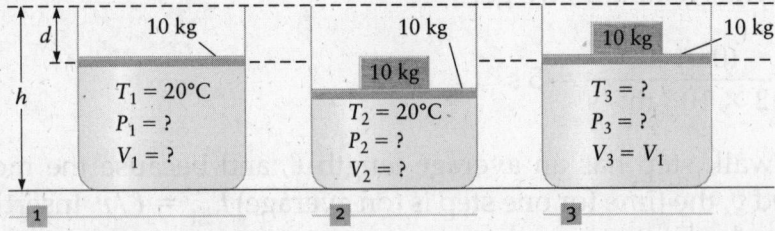

Identify the relationships

(a) Since the piston isn't accelerating, the pressure of the air must be equal to the pressure above the piston plus the added pressure due to the weight of the piston. The pressure above the piston is atmospheric pressure, P_0. Therefore,

$$P_1 = P_0 + P_{piston}$$

The pressure due to the piston is equal to the force of gravity on the piston divided by the area of the piston,

$$P_{piston} = \frac{F_{grav}}{A} = \frac{m_{piston} g}{A}$$

Combining these expressions, we have,

$$P_1 = P_0 + \frac{m_{piston}g}{A}$$

Applying the ideal gas law, we have,

$$\frac{P_1 V_1}{T_1} = \frac{P_0 V_0}{T_0} = N k_B$$

Rearranging this expression to solve for V_1 and substituting in our expression for P_1, we have,

$$V_1 = \frac{P_0 V_0 T_1}{P_1 T_0} = \frac{P_0 V_0 T_1}{\left(P_0 + \dfrac{m_{piston}g}{A}\right) T_0}$$

Volume is equal to the area times the height of the piston and $T_1 = T_0$; therefore,

$$V_1 = \frac{P_0 V_0 \cancel{T_1}}{\left(P_0 + \dfrac{m_{piston}g}{A}\right) \cancel{T_0}}$$

$$\cancel{A}(h - d) = \frac{P_0 \cancel{A} h}{\left(P_0 + \dfrac{m_{piston}g}{A}\right)}$$

$$d = h - \frac{P_0 h}{\left(P_0 + \dfrac{m_{piston}g}{A}\right)}$$

$$d = h\left[1 - \frac{P_0}{\left(P_0 + \dfrac{m_{piston}g}{A}\right)}\right] = h\left(\frac{P_0 + \dfrac{m_{piston}g}{A} - P_0}{P_0 + \dfrac{m_{piston}g}{A}}\right)$$

$$= h\left(\frac{\dfrac{m_{piston}g}{A}}{\dfrac{P_0 A + m_{piston}g}{A}}\right) = h\left(\frac{m_{piston}g}{P_0 A + m_{piston}g}\right)$$

(b) To determine the temperature T_3 we simply need to compare configuration (3) with (1) and apply the ideal gas law,

$$\frac{P_3 V_3}{T_3} = \frac{P_1 V_1}{T_1}$$

Now, we know that $V_3 = V_1$, $P_1 = P_0 + \dfrac{m_{piston}g}{A}$, and P_3 must equal P_1 plus the added pressure to compensate for the additional mass,

$$P_3 = P_1 + \frac{m_{mass}g}{A}$$

$$= P_0 + \frac{(m_{piston} + m_{mass})g}{A}$$

Putting this all together, we have,

$$\frac{P_3 V_3}{T_3} = \frac{P_1 V_1}{T_1} \rightarrow T_3 = \frac{P_3 T_1}{P_1}$$

$$T_3 = \frac{P_0 + \dfrac{(m_{piston} + m_{mass})g}{A}}{P_0 + \dfrac{m_{piston}g}{A}} T_1$$

Simplifying this expression we have,

$$T_3 = \left[\frac{P_0 A + (m_{piston} + m_{mass})g}{P_0 A + m_{piston}g}\right] T_1$$

Solve

(a) Substituting in the numerical values of $h = 0.5\,\text{m}$, $A = \pi r^2 = \pi(0.10\,\text{m})^2 = 0.0314\,\text{m}^2$, $g = 9.8\,\text{m/s}^2$, $P_0 = 1.01 \times 10^5\,\text{Pa}$, and $m_{piston} = 10\,\text{kg}$, we have,

$$d = \left(\frac{m_{piston}g}{P_0 A + m_{piston}g}\right)(0.5\,\text{m})$$

$$= \left[\frac{(10\,\text{kg})(9.8\,\text{m/s}^2)}{(1.01 \times 10^5\,\text{Pa})(0.0314\,\text{m}^2) + (10\,\text{kg})(9.8\,\text{m/s}^2)}\right](0.5\,\text{m})$$

$$= 0.015\,\text{m} = \boxed{1.5\,\text{cm}}$$

(b) Substituting in our numerical values, we have,

$$T_3 = \left[\frac{P_0 A + (m_{piston} + m_{mass})g}{P_0 A + m_{piston}g}\right] T_1$$

$$= \left[\frac{(1.01 \times 10^5\,\text{Pa})(0.0314\,\text{m}^2) + (10\,\text{kg} + 10\,\text{kg})\,9.8\,\text{m/s}^2}{(1.01 \times 10^5\,\text{Pa})(0.0314\,\text{m}^2) + (10\,\text{kg})(9.8\,\text{m/s}^2)}\right](20°\text{C})$$

$$= \boxed{20.6°\text{C}}$$

What does it mean?

Notice that the temperature increase required was quite small since the air had only been compressed a small amount by the weight of the piston and the additional mass.

CP 15.2 Nitrogen in Our Atmosphere

Suppose a nitrogen molecule in the air was traveling at the average speed for a nitrogen molecule at room temperature. Now, suppose that the molecule is traveling vertically upward. If the nitrogen molecule was to travel without colliding with any other atoms or molecules, to what maximum height above the surface of the Earth would it travel? Do not assume that the acceleration of gravity is constant.

Solution

Recognize the principle

From our understanding of the relationship between the temperature of a gas and the average speed of an atom or molecule within the gas, we can determine the speed of the nitrogen molecule. From this speed we can apply the principle of conservation of energy to the molecule to determine the height above the surface of the Earth it will travel. We'll need to be sure to use our general expression for the gravitational potential energy of the molecule-Earth configuration.

Sketch the problem

A simple sketch will help us visualize the problem.

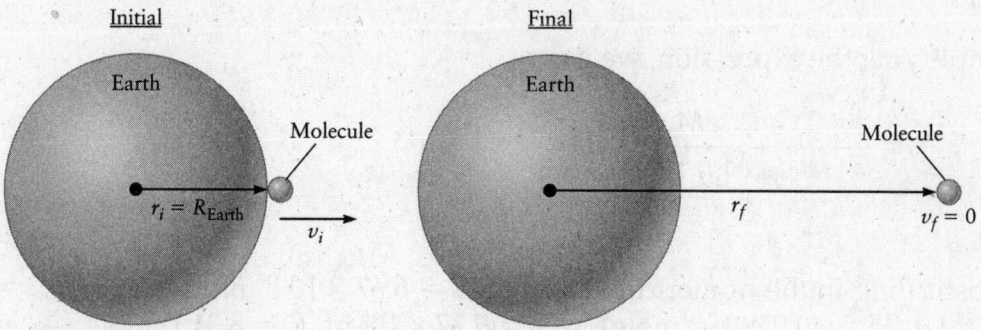

Identify the relationships

We must first determine the speed of the nitrogen molecule. The average speed of an atom or molecule as a function of temperature is given by the expression,

$$v = \sqrt{\frac{3k_B T}{m}} = \sqrt{\frac{3RT}{M}}$$

where we have made use of the fact that $k_B = R/N_A$ and $mN_A = M$. We can now use this speed as the initial speed of the nitrogen molecule as it is "launched" vertically upward from the surface of the Earth.

Initially the nitrogen molecule has kinetic energy determined by the speed we've already found. It also has gravitational potential energy. When the nitrogen molecule reaches its maximum height, it momentarily comes to rest; therefore, it has no kinetic energy. It does, however, have gravitational potential energy. Recall that the gravitational potential energy associated with the configuration of two masses is given by the expression,

$$PE_{grav} = -\frac{Gm_1 m_2}{r}$$

Applying the principle of conservation of energy, we have,

$$E_i = E_f$$

$$KE_i + PE_{grav,i} = KE_f + PE_{grav,f}$$

$$\frac{1}{2} m_{N_2} v_i^2 - \frac{G m_{N_2} m_{Earth}}{r_i} = 0 - \frac{G m_{N_2} m_{Earth}}{r_f}$$

$$\frac{1}{2} v_i^2 - \frac{G m_{Earth}}{r_i} = -\frac{G m_{Earth}}{r_f}$$

Rearranging this expression to solve for r_f and substituting in for $r_i = R_{Earth}$ and v_i from our previous expression, we have,

$$\frac{1}{2} v_i^2 - \frac{G m_{Earth}}{r_i} = -\frac{G m_{Earth}}{r_f}$$

$$\rightarrow r_f = \frac{-G m_{Earth}}{\frac{1}{2} v_i^2 - \frac{G m_{Earth}}{r_i}} = \frac{-G m_{Earth}}{\frac{3RT}{2 M_{N_2}} - \frac{G M_{Earth}}{R_{Earth}}}$$

Simplifying the expression, we have,

$$r_f = \frac{2 G m_{Earth} M_{N_2} R_{Earth}}{2 G m_{Earth} M_{N_2} - 3 RT R_{Earth}}$$

Solve

Substituting in the numerical values of $G = 6.67 \times 10^{-11}$ Nm²/kg², $m_{Earth} = 5.98 \times 10^{-24}$ kg, $M_{N_2} = 0.028$ kg/mole, $R_{Earth} = 6.37 \times 10^6$ m, $R = 8.31$ J/mole $\cdot$ K, and $T = 298$ K, we have,

$$r_f = \frac{2 G m_{Earth} M_{N_2} R_{Earth}}{2 G m_{Earth} M_{N_2} - 3 RT R_{Earth}}$$

$$= \frac{2(6.67 \times 10^{-11}\,\text{N} \cdot \text{m}^2/\text{kg}^2)(5.98 \times 10^{24}\,\text{kg})(0.028\,\text{kg/mole})(6.37 \times 10^6\,\text{m})}{2(6.67 \times 10^{-11}\,\text{N} \cdot \text{m}^2/\text{kg}^2)(5.98 \times 10^{24}\,\text{kg})(0.028\,\text{kg/mole}) - 3(8.31\,\text{J/mole} \cdot \text{K})(298\,\text{K})(6.37 \times 10^6\,\text{m})}$$

$$= \boxed{6.38 \times 10^6\,\text{m}}$$

Therefore, the maximum height above the surface of the Earth which the nitrogen molecule travels is,

$$h = 6.38 \times 10^6\,\text{m} - 6.37 \times 10^6\,\text{m} = \boxed{1.0 \times 10^4\,\text{m}}$$

What does it mean?

The maximum height the nitrogen molecule will travel is about 10 km $\approx$ 6 mi above the surface of the Earth. This simple calculation tells us that the air must be much thinner at high altitudes!

Part E. MCAT Review Problems and Solutions

PROBLEMS

1. An ideal gas is confined to a container with a constant volume. The number of moles is held constant. If the absolute temperature of the gas is tripled, by what factor will the pressure change?

   (a) 1/6
   (b) 1/3
   (c) 3.0
   (d) 6.0

2. Suppose the number of moles and the temperature of an ideal gas are held constant. If the volume is increased to 5 times its original value, by what factor will the pressure change?

   (a) 0.2
   (b) 1.0
   (c) 5.0
   (d) 25.0

3. How many atoms are present in a 300-g piece of pure iron? (The atomic mass of iron is 56 g/mol and $N_A = 6.02 \times 10^{23}$/mol.)

   (a) 3.2×10^{19}
   (b) 3.2×10^{22}
   (c) 3.2×10^{24}
   (d) 3.2×10^{28}

4. Two moles of an ideal gas are at a pressure of 3.0 atm and a temperature of 10°C. The gas is then heated to 150°C while the volume is held constant. What is the final pressure?

   (a) 4.5 atm
   (b) 1.8 atm
   (c) 0.14 atm
   (d) 1.0 atm

5. The ideal gas law treats gasses as consisting of

   (a) chemicals.
   (b) molecules.
   (c) bubbles.
   (d) empty space.

6. Which of the following graphs best represents the relationship between the pressure and volume of an ideal gas if its temperature and number of moles are held constant?

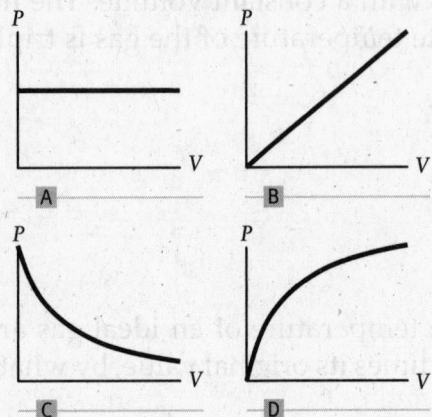

(a) A
(b) B
(c) C
(d) D

7. Which of the following graphs best represents the relationship between the pressure and temperature of an ideal gas if its volume and number of moles are held constant?

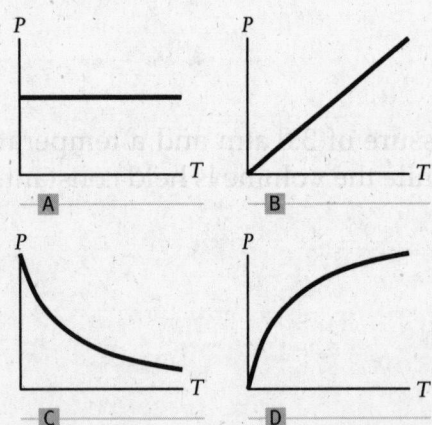

(a) A
(b) B
(c) C
(d) D

8. Two identical containers each contain 10 moles of an ideal gas. The temperature is the same for both gases. Container A holds helium (molecular mass = 4 g/mole) and container B holds oxygen (molecular mass = 16 g/mole). What is the relationship between the pressures in each container?

   (a) Container A has 4 times the pressure of container B.

   (b) Container A has 2 times the pressure of container B.

   (c) Both containers have the same pressure.

   (d) There is not enough information to determine the relationship.

9. If the temperature of an ideal gas in a container is increased,

   (a) the average speed of the molecules of the gas will increase.

   (b) the average speed of the molecules of the gas will decrease.

   (c) the distance between the molecules in the gas will increase.

   (d) the distance between the molecules in the gas will decrease.

10. Suppose you double the length of a box containing an ideal gas. The time it takes a molecule to diffuse across the length of the box will

   (a) also double.

   (b) increase by a factor of 4.

   (c) not change.

   (d) decrease by a factor of 2.

SOLUTIONS

1. *MCAT Strategies*

To answer this problem we simply need to know the ideal gas law, $PV = nRT$. If the number of moles and the volume of an ideal gas are held constant, then we can write the expression,

$$\frac{P}{T} = \frac{nR}{V} = \text{constant}$$

We can see from this expression that if we triple the absolute temperature of the gas, then the pressure must also triple so that the ratio remains constant. So, the correct answer is (c).

2. *MCAT Strategies*

Once again, solving this problem requires us to manipulate the ideal gas law. If the number of moles and the temperature of a gas are held constant, then we can write the expression,

$$PV = nRT = \text{constant}$$

We can see from this expression that the volume is increased to five times its original value, then the pressure must decrease to 1/5th = 0.2 times its initial value. So, the correct answer is (a).

3. MCAT Strategies

Since these answers are all similar, let's proceed with our usual problem-solving strategy.

Recognize the principle

To solve this problem we'll need to know the relationship between the number of particles and the number of moles of a substance, and know the meaning of atomic mass. We may also need to make some approximations as we do the calculations.

Sketch the problem

No sketch required.

Identify the relationships

The atomic mass is the mass of one mole of atoms. To calculate the mass of one atom we divide the atomic mass by Avogadro's number,

$$m_{iron} = \frac{M_{iron}}{N_A}$$

The total number of iron atoms in the piece of iron is simply the mass of the iron divided by the mass of one iron atom,

$$N = \frac{300\ g}{m_{iron}}$$

Combining these two expressions, we have,

$$N = \frac{300\ g}{m_{iron}} = \frac{300\ g}{M_{iron}} N_A$$

Solve

Substituting in the numerical values, we have,

$$N = \left(\frac{300\ g}{56\ g/mole}\right)(6.02 \times 10^{23}\ atoms/mole)$$

$$\approx \left(\frac{300 \times 6}{60}\right) \times 10^{23}\ atoms$$

$$\approx 3.0 \times 10^{24}\ atoms$$

So, the correct answer is (c).

What does it mean?

Notice the approximations we made ($6.02 \approx 6$ and $56 \approx 60$) helped to simplify the calculations and allowed us to zero in on the correct answer.

4. MCAT Strategies

Since we're holding the number of moles and the volume constant, we know that increasing the temperature of the gas will result in an increase in pressure. This is seen by examining the ideal gas law by writing it in the following way,

$$\frac{P}{T} = \frac{nR}{V} = constant$$

Since the initial pressure of the gas was 3 atm, the final pressure must be greater than 3 atm. The only answer which is greater than 3 atm is (a). So, the correct answer is (a).

5. MCAT Strategies

This question is simply testing our knowledge of the ideal gas law and its limitations. An ideal gas is a dilute gas of atoms or molecules so the correct answer is (b).

6. MCAT Strategies

Applying the ideal gas law we see that if the temperature and number of moles are held constant, then we have the following expression,

$$PV = nRT = \text{constant}$$

In other words,

$$P \propto \frac{1}{V}$$

So as the volume increases, the pressure decreases. The only graph which illustrates this relationship is (c). So, the correct answer is (c).

7. MCAT Strategies

Once again, by examining the ideal gas law for the situation where the volume and number of moles are held constant, we have the following expression,

$$\frac{P}{T} = \frac{nR}{V} = \text{constant}$$

In other words,

$$P \alpha T$$

So as the temperature increases, the pressure increases as well. The only graph which illustrates this relationship is (b). So, the correct answer is (b).

8. MCAT Strategies

Once again, let's look carefully at the ideal gas law,

$$PV = nRT$$

Rearranging this expression to solve for the pressure, we have,

$$P = \frac{nRT}{V}$$

Now, in the statement of the problem the number of moles, the temperature, and the volume of the two containers are all the same. Therefore, the two gases must be at the same pressure. So, the correct answer is (c).

9. MCAT Strategies

Even if we can't remember the exact expression, we know that the average speed of the molecules in an ideal gas is related to the temperature of the gas. As you increase the temperature of the gas, the average speed of the molecules will also increase. It may also be true that the distance between the molecules increases or decreases depending on the nature of the container; however, we cannot definitively conclude either (c) or (d) as true without additional information. So, the correct answer is (a).

10. *MCAT Strategies*

To answer this question we must know the relationship between the diffusion time and average net distance of the molecule,

$$\Delta r = \sqrt{Dt}$$

$$\rightarrow t = \frac{(\Delta r)^2}{D}$$

where Δr is the average net distance the molecules move, D is the diffusion constant, and t is the time. We see from this expression that if you double Δr, then t will increase by a factor of 4. So, the correct answer is (b).

16 Thermodynamics

Part A. Summary of Key Concepts and Problem-Solving Strategies

KEY CONCEPTS

Thermodynamics is about the properties of systems

Thermodynamics is concerned with the behavior of systems containing many particles such as a liquid, a gas, or a solid. Such systems interact with their environment, and the laws of thermodynamics describe these interactions.

Zeroth law of thermodynamics

If two systems A and B are both in thermal equilibrium with a third system, then A and B are in thermal equilibrium with each other.

First law of thermodynamics

If an amount of heat Q flows *into* a system from its environment and an amount of work W is done *by* the system on its environment, the internal energy of the system changes by an amount

$$\Delta U = U_f - U_i = Q - W$$

This is a statement about the conservation of energy as it applies to a thermodynamic system.

Second law of thermodynamics

The second law of thermodynamics can be stated in several different ways.

- *Second law of thermodynamics (formulation involving heat flow):* Heat flows spontaneously from a warm body to a colder one. It is not possible for heat to flow spontaneously from a cold body to a warmer one.

- *Second law of thermodynamics (formulation involving heat engines):* The efficiency of a reversible heat engine operating between reservoirs at temperatures T_H and T_C is

$$e = 1 - \frac{T_C}{T_H}$$

  No heat engine can have a greater efficiency.

- *Second law of thermodynamics (entropy version):* In any thermodynamic process the total change in entropy of the universe must be greater than or equal to zero, $\Delta S_{universe} \geq 0$.

Third law of thermodynamics

It is impossible for the temperature of a system to reach absolute zero.

APPLICATIONS

Thermodynamic processes

A *thermodynamic process* takes a system from some initial state with an initial temperature T_i, pressure P_i, and volume V_i to a final state described by T_f, P_f, and V_f. Such a process can be described as a path in a *P-V diagram*. The work done by the system on its environment is the area under this P-V curve. An *isothermal process* is one in which the temperature is constant ($T_i = T_f$). In an *adiabatic process*, there is no heat flow into or out of the system ($Q = 0$). An *isochoric process* is one in which the volume remains constant, and in an *isobaric process* the pressure does not change.

Heat engines

A *heat engine* takes heat energy from a hot reservoir, expels heat energy to a cold reservoir, and does an amount of work on its environment. The *efficiency of a heat engine* is defined as

$$e = \frac{W}{Q_H} = \frac{Q_H - Q_C}{Q_H}$$

The efficiency of a thermodynamically reversible heat engine operating between two temperatures T_C and T_H is given by the laws of thermodynamics as

$$e = 1 - \frac{Q_C}{Q_H} = 1 - \frac{T_C}{T_H}$$

No heat engine can have a greater efficiency.

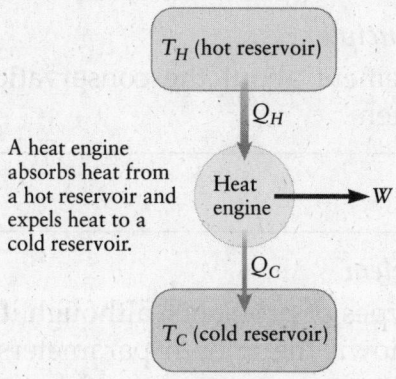

A heat engine absorbs heat from a hot reservoir and expels heat to a cold reservoir.

Refrigerators

In a *refrigerator,* work is done by the environment on the system, which then extracts heat from a cold reservoir and expels heat to a hot reservoir. *The efficiency of a refrigerator* is $e_{refrig} = Q_C/W$. A thermodynamically reversible refrigerator has an efficiency of

$$e_{refrig} = \frac{T_C/T_H}{1 - T_C/T_H}$$

No refrigerator can have a greater efficiency.

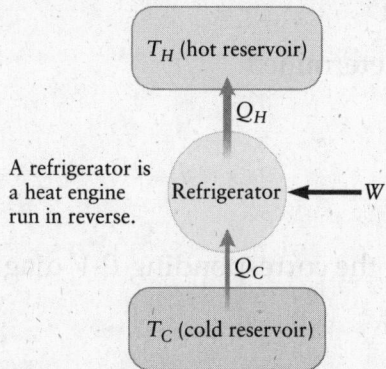

A refrigerator is a heat engine run in reverse.

PROBLEM-SOLVING STRATEGIES

One of the key principles discussed in this chapter is the first law of thermodynamics. This law is a statement about the conservation of energy as it applies to a thermodynamic system. As we have seen before, the principle of conservation of energy is an extremely powerful tool which proves useful in the study of systems which interact by exchanging heat energy. With the first law of thermodynamics we have generalized the concept to include work done by the system. As you tackle problems involving the first law of thermodynamics, the following schematic will help guide you through the thought process involved in applying this principle.

Problem Solving: The First Law of Thermodynamics

Recognize the principle

The first law of thermodynamics is a statement about the conservation of energy as it applies to a thermodynamic system.

Sketch the problem

Usually a sketch is not necessary for these types of problems, although it may be useful to sketch the system and write down the known parameters and given information. These can include temperature, pressure, volume, heat, work, and internal energy.

Identify the relationships

The first law of thermodynamics is given by the expression,

$$\Delta U = U_f - U_i = Q - W$$

Positive Q is heat flow into the system while negative Q is heat flow out of the system. Similarly, positive W is work done by the system on its environment, while negative W is work done on the system.

For many processes, variables can easily be determined:

Adiabatic: $Q = 0$

Isobaric (constant pressure): $W = P\Delta V$

Isochoric (constant volume): $W = 0$

Cyclic: W = area enclosed by the path on the corresponding P-V diagram.

Isothermal (ideal gas): $W = nRT\ln\left(V_f/V_i\right)$

Solve

Using the expressions above and the given information, we can solve for the unknowns.

What does it mean?

Always *consider what your answer means*, and check that it makes sense.

Part B. Frequently Asked Questions

1. *When applying the first law of thermodynamics, how do we know if the work is positive or negative?*

As you may recall, we were very specific in how we defined work. Positive work is work done by the system on the environment. So, if the system is doing work on the environment, then W is positive. If the environment is doing work on the system, then W is negative. For all of the thermodynamic processes we've discussed, work involved the change in volume of our system. If the volume of our system increases during a process, then our system is doing work on the environment and thus W is positive. If the volume of our system decreases, then the environment is doing work on our system and thus W is negative.

2. *There are so many processes discussed in this chapter that I have trouble remembering what they all mean. What's the difference between isothermal, isobaric, isochoric, and adiabatic?*

The prefix "iso" is from the Greek "isos" meaning equal or constant. Also, it's easy to remember that thermal relates to temperature, baric relates to pressure (as in barometric pressure), and choric is another name for volumetric which relates to volume. With this in mind we have the following:

Isothermal → Constant Temperature
Isobaric → Constant Pressure
Isochoric → Constant Volume

The term adiabatic is from the Greek word for impassible, referring here to the inability of heat to pass. So, an adiabatic process is one in which no heat enters or exits the system,

Adiabatic → $Q = 0$

By remembering and understanding these definitions, we are able to easily apply the first law of thermodynamics to these processes and solve for unknowns.

3. *I know that the area under the curve on a P-V diagram is the work done by the system, but how do we know if it's positive or negative?*

Well, as was described in FAQ #1, if the volume of the system increases then the work will be positive, and if the volume of the system decreases then the work will be negative. The following figures help to illustrate this point:

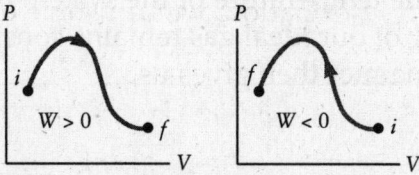

4. *What is the meaning of internal energy? Isn't it just the average kinetic energy of the molecules in a system?*

As we initially defined in Chapter 14, the internal energy of a system is the energy associated with all the particles of a system. This energy includes both the kinetic energy of the particles and the potential energy associated with the chemical bonds or interactions between the particles. This can include vibrational motions which we can model as a spring with spring-potential energy. Though it has not been discussed thus far, if molecules are free to rotate, this rotational energy contributes to the internal energy of the system as well. If the entire system is moving as a whole, we do not include this kinetic energy in the internal energy.

5. *When determining Q, do we need to consider only the transfer of heat by direct contact?*

No. The heat transferred into or out of the system can happen in a number of ways including by direct contact. Recall from Chapter 14 the heat flow can occur through direct contact, by convection, and through radiative transfer. All these forms of heat transfer must be included in Q.

6. *In other textbooks I've seen the first law of thermodynamics written as $\Delta U = Q + W$. Which way is correct?*

Actually, both equations are correct. Remember, the first law of thermodynamics is really a statement about the conservation of energy. In this textbook we defined positive work as work done by the system on the environment. By defining work this way we see that if positive work is performed, the result will be a decrease in the internal energy of the system. On the other hand, if negative work is performed, then the internal energy of the system will increase. Applying the principle of conservation of energy, we get the form of the first law of thermodynamics used in this textbook,

$$\Delta U = Q - W$$

It is also quite common to define positive work as work done on the system. By defining it this way we see that if positive work is performed, the result will be an increase in the internal energy of the system. If negative work is performed, then the internal energy of the system will decrease. Applying the principle of conservation of energy, the first law of thermodynamics would be,

$$\Delta U = Q + W$$

Both expressions are valid, but we must be careful in how we define work!

7. *For an ideal gas, since the temperature doesn't change for an isothermal process, doesn't that mean that $Q = 0$?*

No. An isothermal process is one in which the temperature of the system remains constant. This means that the internal energy of our ideal gas remains constant, so $\Delta U = 0$. Applying the first law of thermodynamics then gives us,

$$\Delta U = Q - W$$
$$0 = Q - W \rightarrow Q = W$$

So, the heat flow Q is equal to the work done by the system. The process for which $Q = 0$ is called an adiabatic process.

8. *Is there any real process which is reversible?*

No. Although all real processes are irreversible, there are some processes which are almost reversible. If a real process is performed very slowly so that at all times the system is nearly in equilibrium, then this process is very nearly reversible. For example, consider a gas which expands as it slowly absorbs heat from the environment. This process could be reversed so that the gas contracts as heat energy slowly flows out into the environment. Of course, in any real process there are dissipative effects such as friction and turbulence which convert some of the energy into forms which cannot be recovered. Also, nearly reversible processes are inherently slow, making them impractical for use in a real system.

9. *Shouldn't the maximum efficiency of a heat engine depend on many factors, not just the temperatures of the hot and cold reservoirs?*

No. In general, the efficiency of a heat engine is given by the expression,

$$e = 1 - \frac{Q_C}{Q_H}$$

Certainly Q_C and Q_H can depend on many factors in addition to the temperatures of the hot and cold reservoirs. However, the maximum possible efficiency for a heat engine occurs when the engine is operated as a Carnot engine. Carnot showed by using only reversible processes that the following must be true,

$$\frac{Q_C}{Q_H} = \frac{T_C}{T_H}$$

Therefore, the efficiency of a Carnot engine is given by the expression,

$$e = 1 - \frac{T_C}{T_H}$$

Since the Carnot engine has the maximum possible efficiency for a heat engine operating between the same two temperature extremes, the maximum efficiency of a heat engine only depends on the temperatures of the hot and cold reservoirs.

10. *Suppose that two systems are brought into thermal contact and heat flows from one system to another system. If the systems are isolated from their surroundings, since the entropy of the first system decreases by ΔS, does that mean that the entropy of the second system increases by ΔS?*

No. In fact, the entropy increase of the hotter system is greater than the entropy decrease of the colder system. So, the total entropy change is greater than zero. To illustrate this let's consider the following schematic:

Now, the Q flowing out of the hotter system is equal to the Q flowing into the colder system, but $T_H > T_C$. Therefore, $|\Delta S_C| > |\Delta S_H|$ resulting in $\Delta S_{total} > 0$.

Part C. Selection of End-of-Chapter Answers and Solutions

QUESTIONS

Q16.8 Consider a thermodynamic process in which an ice cube melts to become a liquid (water). Is the change of the internal energy of the H_2O positive or negative? Is the change of the entropy of the H_2O positive or negative?

Answer

The internal energy of the H_2O increases when the ice melts to liquid water. Since heat is transferred to the ice at constant temperature (273 K), the change in entropy of the water is positive. This is clear when we think in terms of disorder. Since H_2O in a liquid state (water) is more disordered than when it is solid (ice), the entropy of the water is greater than the ice.

Q16.18 Explain why heat pumps that use air as the cold reservoir are most efficient in mild climates. (These are called "air-to-air" heat pumps.) Why does a heat pump that uses water from deep underground as the cold reservoir have a thermodynamic advantage over air-to-air heat pumps when the weather is very cold?

Answer

The efficiency of a heat pump is given by the expression,

$$e_{pump} = \frac{1}{\left(1 - \frac{T_C}{T_H}\right)}$$

This will be largest when T_C and T_H are closest in value. Thus, a heat pump will work best when its cold reservoir is close to the temperature inside the house. In temperate climates, the air doesn't get too cold, so this is a perfectly effective cold reservoir. However, in harsh climates, the air can get very cold, lowering the efficiency. Water pumped from deep underground will tend to be warmer than the air, making the pump more efficient.

PROBLEMS

P16.3 The temperature of an ideal monatomic gas increases by 100 K while 3000 J of heat are added to the gas. If the gas contains 4.0 moles, how much work does it do on its surroundings?

Solution

Recognize the principle

The internal energy increase can be determined by the rise in temperature. Any heat not involved in this increase must do work on the surroundings.

Sketch the problem

No sketch required.

Identify the relationships

The change in internal energy (ΔU) for a monatomic ideal gas with a change in temperature is given by the expression,

$$\Delta U = \frac{3}{2} nR\Delta T$$

The first law of thermodynamics also defines this change as the difference between the heat transferred to the system (Q) and the work done by the system (W),

$$\Delta U = Q - W$$

Solve

Setting these two expressions equal and solving for the work done yields:

$$Q - W = \frac{3}{2} nR\Delta T \quad \rightarrow \quad W = Q - \frac{3}{2} nR\Delta T$$

Then inserting the numerical values, we have,

$$W = 3000 \text{ J} - \frac{3}{2}(4 \text{ mol})(8.31 \text{ J/mol} \cdot \text{K})(100 \text{ K}) = -1986 \text{ J} \approx \boxed{-2000 \text{ J}}$$

What does it mean?

The negative work implies that work must be done on this system rather than the system doing work on its surroundings. This means that in order to see this temperature change, the gas must also have 2000 J of work done on it in addition to the 3000 J of heat transferred to it. This could be done, for instance, by compressing the gas to a smaller volume as heat is transferred to it.

P16.17 Consider a balloon of volume 2.0 m³ that contains 3.5 moles of helium gas at a temperature of 300 K. If the gas is compressed adiabatically so as to have a final temperature of 400 K, what is the work done by the gas?

Solution

Recognize the principle

In an adiabatic process, no heat enters or leaves the system. According to the first law of thermodynamics, the change in internal energy is therefore equal in magnitude but opposite in sign to the work done.

Sketch the problem

No sketch required.

Identify the relationships

The first law of thermodynamics tells us

$$\Delta U = Q - W$$

and, for a monatomic ideal gas,

$$\Delta U = \frac{3}{2} nR\Delta T$$

Solve

For an adiabatic process, $Q = 0$ and

$$\Delta U = -W$$

Therefore,

$$W = -\Delta U = -\frac{3}{2} nR\Delta T$$

Substituting in the numerical values, we have,

$$W = -\frac{3}{2} nR\Delta T = -(1.5)(3.5 \text{ mole})(8.31 \text{ J/mole} \cdot \text{K})(400 \text{ K} - 300 \text{ K})$$

$$= \boxed{-4400 \text{ J}}$$

What does it mean?

The work done by the gas is negative, as it should be since this is a compression and work is done on the gas.

P16.23 Figure P16.23 shows a system that is compressed from an initial state *i* to a final state *f*. If the internal energy of the system decreases by 5000 J, approximately how much heat was added to the system?

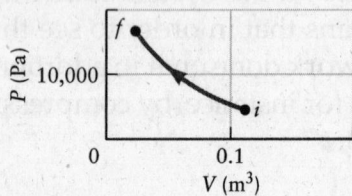

Figure P16.23

Solution

Recognize the principle

The heat transferred, internal energy, and work done on the system are related by the first law of thermodynamics. The work done on the system can be found by approximating the area under the *P-V* curve.

Sketch the problem

We can estimate the area under the *P-V* curve by approximating the curve as a straight line:

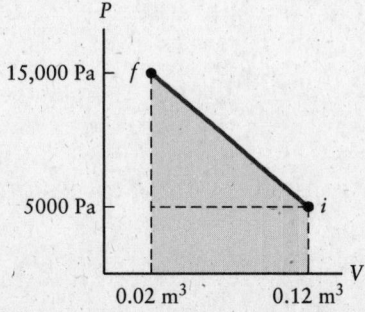

Identify the relationships

We can approximate this area, which represents the work, as the combination of a square and a triangle, as shown in the sketch. The first law of thermodynamics then relates the change in internal energy (ΔU) to the added heat (Q) and the work done (W) according to the expression,

$$\Delta U = Q - W$$

Solve

We approximate the work done on the system as:

$$W = -[(\text{area of the triangle}) + (\text{area of the rectangle})]$$
$$= -\left[\frac{1}{2}(0.12 \text{ m}^3 - 0.02 \text{ m}^3)(15000 \text{ Pa} - 5000 \text{ Pa}) + (0.12 \text{ m}^3 - 0.02 \text{ m}^3)\,(5000 \text{ Pa})\right]$$
$$= -[500 \text{ J} + 500 \text{ J}] = -1000 \text{ J}$$

Then solving the first law of thermodynamics for the heat transferred, and inserting the numerical value found for the work and the change in internal energy given in the problem, we have

$$Q = \Delta U + W = -5000 \text{ J} - 1000 \text{ J} = \boxed{-6000 \text{ J}}$$

What does it mean?

Approximately 6000 J of heat energy is transferred from the system. Since the gas is being compressed, work is done on the system, and this would tend to increase the internal energy. If the internal energy actually decreases, more heat energy must be transferred out of the system.

P16.31   Assume that the gasoline engine in your car is a heat engine operating between a hot reservoir at 800 K and a cold reservoir at 280 K and that your engine produces a peak power output of 250 hp (horsepower). If the temperature of the hot reservoir is increased to 900 K, what is the theoretical peak power output?

Solution

Recognize the principle

To estimate peak power outputs, we treat the car engine as a Carnot engine. The work done by a heat engine depends on the temperatures of the two reservoirs. Based on the efficiencies calculated from the temperature ratios, we can also find the ratio of the power outputs.

Sketch the problem

No sketch required.

Identify the relationships

The maximum efficiency for an ideal heat engine is given by the expression,

$$e = 1 - \frac{T_C}{T_H}$$

The efficiency can also be written in the form,

$$e = \frac{W}{Q_H}$$

Solve

When the hot reservoir is at 800 K, the engine's maximum efficiency is

$$e_{800} = 1 - \frac{280 \text{ K}}{800 \text{ K}} = 0.65$$

But at 900 K, the maximum efficiency is

$$e_{900} = 1 - \frac{280 \text{ K}}{900 \text{ K}} = 0.69$$

If we then create a ratio of these two efficiencies and use our definition of efficiency in terms of the work done and heat removed from the hot reservoir, we have,

$$\frac{e_{800}}{e_{900}} = \frac{0.65}{0.69} = \frac{\dfrac{W_{800}}{Q_H}}{\dfrac{W_{900}}{Q_H}} = \frac{W_{800}}{W_{900}}$$

And, since the work done is just the power multiplied by time, the ratio must also hold for power. That is,

$$\frac{e_{800}}{e_{900}} = \frac{0.65}{0.69} = \frac{\dfrac{W_{800}}{t}}{\dfrac{W_{900}}{t}} = \frac{P_{800}}{P_{900}}$$

Solving this expression for the power at 900 K, and inserting the numerical values, we have,

$$P_{900} = P_{800}\frac{e_{900}}{e_{800}} = 250 \text{ hp}\left(\frac{0.69}{0.65}\right) = \boxed{270 \text{ hp}}$$

What does it mean?

This higher temperature reservoir increases the power output by about 6%. Increasing engine temperature can be used to increase efficiency, but the power gains are limited by the need for materials that can safely withstand higher temperatures.

P16.44 A heat engine absorbs 8000 J of heat from a hot reservoir and produces 5500 J of work. What is the efficiency of the engine?

Solution

Recognize the principle

The efficiency can be found from the ratio of the work done to the heat transferred from the hot reservoir.

Sketch the problem

No sketch required.

Identify the relationships

The efficiency of a heat engine is given by the expression,

$$e = \frac{W}{Q_H}$$

Solve

Substituting in the numerical values, we have,

$$e = \frac{W}{Q_H} = \frac{5500 \text{ J}}{8000 \text{ J}} = \boxed{0.69}$$

What does it mean?

This heat engine converts about 69% of the heat taken from the hot reservoir into useful work.

P16.45 Suppose the refrigerator in your house is a reversible refrigerator. Estimate how much heat is extracted from the fresh-food compartment when 1000 J of work are done. *Hint*: You will first need to identify the hot and cold reservoirs and estimate their temperatures. Ignore the freezer compartment.

Solution

Recognize the principle

A refrigerator extracts heat from a cold reservoir (the refrigerator) and deposits heat to the hot reservoir (the room). For a reversible refrigerator, the ratio of the heat removed from the cold reservoir to that expelled to the hot reservoir is the same as the ratio of absolute temperatures.

Sketch the problem

No sketch required.

Identify the relationships

For a refrigerator, the work done and heats absorbed/expelled are related by the expression,

$$W + Q_C = Q_H$$

The ratio of temperatures and heats is given by the expression,

$$\frac{Q_C}{Q_H} = \frac{T_C}{T_H}$$

Solve

We can estimate that the cold reservoir (refrigerator) is about 3°C (276 K), while the hot reservoir (room) is about 20°C (293 K). Using our estimated temperatures, we can then find the heat extracted from the fresh-food compartment in terms of the heat deposited into the hot reservoir,

$$\frac{Q_C}{Q_H} = \frac{276 \text{ K}}{293 \text{ K}}$$
$$Q_H = (1.06)Q_C$$

We can then substitute this definition into our work/heat equation, along with the given work, and solve for the heat removed from the cold reservoir.

$$1000 \text{ J} + Q_C = 1.06 \, Q_C$$

$$Q_C = 16200 \text{ J} \approx \boxed{20{,}000 \text{ J}}$$

What does it mean?

About 1000 J of work by a compressor on the back of a refrigerator can remove about 16200 J of energy from the fresh-food compartment. A total of 17200 J of heat would then be dumped into the room.

P16.47   Suppose 7000 J of heat flows out of a house that is at 293 K to the outside environment at 273 K. Find the change in entropy of the house, the change in the entropy of the environment, and the total change in entropy of the universe. Is this process reversible or irreversible?

Solution

Recognize the principle

Entropy is calculated from the amount of heat transferred at a given temperature.

Sketch the problem

No sketch required.

Identify the relationships

The change in entropy in each case is given by the expression,

$$\Delta S = \frac{Q}{T}$$

We can use this expression to find the change in entropy for each heat transition. The sum of these changes in entropy is then the change in entropy of the universe. If this total change is greater than zero, then the process is irreversible.

Solve

The change in entropy of the house is:

$$\Delta S_{\text{house}} = \frac{\Delta Q}{T_{\text{inside}}} = \frac{-7000 \text{ J}}{293 \text{ K}} = -23.89 \text{ J/K} \approx \boxed{-24 \text{ J/K}}$$

The change in entropy of the environment is:

$$\Delta S_{\text{environment}} = \frac{\Delta Q}{T_{\text{outside}}} = \frac{7000 \text{ J}}{273 \text{ K}} = 25.64 \text{ J/K} \approx \boxed{+26 \text{ J/K}}$$

The change in entropy of the universe is then:

$$\Delta S_{\text{universe}} = \Delta S_{\text{house}} + \Delta S_{\text{environment}} = -23.89 \text{ J/K} + 25.64 \text{ J/K} = 1.75 \text{ J/K}$$

$$\approx \boxed{+1.8 \text{ J/K}}$$

What does it mean?

Since the total entropy change is greater than zero, this is an ⟨irreversible process⟩.

P16.54   Figure P16.54 shows the *P-V* diagrams for a variety of different processes involving an ideal gas. (a) Which diagrams describe an expansion? (b) Which diagrams describe a compression? (c) Which diagrams might describe an adiabatic process? (d) Which diagrams might describe an isothermal process?

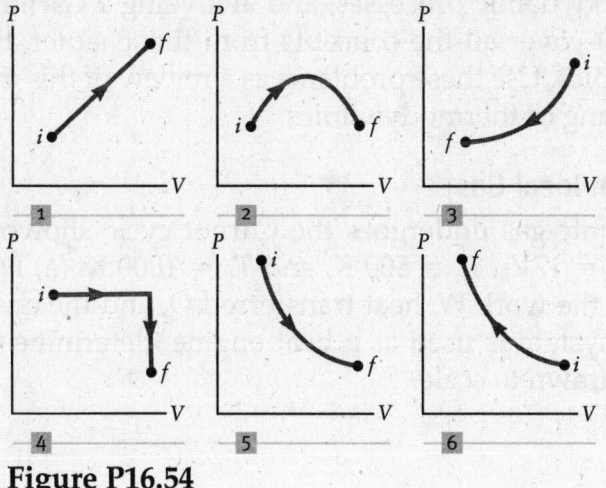

Figure P16.54

Solution

Recognize the principle

By analyzing the relationship between *P* and *V* in the diagrams, and using our ideal gas equation $PV = nRT$, we can determine the likely process.

Sketch the problem

The sketch is Figure P16.54, provided in the problem.

Identify the relationships and Solve

(a) An expansion is any process in which the volume increases. For processes (1), (2), (4), and (5), the volume is increasing, so they represent an expansion.

(b) A compression is any process in which the volume decreases. For processes (3) and (6), the volume is decreasing, so they represent a compression.

(c) The expression for an adiabatic process is PV^γ = constant, and is represented on a *P-V* diagram as a negative slope, upward curving line. Figures (5) and (6) are shaped like this, therefore could represent an adiabatic process.

(d) The expression for an isothermal process is $PV = nRT$ = constant, therefore $P \propto 1/V$. This is represented on a *P-V* diagram as a negative slope, upward curving line. Figures (5) and (6) are shaped like this, therefore could represent an isothermal process.

What does it mean?

The shape of a *P-V* curve provides us with information on the process.

Part D. Additional Worked Examples and Capstone Problems

The following worked example and capstone problems provide you with additional practice applying the first law of thermodynamics, improving your understanding of the various thermodynamic processes, and analyzing a Carnot cycle. Though these problems do not cover all the concepts from this chapter, they do focus on several of the main topics. Use these problems as a review of this material and to gauge your understanding of thermodynamics.

WE 16.1 The Carnot Cycle for an Ideal Gas

One mole of an ideal monatomic gas undergoes the Carnot cycle shown in the figure. $V_A = 3V_B$, $V_C = 6V_B$, $V_D = 17V_B$, $T_A = 500$ K, and $T_B = 1000$ K. (a) For each process in the cycle determine the work W, heat transferred Q, and the change in internal of the gas. (b) If this system is used as a heat engine, determine its efficiency. *Note*: The figure is not drawn to scale.

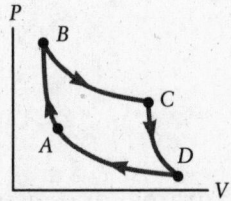

Solution

Recognize the principle

A Carnot cycle is composed of two adiabatic processes and two isothermal processes. We can combine the first law of thermodynamics with the ideal gas law to determine the work W, heat Q, and change in internal energy ΔU for each process.

Sketch the problem

Let's begin by redrawing our cycle and labeling the processes.

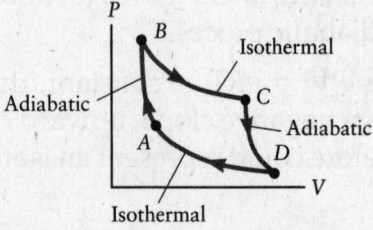

Identify the relationships

(a) Since we are given the temperature of our system at points A and B and the volume at point A, we can apply the ideal gas law to relate the temperature, pressure, and volume for all points.

$$\frac{PV}{T} = nR = \text{constant}$$

$$\rightarrow \frac{P_A V_A}{T_A} = \frac{P_B V_B}{T_B} = \frac{P_C V_C}{T_C} = \frac{P_D V_D}{T_D}$$

Also, since for an isothermal process $T = \text{constant}$, we have that $T_B = T_C$ and $T_D = T_A$. Combining these expressions with the first law of thermodynamics we can tackle this problem.

Process A to B

This is an adiabatic process so $Q = 0$. Applying the first law of thermodynamics then gives us,

$$W = -\Delta U$$

For an ideal monatomic gas the internal energy is given by the expression,

$$U = \frac{3}{2} N k_B T = \frac{3}{2} nRT$$

Therefore, the change in the internal energy is given by,

$$\Delta U = \frac{3}{2} nR\Delta T = \frac{3}{2} nR(T_B - T_A)$$

So, in going from A to B we have,

$$Q = 0$$
$$\Delta U = \frac{3}{2} nR(T_B - T_A)$$
$$W = -\frac{3}{2} nR(T_B - T_A)$$

Process B to C

This is an isothermal process so $\Delta T = 0$. Since this is an ideal monatomic gas, this also means that,

$$\Delta U = 0$$

Applying the first law of thermodynamics then gives us,

$$Q = W$$

For an ideal gas, the work done in an isothermal process is given by the expression,

$$W = nRT \ln\left(\frac{V_f}{V_i}\right) = nRT \ln\left(\frac{V_C}{V_B}\right)$$

Since $V_C = 6V_B$ we have,

$$W = nRT_B \ln(6)$$

So, in going from B to C we have,

$$Q = nRT_B \ln(6)$$
$$\Delta U = 0$$
$$W = nRT_B \ln(6)$$

Process C to D

This is an adiabatic process so $Q = 0$. Applying the first law of thermodynamics then gives us,

$$W = -\Delta U$$

For an ideal monatomic gas the internal energy is given by the expression,

$$U = \frac{3}{2} N k_B T = \frac{3}{2} nRT$$

Therefore, the change in the internal energy is given by

$$\Delta U = \frac{3}{2} nR \Delta T = \frac{3}{2} nR(T_D - T_C)$$

But, $T_D = T_A$ and $T_C = T_B$, therefore,

$$\Delta U = \frac{3}{2} nR(T_A - T_B)$$

So, in going from C to D, we have,

$$Q = 0$$
$$\Delta U = \frac{3}{2} nR(T_A - T_B)$$
$$W = -\frac{3}{2} nR(T_A - T_B)$$

Process D to A

This is an isothermal process so $\Delta T = 0$. Since this is an ideal monatomic gas, this also means that,

$$\Delta U = 0$$

Applying the first law of thermodynamics then gives us,

$$Q = W$$

For an ideal gas, the work done in an isothermal process is given by the expression,

$$W = nRT \ln\left(\frac{V_f}{V_i}\right) = nRT \ln\left(\frac{V_A}{V_D}\right)$$

Since $V_A = 3V_B$ and $V_D = 17 V_B$, we have,

$$W = nRT_A \ln\left(\frac{3}{17}\right)$$

So, in going from D to A, we have,

$$Q = nRT_A \ln\left(\frac{3}{17}\right)$$
$$\Delta U = 0$$
$$W = nRT_A \ln\left(\frac{3}{17}\right)$$

(b) Since the lowest temperature is along the isothermal process from D to A and the highest temperature along B to C, the efficiency of the engine is given by the expression,

$$e = 1 - \frac{T_{\text{cold}}}{T_{\text{hot}}} = 1 - \frac{T_A}{T_B}$$

Solve

(a) Substituting in the numerical values, we have,

Process A to B

$$Q = \boxed{0}$$

$$\Delta U = \frac{3}{2}nR(T_B - T_A) = \frac{3}{2}(1 \text{ mole})(8.31 \text{ J/mole} \cdot \text{K})(500 \text{ K}) = \boxed{6.23 \times 10^3 \text{ J}}$$

$$W = -\frac{3}{2}nR(T_B - T_A) = \boxed{-6.23 \times 10^3 \text{ J}}$$

Process B to C

$$Q = nRT_B \ln(6) = (1 \text{ mole})(8.31 \text{ J/mole} \cdot \text{K})(1000 \text{ K})\ln(6) = \boxed{1.49 \times 10^4 \text{ J}}$$

$$\Delta U = \boxed{0}$$

$$W = nRT_B \ln(6) = \boxed{1.49 \times 10^4 \text{ J}}$$

Process C to D

$$Q = \boxed{0}$$

$$\Delta U = \frac{3}{2}nR(T_A - T_B) = \frac{3}{2}(1 \text{ mole})(8.31 \text{ J/mole} \cdot \text{K})(-500 \text{ K}) = \boxed{-6.23 \times 10^3 \text{ J}}$$

$$W = -\frac{3}{2}nR(T_A - T_B) = \boxed{6.23 \times 10^3 \text{ J}}$$

Process D to A

$$Q = nRT_A \ln\left(\frac{3}{17}\right) = (1 \text{ mol})(8.31 \text{ J/mole} \cdot \text{K})(500 \text{ K}) \ln\left(\frac{3}{17}\right) = \boxed{-7.21 \times 10^3 \text{ J}}$$

$$\Delta U = 0$$

$$W = nRT_A \ln\left(\frac{3}{17}\right) = \boxed{-7.21 \times 10^3 \text{ J}}$$

(b) Substituting in the numerical values we have,

$$e = 1 - \frac{T_A}{T_B} = 1 - \frac{500 \text{ K}}{1000 \text{ K}} = \boxed{0.50}$$

What does it mean?

There are a couple of important things to note. First, notice that the total change in the internal energy of the gas is zero. This must be the case for a cyclic process since the system ends up back where it started. Second, if we add up the total work for the cycle and divide it by the heat entering the system from B to C (this is Q_H), we would get an efficiency of 0.5 which agrees with our result in part (b).

CP 16.1 The Cyclic Process

Suppose that 1.00 mole of an ideal monatomic gas is taken through the cycle shown. (a) What is the net work done by the gas during one cycle? (b) Determine the net heat flow into the gas during one cycle. (c) If this system is used as a heat engine, calculate its efficiency. (d) Determine the ideal efficiency of a heat engine operated between the same two temperature extremes.

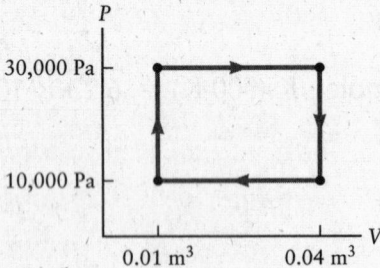

Solution

Recognize the principle

We can combine the first law of thermodynamics with the ideal gas law to determine the work W, heat Q, and change in internal energy ΔU for each process. However, the net work can be determined from the area enclosed by the cycle on the P-V diagram and since the change in internal energy is zero for a complete cycle, then $Q_{net} = W_{net}$.

Sketch the problem

Let's begin by redrawing the diagram to include where the heat flows into and out of the system. We'll also label the points in the figure. Heat flows into the system during the isochoric pressure increase and during the isobaric expansion, and out of the system during the other two processes. Therefore, we have the following:

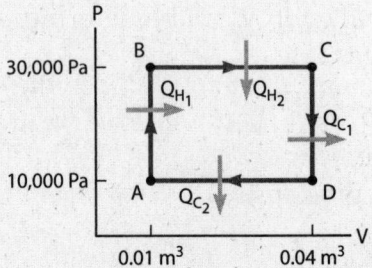

Identify the relationships

(a) The net work done by the gas during one cycle is simply the area enclosed by the curve,

$$W_{net} = \text{area enclosed} = \Delta P \Delta V = (P_B - P_A)(V_D - V_A)$$

(b) Since the change in internal energy of the gas is zero for one complete cycle, the net heat flow is the net work,

$$Q_{net} = W_{net} = \Delta P \Delta V = (P_B - P_A)(V_D - V_A)$$

(c) To determine the efficiency of the heat engine we'll need to determine the total Q_H, where $Q_H = Q_{H_1} + Q_{H_2}$. To find this we can calculate the heat flow into the system during the process from A to B and add that to the heat flow into the system during the process from B to C.

Process A to B

During this process the work is zero since the volume does not change. Applying the first law of thermodynamics, we have,

$$\Delta U = Q - W = Q - 0 \rightarrow Q = \Delta U$$

Since our system is a monatomic ideal gas, the change in internal energy is given by,

$$\Delta U = \frac{3}{2} N k_B \Delta T = \frac{3}{2} n R \Delta T$$

From the ideal gas law, we have,

$$PV = nRT \rightarrow \Delta(PV) = nR\Delta T$$

Combining these expressions, we have,

$$Q = \Delta U = \frac{3}{2} n R \Delta T = \frac{3}{2} \Delta(PV) = \frac{3}{2} V \Delta P$$

Writing this in terms of our variables, we have,

$$Q = Q_{H_1} = \frac{3}{2} V_A (P_B - P_A)$$

Process B to C

During this process the work is not zero since the volume is changing, but can be calculated as follows,

$$W = P \Delta V = P_B (V_C - V_B)$$

As in the previous process, the change in internal energy can be determined from the change in temperature and using the ideal gas law,

$$\Delta U = \frac{3}{2} n R \Delta T = \frac{3}{2} \Delta(PV) = \frac{3}{2} P \Delta V = \frac{3}{2} P_B (V_C - V_B)$$

Applying the first law of thermodynamics then gives us,

$$Q = Q_{H_2} = \Delta U + W = \frac{3}{2} P_B (V_C - V_B) + P_B (V_C - V_B) = \frac{5}{2} P_B (V_C - V_B)$$

So the heat flow into the system is the sum of the two,

$$Q_H = \frac{3}{2} V_A (P_B - P_A) + \frac{5}{2} P_B (V_C - V_B)$$

And the efficiency of the heat engine is given by,

$$e = \frac{W}{Q_H} = \frac{(P_B - P_A)(V_D - V_A)}{\frac{3}{2} V_A (P_B - P_A) + \frac{5}{2} P_B (V_C - V_B)}$$

(d) To determine the maximum efficiency of a heat engine operating between the same two temperature extremes, we need to know the two temperature extremes. From our P-V diagram, we see that the highest temperature occurs at point C where PV is a maximum. The lowest temperature occurs at point A where PV is a minimum. So, using the ideal gas law and our expression for the maximum efficiency of a heat engine, we have,

$$e_{max} = 1 - \frac{T_{cold}}{T_{hot}} = 1 - \frac{T_A}{T_C} = 1 - \frac{P_A V_A}{P_C V_C}$$

Solve

(a), (b), (c), and (d) Substituting in the numerical values, we have,

$$W_{net} = (P_B - P_A)(V_D - V_A) = (20,000 \text{ Pa})(0.03 \text{ m}^3) = \boxed{600 \text{ J}}$$

$$Q_{net} = W_{net} = \boxed{600 \text{ J}}$$

$$e = \frac{(P_B - P_A)(V_D - V_A)}{\frac{3}{2} V_A(P_B - P_A) + \frac{5}{2} P_B (V_C - V_B)}$$

$$= \frac{600 \text{ J}}{\frac{3}{2}(0.01 \text{ m}^3)(20,000 \text{ Pa}) + \frac{5}{2}(30,000 \text{ Pa})(0.03 \text{ m}^3)}$$

$$= \boxed{0.24}$$

$$e_{max} = 1 - \frac{P_A V_A}{P_C V_C} = 1 - \frac{(10,000 \text{ Pa})(0.01 \text{ m}^3)}{(30,000 \text{ Pa})(0.04 \text{ m}^3)} = \boxed{0.92}$$

What does it mean?

Comparing 24% efficiency with the ideal efficiency of 92%, we see that a rectangular cycle is not very efficient!

CP 16.2 Freezing Water

Suppose that 200 g of water at 20°C are placed into a freezer which is at −10°C. (a) If the refrigeration efficiency of the freezer is 5.00, determine the work done by the refrigerator during the process of freezing the water to a temperature of −10°C. (b) How much energy is expelled into the room during this process?

Solution

Recognize the principle

In order to determine the work done by the refrigerator we'll first need to calculate the energy removed from the water during the entire process. This energy is Q_C, and combined with our expression for the refrigeration efficiency we can find the work. Once we have both Q_C and W, we can easily calculate Q_H, which is the energy expelled into the room.

Sketch the problem

There is really no sketch required for this problem although the schematic of a heat pump running in reverse will help remind us of the refrigeration process.

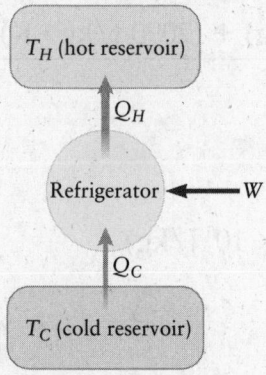

Identify the relationships

(a) We begin by determining the energy removed from the water as it cools, freezes, and then further cools during the process. The total energy removed is given by the expression,

$$Q_C = Q_{\text{cool water}} + Q_{\text{freeze water}} + Q_{\text{cool ice}}$$

Substituting in our expressions for specific heat and latent heat from Chapter 14, we have the following,

$$Q_C = mc_{\text{water}} \Delta T_{\text{water}} + mL_{\text{fusion}} + mc_{\text{ice}} \Delta T_{\text{ice}}$$

Rearranging our expression for the refrigeration efficiency, we have,

$$e = \frac{Q_C}{W} \rightarrow W = \frac{Q_C}{e}$$

Combining these expressions, we find,

$$W = \frac{Q_C}{e} = \frac{m(c_{\text{water}} \Delta T_{\text{water}} + L_{\text{fusion}} + c_{\text{ice}} \Delta T_{\text{ice}})}{e}$$

(b) From our schematic diagram we see that the energy expelled into the room is given by the expression,

$$Q_H = Q_C + W = m(c_{water}\Delta T_{water} + L_{fusion} + c_{ice}\Delta T_{ice})$$

$$+ \frac{m(c_{water}\Delta T_{water} + L_{fusion} + c_{ice}\Delta T_{ice})}{e}$$

$$= \left(1 + \frac{1}{e}\right)m(c_{water}\Delta T_{water} + L_{fusion} + c_{ice}\Delta T_{ice})$$

Solve

(a) and (b) Substituting in the numerical values, we have,

$$W = \frac{m(c_{water}\Delta T_{water} + L_{fusion} + c_{ice}\Delta T_{ice})}{e}$$

$$= \frac{0.2\text{ kg}\left[(4186\text{ J/kg}\cdot\text{K})(20\text{ K}) + (3.34\times10^5\text{ J/kg}) + (2090\text{ J/kg}\cdot\text{K})(10\text{ K})\right]}{5.00}$$

$$= \boxed{1.75\times10^4\text{ J}}$$

$$Q_H = \left(1 + \frac{1}{e}\right)m(c_{water}\Delta T_{water} + L_{fusion} + c_{ice}\Delta T_{ice})$$

$$= \left(1 + \frac{1}{5}\right)(0.2\text{ kg})\left[(4186\text{ J/kg}\cdot\text{K})(20\text{ K}) + (3.34\times10^5\text{ J/kg})\right.$$

$$\left. + (2090\text{ J/kg}\cdot\text{K})(10\text{ K})\right]$$

$$= \boxed{1.05\times10^5\text{ J}}$$

What does it mean?

So, the freezer extracts 8.77×10^4 J of energy from the water during this process while consuming 1.75×10^4 J of electrical energy and dumping 1.05×10^5 J of heat energy into the room.

Part E. MCAT Review Problems and Solutions

PROBLEMS

1. For the process described in the P-V diagram shown, the work done by the gas is given by the expression,

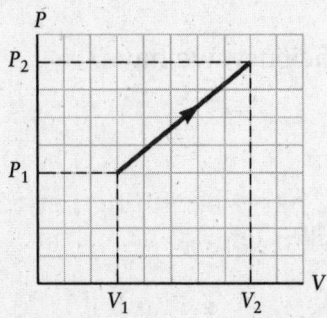

(a) $\frac{1}{2}(P_1 + P_2)(V_1 - V_2)$

(b) $\frac{1}{2}(P_2 + P_1)(V_2 - V_1)$

(c) $(P_2 - P_1)(V_2 - V_1)$

(d) $(P_1 + P_2)(V_1 + V_2)$

2. A 2.0-mol ideal gas system is maintained at a constant volume of 4.0 L. If 100 J of heat flows into the system, what is the work done by the system?

(a) 0 J

(b) 20 J

(c) 100 J

(d) 400 J

3. An ideal gas is taken through the process shown. If the change in the internal energy of the gas is $\frac{3}{2} V_0 (P_1 - P_0)$, how much heat flows into the system?

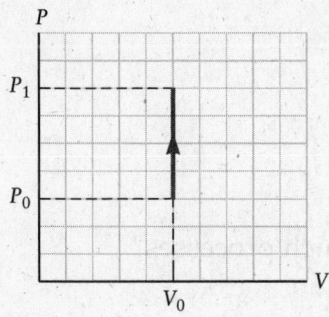

(a) 0

(b) $V_0 (P_1 - P_0)$

(c) $\frac{3}{2} V_0 (P_1 - P_0)$

(d) $\frac{5}{2} V_0 (P_1 - P_0)$

4. A heat engine expels 3000 J of heat while performing 1500 J of useful work. What is the efficiency of this heat engine?

(a) 15%

(b) 33%

(c) 50%

(d) 60%

5. The maximum theoretical efficiency of a heat engine operating between hot and cold reservoirs depends on which of the following?

(a) The temperature of the hot reservoir only.

(b) The temperature of the cold reservoir only.

(c) The temperature of both the hot and cold reservoirs.

(d) None of the above.

6. Which of the following figures best represents a Carnot cycle?

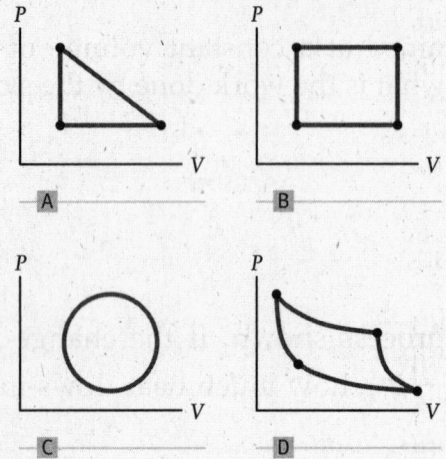

(a) A
(b) B
(c) C
(d) D

7. A Carnot cycle consists of a combination of which processes?
(a) Isobaric and isothermal processes
(b) Isochoric and adiabatic processes
(c) Isothermal and adiabatic processes
(d) Adiabatic and isobaric processes

8. Suppose that a piece of ice at 0°C melts by absorbing 50,000 J of heat energy. What is the change in entropy of the ice during this process?
(a) −50,000 J/K
(b) 0
(c) 183 J/K
(d) Infinite

9. For an irreversible process, which of the following statements is true?
(a) The entropy change of the universe remains constant.
(b) The entropy change of the universe decreases.
(c) The entropy change of the universe increases.
(d) There is not enough information to make a definitive statement about the entropy.

10. Which of the following is a statement of the third law of thermodynamics?

   (a) In any thermodynamic process the change in entropy of the universe must be greater than or equal to zero.

   (b) It is impossible for the temperature of a system to reach absolute zero.

   (c) If two systems *A* and *B* are in thermal equilibrium with a third system, then *A* and *B* are in thermal equilibrium with each other.

   (d) None of the above.

SOLUTIONS

1. *MCAT Strategies*

Since the answers are all similar, let's proceed with our usual problem-solving strategy.

Recognize the principle

We know that the work done during a thermodynamic process is the area under the curve on the P-V diagram.

Sketch the problem

The sketch is provided in the question.

Identify the relationships and Solve

From the diagram we see that the area under the curve can be divided into a triangle and a rectangle. The area of a triangle is ½ (base)(height). So, the total area under the curve is,

$$W = \text{total area} = (\text{area of triangle}) + (\text{area of rectangle})$$

$$= \frac{1}{2}(V_2 - V_1)(P_2 - P_1) + P_1(V_2 - V_1)$$

$$= \frac{1}{2}(V_2 - V_1)(P_2 + P_1)$$

So, the correct answer is (b).

What does it mean?

Notice that since we knew that the work had to be positive, we could have immediately eliminated answer (a).

2. *MCAT Strategies*

This problem is just testing our understanding of the work done on a gas. In order for work to be done, the volume of the gas must change. Since it is stated in the problem that the volume is constant, the work must be zero. So, the correct answer is (a).

3. MCAT Strategies

Once again this problem is testing our understanding of the work done on a gas as well as our understanding of the first law of thermodynamics. Since the process involves no change in volume, the work must be zero. Now, applying the first law of thermodynamics tells us that,

$$\Delta U = Q - W = Q$$

Therefore, the heat flow into the system is the same as the change in the internal energy. So, the correct answer is (c).

4. MCAT Strategies

Since it is not obvious which answer is correct, let's proceed with our usual problem-solving strategy.

Recognize the principle

We know that a heat engine takes in heat energy Q_H, does useful work W, and expels energy Q_C. Also, the efficiency of a heat engine is the benefit divided by the cost.

Sketch the problem

A simple schematic will help us remember the important expressions.

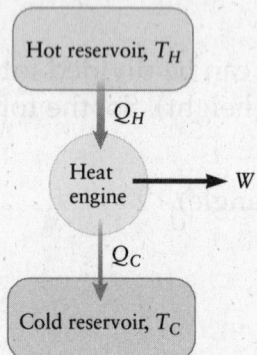

Identify the relationships

From the diagram we see that the energy input Q_H is divided into work W and expels energy Q_C so that,

$$Q_H = W + Q_C$$

The benefit of a heat engine is the useful work while the cost is the energy input. Therefore, our expression for the efficiency becomes,

$$e = \frac{\text{benefit}}{\text{cost}} = \frac{W}{Q_H} = \frac{W}{W + Q_C}$$

Solve

Substituting in the numerical numbers, we find,

$$e = \frac{W}{W + Q_C} = \frac{1500 \text{ J}}{1500 \text{ J} + 3000 \text{ J}} = \frac{1500 \text{ J}}{4500 \text{ J}} = \frac{1}{3} = 33\%$$

So, the correct answer is (b).

What does it mean?

Notice that we did not need to have memorized the expressions for the efficiency and how the energies were related in order to solve this problem. The expressions were derived from a basic understanding of the heat engine and sketching the process.

5. MCAT Strategies

This problem is simply testing our understanding of the maximum efficiency at which a heat engine can operate. Recall that a Carnot engine is the most efficient engine operating between a hot and cold reservoir. What governs the efficiency of a Carnot engine is the temperatures of the reservoirs. So, the correct answer is (c).

6. MCAT Strategies

This problem is testing our understanding of a Carnot cycle. Recall that a Carnot cycle is composed of two adiabatic processes and two isothermal processes. These are best represented by figure D, so the correct answer is (d).

7. MCAT Strategies

As was stated in the last problem, a Carnot cycle is composed of two isothermal and two adiabatic processes. So, the correct answer is (c).

8. MCAT Strategies

Since this process occurs at a constant temperature (0°C = 273 K), the change in entropy of the ice is given by the expression,

$$\Delta S = \frac{Q}{T}$$

The heat energy is given in the problem as 50,000 J and the temperature is 273 K. Without doing the calculation we see that answers (a), (b), and (d) are clearly wrong, so the correct answer is (c). Notice that if we had forgotten to convert our temperature to Kelvin, we would have incorrectly concluded that (d) was the answer.

9. MCAT Strategies

This problem is simply testing our understanding of the second law of thermodynamics. The second law states that during any process the entropy change of the universe must be greater than or equal to zero. For a reversible process the change in entropy of the universe is zero, but for an irreversible process the entropy of the universe increases. So, the correct answer is (c).

10. *MCAT Strategies*

This problem is simply testing our understanding of the third law of thermodynamics. The third law is a statement about the inability of the temperature of a system to reach absolute zero. So, the correct answer is (b).